Disease Evolution

Models, Concepts, and Data Analyses

DIMACS
Series in Discrete Mathematics
and Theoretical Computer Science

Volume 71

Disease Evolution
Models, Concepts, and Data Analyses

Zhilan Feng
Ulf Dieckmann
Simon Levin
Editors

Center for Discrete Mathematics
and Theoretical Computer Science
A consortium of Rutgers University, Princeton University,
AT&T Labs–Research, Bell Labs (Lucent Technologies),
NEC Laboratories America, and Telcordia Technologies
(with partners at Avaya Labs, HP Labs, IBM Research,
Microsoft Research, and Stevens Institute of Technology)

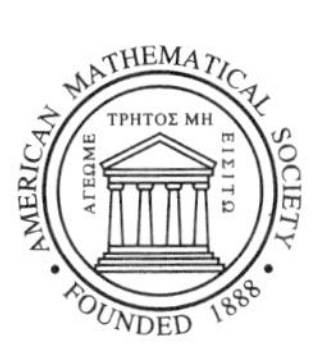

American Mathematical Society

2000 *Mathematics Subject Classification.* Primary 34C60, 35Q80, 62P10, 92B05, 92C50, 92C60, 92D10, 92D15, 92D25, 92D30.

Library of Congress Cataloging-in-Publication Data

Disease evolution : models, concepts, and data analyses/Zhilan Feng, Ulf Dieckmann, Simon Levin, editors.
p. cm. — (DIMACS series in discrete mathematics and theoretical computer science; 71)
Includes bibliographical references
ISBN 0-8218-3753-2 (alk. paper)
1. Communicable diseases–Epidemiology. 2. Molecular epidemiology. 3. Evolutionary genetics. I. Feng, Zhilan, 1959- II. Dieckmann, Ulf, 1966- III. Levin, Simon A. IV. Series.

RA652.2.M3.D57 2006
614.4—dc22 2006042973

∞ The paper used in this book is acid-free and falls within the guidelines
established to ensure permanence and durability.
Visit the AMS home page at http://www.ams.org/

10 9 8 7 6 5 4 3 2 1 11 10 09 08 07 06

Contents

Foreword

This volume has its genesis in the activities of the DIMACS working group on Genetics and Evolution of Pathogens. This working group held a meeting on November 24-25, 2003 at Rutgers University, and we would like to express our appreciation to Zhilan Feng for organizing and planning this successful conference. The volume represents an expansion of the efforts of this working group, and contains papers from experts in the field who were unable to attend this initial meeting. We thank the three editors Zhilan Feng, Ulf Dieckmann and Simon Levin for their efforts in the organization of the volume, and we thank Bruce Levin for his insightful Preface and also the various authors who contributed to the volume.

The meeting was part of the 2002-2007 Special Focus on Computational and Mathematical Epidemiology, and was organized by one of a number of special focus research groups called "working groups" as part of the special focus. We extend our thanks to Martin Farach-Colton, Sunetra Gupta, Donald Hoover, David Krakauer, Simon Levin, Marc Lipsitch, David Madigan, Megan Murray, S. Muthukrishnan, David Ozonoff, Fred Roberts, Burton Singer and Daniel Wartenberg for their work as special focus organizers.

The meeting brought together researchers who approach the study of epidemiology from a variety of disciplines, some applied and some theoretical. These included computer scientists, mathematicians, statisticians, and biologists together with both descriptive and analytical epidemiologists. The goal of the working group as well as of this volume is the exploration of cross-disciplinary approaches to the study of topics related to disease evolution, and how they apply to the study of specific diseases.

DIMACS gratefully acknowledges the generous support that makes these programs possible. Special thanks go to the National Science Foundation, the James S. McDonnell Foundation, the Burroughs-Wellcome Fund, the Purdue University Mathematics Department and to DIMACS partners at Rutgers, Princeton, AT&T Labs - Research, Bell Labs, NEC Laboratories America, and Telcordia Technologies, and affiliate partners Avaya Labs, HP Labs, IBM Research, and Microsoft Research.

Fred S. Roberts
Director

Robert Tarjan
Co-Director for Princeton

Preface

Infections by microparasites (bacteria, viruses, protozoa and single celled fungi) are the primary source of human mortality in the underdeveloped world. And, despite all of the improvement in public health, hygiene, nutrition, living conditions and medical intervention over the past century, infections continue to be a major cause of morbidity and mortality in the developed world as well. Indeed, if we include people compromised by age, cancers and other diseases with immune–suppressing effects and/or treatments, coronary artery, diabetes, and other non-contagious and degenerative diseases, bacterial infections (often acquired in hospitals) may well be the major immediate cause of death even in overdeveloped countries.

Traditionally, the study of infectious diseases and their prevention and treatment has been the purview of epidemiologists, microbiologists, immunologists and clinicians – people who generally have little background in or appreciation for mathematics beyond statistics, if that. While the importance of quantitative reasoning for studies of the epidemiology of infectious diseases has been recognized for some time, this has been less so for investigations of the course of infections and their treatment within individual patients. For the most part, protocols for preventing the spread of infections in hospitals and communities and for the treatment of individual patients are based on qualitative considerations, experience and intuition, with money being the primary quantitative element in their design and implementation.

During the past two decades, studies of the epidemiology, evolution and within-host biology of infectious diseases and the development of methods for their prevention and treatment have been increasingly infiltrated by quantitative methods beyond statistics. A number of applied and not-so-applied mathematicians, mathematically trained and oriented epidemiologists, microbiologists, immunologists, ecologists, population and evolutionary biologists and even real doctors have been using mathematical and numerical models (computer simulations) to study the epidemiology, evolution and within-host dynamics of infectious diseases and to develop and evaluate protocols for their prevention and treatment. This collection is an impressive sampling of the nature and diversity of this epidemic of mathematical and numerical modelling for the studies of infectious diseases. It illustrates some of the delicious problems and opportunities for mathematicians and mathematical biologists that infectious diseases pose – problems that have the virtues of being important to human health and well-being and, at the same time, being challenging and intriguing even from the precious heights of academe.

For generality, tractability and the aesthetic appeal of closed-form mathematical analysis, traditional models of the epidemiology and evolution of infectious diseases have been deterministic and give little or no consideration to the spatial, temporal and other heterogeneities of human and other host communities and those of the microbes that infect them. In recent years, increasing numbers of modelers have been confronting these inconveniences and the unfortunate finiteness of the real world and exploring how they affect the inferences about the epidemiology and evolution of infectious diseases drawn from simpler models. Three of the chapters in this collection illustrate this trend. Mike Boots and his collaborators consider the effects of the spatial structure of host populations on the evolution of the virulence of the microparasites that infect them. The two chapters by Wayne Getz and his cohorts examine, in a pedagogically useful as well as scholarly way, the consequences of spatial structure and stochastic processes on the spread of infectious diseases. Their chapters illustrate the utility of modelling to understanding the ascent and spread of emerging and reemerging diseases like SARS and tuberculosis, predicting their emergence and evaluating methods to control their dissemination.

In a commentary with a title that should appeal to this audience, "In Theory", Sidney Brenner referred to molecular biology as the "great leveler" and suggested that for many it has made thinking unnecessary, a position I do not challenge. On the other side, the ease with which data and particularly those on the nucleotide sequences of DNA can be and have been gathered has also provided an opportunity for the quantitative study of evolutionary history through phylogenies. God is no longer the only one who can make a tree. Phylogenies generated from DNA sequence data – molecular phylogenies – have been the largest growth industry in the evolution business for the past decade. And, the development (if not always the application) of methods for generating, analyzing and interpreting these trees is an activity that requires serious thinking. The chapter by Charleston and Galvani is a fine example of this thinking applied to ascertaining the evolutionary relationship between interacting organisms like parasites and their hosts. "Co-phylogenetic" methods of the sort they are developing are of practical as well as academic interest. They can be used to determine the origins (original hosts) of newly emerging infectious diseases like HIV/AIDS and in that way better understand the conditions responsible for their emergence.

A prominent approach to drawing inferences about the nature and direction of evolution of parasites and their virulence has been to study their ecology (population dynamics and demography) within individual hosts or communities of hosts. In this perspective, the fitness of the microparasites is proportional to their reproductive number, R_0 – the number of secondary infections in a largely (or better yet, wholly) uninfected population of cells or tissues in an infected host or among individuals in a community of hosts. While this ecological approach to evolutionary inferences is explicit about nature and functional form of the selection pressures responsible for evolution, it does not consider the genetic basis of the variability upon which that selection is operating. In different ways and with different foci, two chapters consider ways to meld the ecological approach to the study of the evolution of microparasites and their virulence with those of population and quantitative genetics in situations where the nature of inheritance is explicit but where the ecological basis of selection is not. In their chapter, Troy Day and Sylvain Gandon consider how to apply classy population genetic approaches, like the Price equation, to studies of the evolution of microparasites and their virulence in communities of hosts. In his chapter, John Kelly uses a combination of ecological and population genetic methods to explore the contribution of tissue heterogeneity to the evolution of viruses in within infected hosts.

The contribution of the heterogeneity of the within-host habitat to the population and evolutionary dynamics of microparasites is also the focus of three other chapters in this collection. While modelers as well as experimentalists have the convenience of separately studying microparasite ecology and evolution within infected host and in communities of hosts, in the real world microbes have no choice but to deal with both of these elements of their ecology and evolution. Although their models focus primarily on the within-host population and evolutionary dynamics of viruses and the contribution of within-host heterogeneity to that evolution, in their chapter Robert Holt and Michael Barfield consider how the within-host biology of microparasites contributes to their evolution in communities of hosts. In the chapter by Zhilan Feng and Libin Rong, the within-host heterogeneity of concern is comprised of the selective environments imposed by treatment with multiple drugs that act at different stages in the microparasite replication cycle. Using an age-structured model, they consider the treatment of HIV/AIDS with reverse-transcriptase – and protease – inhibitors, and how this treatment contributes to the evolution of resistance and rates of viral replication. Antimicrobial chemotherapy, heterogeneity and resistance are also the subjects of the chapter by David Smith and his collaborators. In their case, the drugs are antibacterial (antibiotics) rather than antiviral; the heterogeneity is both spatial and in the extent to which the drugs are employed; and the focus is the epidemiology and evolution of resistance in communities of hosts rather than in individual treated patients.

Investigators studying the evolution of infectious disease are almost invariably adaptationists; they assume that selection in the host, parasite or both populations is responsible for the virulence of the parasite and for maintaining genetic diversity in the parasite population. In their article on the serological diversity of the rhinoviruses responsible for the common cold, William Koppelman and Frederick Adler consider the neutral, null hypothesis alternative – that the 100 or so serotypes of Rhinoviruses responsible for colds are consequences of a high mutation rate and genetic drift rather than immune-mediated selection. Rhinoviruses and the cross-immunity they engender are also stage center in the chapter by Alun L. Lloyd and Dominik Wodarz, but the focus of their investigation of these ubiquitous and annoying, albeit rarely lethal, viruses is chemotherapy and the contribution of the host immune response to the evolution of resistance to the antiviral drugs employed.

This collection can be and I believe should be seen as a testimony to the work of Roy Anderson and Robert May. While they are not the discoverers of infectious diseases (at least I don't think they are) or even the first to use mathematical models to investigate them, their research more than that of any other investigators has been responsible for the renaissance (epidemic) in the use of models for studying infectious diseases and their control. The research reported in almost all of the chapters in this volume have antecedents in Anderson and May's work. While there is no formal dedication to them in the front matter of this volume, that dedication is where it really counts. The contributions of either Robert May and/or Roy Anderson are acknowledged at least once in every chapter and now, appropriately, in this Preface.

Enjoy,
Bruce R. Levin
Atlanta, September 2005

Editors' introduction

The goal of this volume is to show how to use mathematical tools to understand the evolution of infectious diseases. Inspiration for this project comes from work of the DIMACS Working Group on Genetics and Evolution of Pathogens, which is organized under the auspices of DIMACS' Special Focus on Computational and Mathematical Epidemiology.

This volume is divided into two sections: Model Infrastructure and Applications to Specific Diseases. Section I discusses the impact on disease evolution of various factors, including spatial structure, transient dynamics, coupling of within-host and between-host dynamics, heterogeneity in host populations, and drug resistance. Section II is concerned with investigations associated with specific infectious diseases such as rhinovirus, HIV/AIDS, tuberculosis, and malaria.

We thank Bruce Levin for his excellent Preface. We also express our gratitude to members of DIMACS' staff who kindly helped with the support of the workshop and the preparation of this volume. The leadership of Fred Roberts in developing the multi-year epidemiology program has been an inspiration to many researchers, and this volume owes its existence to his efforts. We also thank all the Purdue Mathematics Department for providing technical support. Finally, we thank the authors for their outstanding contributions.

The workshop and the preparation of this volume were partially supported by an NSF grant to DIMACS, and by NSF and James S. McDonnell Foundation grants to ZF.

Zhilan Feng (Purdue University)
Ulf Dieckmann (International Institute for Applied Systems Analysis, Austria)
Simon Levin (Princeton University)

Section I

Model Infrastructure

DIMACS Series in Discrete Mathematics
and Theoretical Computer Science
Volume **71**, 2006

The implications of spatial structure within populations to the evolution of parasites

Mike Boots, Masashi Kamo, and Akira Sasaki

ABSTRACT. It is well understood that the spatial structure inherent in most if not all populations can have important implications to the evolutionary dynamics of a wide range of traits. One of the best developed areas of evolutionary theory focusses on the evolution of parasites and spatially explicit models have illustrated the importance of structure within the host to the selection of the infectious organisms. Here we review this theory and show how approximation techniques may be useful in addressing the problem. We show that transmission can be constrained without trade-offs and there is the possibility of multiple stable states due to interactions with self-generated population structures. Models intermediate between local and the mean-field have demonstrated that these effects are not only the result of extreme local interactions and may therefore be applicable not only to plant populations with rigidly local interactions. Advances in this field have mostly been driven by the use of computer simulation, but moment closure approximations offer an opportunity to develop our analytical understanding of these processes. However, advances in these approximation techniques are required for models with extreme local interactions.

1. Introduction

There is a striking variation in the rate at which pathogenic organisms are able to transmit themselves and the harm that they can cause their hosts. Why do different diseases adopt different strategies to their common fundamental problem of being passed from one host to another? Understanding the processes that underpin and cause this variation remains one of the major challenges of evolutionary theory and it is only by understanding these processes can we hope to predict and manage the evolution of disease virulence effectively in human, agricultural and wildlife populations. The evolutionary theory of infectious disease is well developed and has made a significant contribution to our understanding of what determines the virulence seen in nature. Conventional "wisdom" from the simplest of models predicts that parasites necessarily evolve to become harmless to their hosts and therefore the virulence that we see is a consequence of maladaptation (see [**21**]). At the heart of this older idea were a series of group selection arguments based on the idea that the parasite should evolve for the benefit of the parasite species. Relatively recent

Key words and phrases. Evolution, Parasites, Spatial structure, Moment closures.

work has overturned these traditional ideas and approach; the modern theory of the evolution of parasites is based on more fundamental individual selection arguments focused on the dominance of favoured individual strategies.

Simple classical "mean-field" or homogeneous mixing models (which ignore population structure) where there is no co or super-infection predict that selection will tend to maximise the parasites basic reproductive number (the epidemiological R_0) [**8, 21**]. R_0 is the most important epidemiological characteristic and determines the ability of the disease to spread in a population; it is defined as the average number of secondary infections caused by an average infected host in a susceptible host population (see [**1**]). It is therefore determined by the rate of infection and the duration of the infectious period; the infectious period is governed by the rate that infected individuals either recover or die, including the increased death rate due to infection (virulence). Therefore, in order to maximise R_0, evolution should maximise the transmission rate and minimise virulence and recovery. However it is doubtful that the disease behaviour is completely unconstrained, and we expect there to be a trade-off from the point of view of the parasite between transmission and virulence. Higher transmission can only be 'bought' at the expense of higher virulence as the processes of producing of the necessary amounts of parasite transmission cause damage to the host [**19**]. If transmission is increasingly costly in terms of virulence, models predict the evolution of a finite transmission rate and virulence, otherwise evolution will maximise transmission and virulence; in both cases maximising R_0. This analysis is by no means always applicable to all circumstances. For example, superinfection of parasites [**11, 23, 29, 33**] leads to a higher ESS virulence because the intra-host competition among strains favors a more virulent parasite than that maximizes the basic reproductive number. The virulence evolved in expanding population should also be larger than that in population in demographic plateau [**18**].

In addition, all of the general theory assumes that the host population is completely mixed and that therefore any individual is as likely to infect any one individual as any another. The assumption of homogeneous mixing in host populations ignores the fact that certain individuals are more likely to contact and therefore infect others. One obvious example of this is that on average an infectious individual is more likely to infect those closest to them than those at a distance. Here we will review the literature on the evolution of parasites in spatially structured populations, addressing the importance of population extinctions, of spatial structure intermediate between completely local and completely global and the role of added biological realism. We will also describe the use and limitations of approximation techniques in addressing these problems and point the way forward in the development of the analysis.

2. Within population spatial structure and the evolution of parasites

The inclusion of spatial/social structure into host-parasite models has shown that this more realistic assumption about the structure of host populations has dramatic implications to the evolution of the parasite. A successful approach to examining the role of the spatial structure of individual hosts is by using lattice models (also called probabilistic cellular automata (PCA)) [**6, 15, 30, 32, 34**]. This approach examines the fundamental spatial relationships of individuals within populations and uses biologically realistic and quantifiable parameters. Space within a population is represented as a network of sites that correspond to either individuals or empty space. The states of the sites change probabilistically in response to a set of simple rules that define the ecological characteristics of the system.

The epidemiological dynamics of simple spatially structured host pathogen populations have been shown to be dramatically different to classical ones [**34**]. Using a lattice structured population Sato et al. [**34**] examined the equilibrium states using pair approximation. This technique, which keeps track of the spatial correlation between the nearest neighbor sites, showed that the parasite could drive the host to extinction even in infinitely large populations. The conditional probabilities or local densities (e.g. the probability that a susceptible host is in the nearest neighbor of an empty site) played a critical part in determining the invasibility and the persistence of infected in the population. If the transmission rate is sufficiently large, the host goes to extinction, a result which does not occur in mixed populations. Sato et al. [**34**] showed that the local density of susceptibles around each infected individual remains high enough to allow the local expansion of the infected individual even if the global density of the host approaches zero. In a completely mixed population, in contrast, decreased host density will immediately result in a reduced mean transmission rate, which stops the population from going to extinction. Further work has shown that parasite driven extinction is possible so long as the parasite reduces the host fecundity to some degree [**6**]. With such important differences demography such as this caused by spatial structure and the fact that local spatial structure is particularly important to a rare mutant, it was highly likely that spatial structure would play an important role in the evolution of pathogens.

Haraguchi and Sasaki [**15**] developed a detailed model of the evolution of parasites in a generic host parasite model with simple spatial structure characterized by a regular lattice. They considered a model where each site of the lattice is either empty, occupied by a susceptible, or occupied by an infected. A $L \times L$ regular lattice with a periodic boundary was assumed so that each site has 4 nearest neighbors. The state of the x-th site in the lattice at time t is denoted by $\sigma_x(t) \in \{0, S, I\}$, where the state 0, S, and I indicate respectively that the site is empty, occupied by a susceptible, and occupied by an infected host. When we consider the evolution of parasites, we introduce the state I_j which indicates that the site is occupied by an individual that infected by the j-th strain of parasite.

A continuous time Markov process was defined by specifying the transition probability of each site in a unit time interval. The state of the x-th site changes by

(i) the mortality of a susceptible individual:

$$S \to 0, \quad \text{at rate } d_1 \ (= 1); \tag{2.1}$$

(ii) the mortality of an infected individual:

$$I \to 0, \quad \text{at rate } d_2 \ (= 1 + \alpha); \tag{2.2}$$

(iii) the reproduction of susceptible individuals:

$$0 \to S, \quad \text{at rate } r n_x(S)/z; \tag{2.3}$$

(iv) infection:

$$S \to I, \quad \text{at rate } \beta n_x(I)/z; \tag{2.4}$$

where $n_x(\sigma)$ represents the number of sites with the state σ in the nearest neighbor of the x-th site, and z is the number of nearest neighbor sites ($z = 4$ for a regular lattice). The mortality of a susceptible and an infected are denoted by d_1 and d_2, where time is scaled so that the mortality of a susceptible host d_1 equals 1 (i.e., the mean lifetime of a susceptible is 1); $\alpha = d_2 - d_1$ is the additional mortality (virulence) of an infected individual; r is the reproduction rate of a susceptible of progeny to its nearest neighbor sites; β is the transmission rate of parasite. They assumed that the pathogen sterilizes the host immediately after infection.

Pair approximations of this model are derived as follows. Let $\rho_\sigma(t)$ be the probability that a randomly chosen site has the state σ at time t, and $p_{\sigma\sigma'}$ be the probability that a randomly chosen pair of nearest neighbor sites has state $\sigma\sigma'$. We can define the conditional density $q_{\sigma/\sigma'}$ as the probability that a nearest-neighbor of σ'-site has the state σ. Likewise, $q_{\sigma/\sigma'\sigma''}$ represents the conditional probability that a nearest-neighbor of σ' site in the $\sigma'\sigma''$-pair has the state σ. By definition we should have $\sum_\sigma \rho_\sigma = 1$, $\sum_\sigma q_{\sigma/\sigma'} = 1$ and

$$p_{\sigma\sigma} = \rho_\sigma q_{\sigma/\sigma'}, \tag{2.5}$$

$$p_{\sigma\sigma'} = q_{\sigma'/\sigma\sigma} = q_{\sigma/\sigma'\sigma'}, \text{ if } \sigma \neq \sigma'. \tag{2.6}$$

From the transitions probabilities (2.1)–(2.4), the following differential equations for pair densities are derived:

$$\dot{p}_{00} = -2r(1-\theta)q_{S/00}p_{00} + 2p_{S0} + 2(1+\alpha)p_{I0} \tag{2.7}$$

$$\begin{aligned} \dot{p}_{S0} = &- \left[1 + r\left\{\theta + (1-\theta)q_{S/0S}\right\} + \beta(1-\theta)q_{I/S0}\right] p_{S0} \\ &+ r(1-\theta)q_{S/00}p_{00} + p_{SS} + (1+\alpha)p_{IS} \end{aligned} \tag{2.8}$$

$$\dot{p}_{SS} = -\left\{2 + 2(1-\theta)\beta q_{I/SS}\right\} p_{SS} + 2r\left\{\theta + (1-\theta)q_{S/0S}\right\} p_{S0} \tag{2.9}$$

$$\begin{aligned} \dot{p}_{I0} = &- \left[1 + \alpha + r\left\{\theta + (1-\theta)q_{S/0I}\right\}\right] p_{I0} \\ &+ \beta(1-\theta)q_{I/S0}p_{S0} + p_{IS} + (1+\alpha)p_{II} \end{aligned} \tag{2.10}$$

$$\begin{aligned} \dot{p}_{IS} = &- \left[2 + \alpha + \beta\left\{\theta + (1-\theta)q_{I/SI}\right\}\right] p_{IS} \\ &+ r(1-\theta)q_{S/0I}p_{I0} + \beta(1-\theta)q_{I/SS}p_{SS} \end{aligned} \tag{2.11}$$

$$\dot{p}_{II} = -2(1+\alpha)p_{II} + 2\beta\left\{\theta + (1-\theta)q_{I/SI}\right\} p_{IS} \tag{2.12}$$

where dots represent the time derivative, and $\theta = 1/z$. As pair density dynamics depends on triplet densities (e.g. $q_{I/SS}$), the following approximation for triplet correlation is used to complete the pair approximation:

$$\begin{aligned} q_{I/S\sigma} &= q_{I/S} \text{ for any } \sigma, \text{ and} \\ q_{S/0I} &= q_{S/0}, \\ q_{S/00} &= \varepsilon q_{S/0}, \end{aligned}$$

where ε is determined to be consistent with the threshold reproductive rate for the persistence of host in disease-free population, and is 0.8093 for two-dimensional regular lattice [**16, 34**]. These pair approximations predicted a region of parasite driven host extinction that was confirmed by simulation (see Figure 1). It is this ability to predict new outcomes that may have been missed if simulation is the only form of analysis that justifies the use of these approximation techniques. There are also some important discrepancies between the approximations and the results of the simulations shown in Figure 1. In particular there is a region where we get extinction when the approximations predict an endemic equilibrium. This occurs due to a stochastic extinction process in the finite population size of the simulation in regions where the approximations predict oscillatory dynamics. There is also a region where there is the unexpected loss of the parasite in the full simulations. This discrepancy is likely to be a result of a failure of the approximation since it is insensitive to increasing the population size in the simulation [**15**]. The most likely cause of this failure is that the loss of the parasite is due to higher correlation effects that are not captured in the pair dynamics of the system.

As discussed in the introduction, in a completely mixed population with the assumptions we have here, it has been shown [**21**] that the parasite trait evolves to maximize the basic reproduction number:

$$R_0 = \frac{\beta}{d_1 + \alpha}.$$

This implies that if the additional mortality α of the infected host and the transmission rate β are independent of each other, β evolves to infinity and α evolves to zero. Within the spatial model, Haraguchi and Sasaki considered a parasite population consisting of two different strains with different transmission rates — a wild type with transmission rate β_I and the additional mortality α_I, and a mutant with β_J and α_J. By modifying the doublet density dynamics described above to include two strains of the parasite (but ignoring double infection), we see that the global density of mutant parasite strain, ρ_J, changes when rare as

$$\frac{d\rho_J}{dt} = (\beta_J q_{S/J} - d_1 - \alpha_J)\rho_J, \tag{2.13}$$

where $q_{S/J} = p_{SJ}/\rho_J$ is the local density of susceptible (S) in the nearest neighbor of a mutant infected host (J). Let $q_{S/J} = q_{S/I} + \Delta q_{S/J(I)}$ and noting $q_{S/I} = (d_1 + \alpha_I)/\beta_I$ at equilibrium of wild type infected population [**34**], the marginal growth rate of the mutant can be written as

$$\lambda(J|I) = \beta_J \left\{ \frac{d_1 + \alpha_I}{\beta_I} - \frac{d_1 + \alpha_J}{\beta_J} + \Delta q_{S/J(I)} \right\} \tag{2.14}$$

Thus if the mutant parasites are surrounded by more susceptibles than the wild type parasites (i.e., if $\Delta q_{S/J(I)} > 0$), the mutant can invade even when it has a lower basic reproductive number. Furthermore since lower transmission rate and lower virulence tend to make the local density of susceptible around an infected increase, equation (2.6) suggests spatially structured populations favour lower transmission rate and lower virulence through the Δq term. Note that the mechanism does not require any group selection argument, although one may be able to interpret these results from the viewpoint of inter-cluster selection.

Simulation results confirmed the evolution of milder parasites with a finite ESS transmission rate even if there was no trade-off with virulence (Figure 2a) or if there was a linear relationship between the two (Figure 2b). Without spatial structure, theory would have predicted the evolution of maximal transmission in both cases. This is a key result. Spatial structure in itself can restrain the evolution of transmission rates leading to milder parasites. They also found that there was the evolution to just below the point at which the parasite would drive its host to extinction.

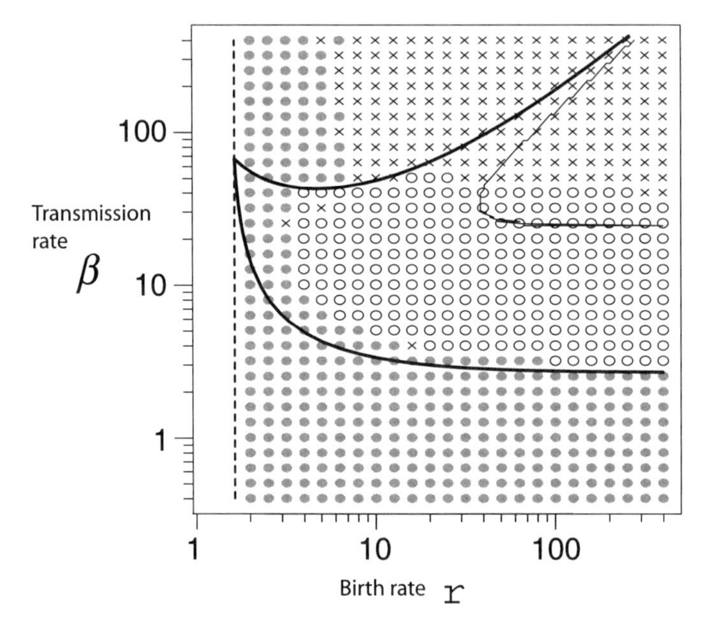

FIGURE 1. The epidemiological equilibrium states in a spatially structured SI lattice-structured population, predicted from both pair approximations and observed in Monte Carlo simulations. The curves represents two threshold transmission rates, β_c and β_f, as a function of host reproduction rate r, predicted from the pair approximation described in the text. If $\beta > \beta_c$ the parasite will drive the host to extinction; if $\beta_f < \beta < \beta_c$ the population is endemic; and if $\beta < \beta_f$ the parasite cannot invade the disease free population. The circles and hatches indicate the most frequently observed equilibrium state in simulation of 5 independent replicates with 100×100 regular lattice. The outcome of each simulation run is classified into endemic (open circle, both susceptibles and infecteds persisted until $t = 50$), disease-free (gray circle, infecteds went to extinction, leaving disease-free population) and parasite-driven host extinction (hatch, susceptibles went to extinction, leading to the extinction of entire population). The thinner curved line shows the region within which the endemic equilibrium is predicted by pair approximation to be locally unstable, leading to limit cycles in the pair density dynamics.

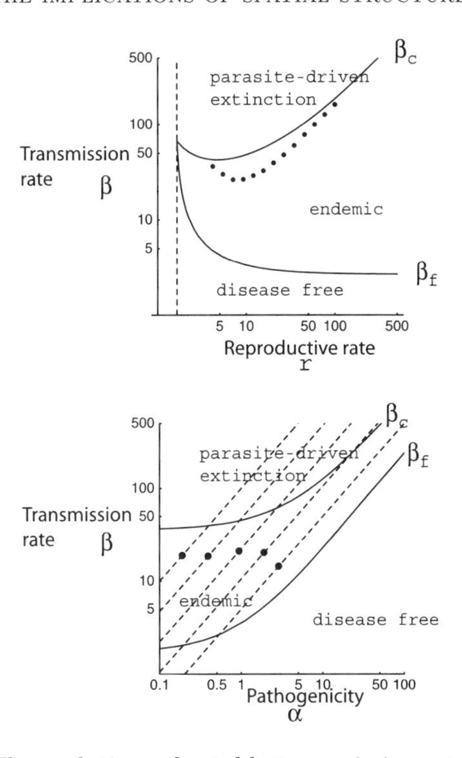

FIGURE 2. The evolutionarily stable transmission rate for the parasite in a spatial SI model at a variety of host reproduction rates. In figure **(a)** we have the results of Monte Carlo simulations for the evolution of transmission rate in a 200×200 lattice when there is no constraint through a trade-off with virulence. There were 101 strains of parasite with equally divided transmission rate from $\beta_{\min} = 5.0$ to $\beta_{\max} = 200.0$. The simulation started with the introduction of the parasite strain with the minimum transmission rate, and allowed to evolve for 500 host generations (i.e., until $t = 500$, as the mean life time of a susceptible host is $1/d_1 = 1$), where a parasite strain can mutate to one of the adjacent strains with the rate 0.5 per host generation. For each value of host reproductive rate (horizontal axis), the long-term average of the population mean transmission rate over last 50 generations are plotted (dots). Curves represents two threshold transmission rates predicted from pair approximation. When the host reproductive rate is smaller than 4, the evolutionary increase of transmission rate leads to the extinction of the parasite; whereas, when it is larger than 100, the evolution leads to the extinction of the host population. Otherwise, the parasite transmission rate stopped increasing fairly below the threshold for population extinction (note logarithmic scale). **(b)** The evolutionarily stable transmission rate β and virulence α, when there is a linear trade-off between them such that $\alpha = c\beta$. The coefficient c of trade-off varied from 0.01 to 0.2, and the pairs of transmission rate and virulence constrained by trade-off are shown by broken lines. For a given trade-off, transmission rate and virulence evolve along the broken lines, and balance at intermediate values (dots) within the region of endemic equilibrium. The size of lattice is 200×200. Parameters: (a) $d_1 = 1.0$, $\alpha = 1.0$; (b) $d_1 = 1.0$, $r = 8.0$.

Rand et al. [**30**] had previously analyzed a model similar to Haraguchi and Sasaki, with the distinction that they assumed that all host mortality is due to parasite virulence. They found that there was a critical transmissibility above which pathogen goes to extinction (leaving disease-free host population), and that the transmission rate evolves towards the criticality. They did not observe parasite driven extinction, perhaps because the parameters that Rand et al. used in their Monte Carlo simulations had low host reproductive rates (Figure 1). Rand et al. [**30**] described the evolution of the transmission rate towards this criticality, which contrasts with the results of Haraguchi and Sasaki who suggested that the ESS transmission rate was strictly below the threshold for population extinction, and that therefore in sufficiently large lattices, population size remains endemic at the evolutionary stable state. In a small population, however, the evolution of transmission rate could lead to the extinction of both the host and the parasite, or the parasite alone, depending on the host reproduction rate. Claesson and DeRoos [**9**] had also examined the evolution of parasites in a spatially structure population. Their model assumed a constant population size and only found an effect on the evolution of parasite virulence once superinfection was assumed.

All this work has shown that parasite evolution is strongly affected by the spatial structure in the host population. It is important to note however that this is not simply the result of the fact that there is the possibility of parasite driven extinction in the models. Another established theoretical scenario in which pathogens may drive their hosts deterministically to extinction is when the transmission process depends on the frequency of infected hosts in the population rather than the density [**14**]. This form of transmission is often proposed for sexually transmitted diseases where mating rate may be independent of density [**14**]. The argument is that even at low population numbers, individuals still find each other to mate and therefore transmission of the disease is dependent on the frequency in the population. Frequency dependent transmission has been suggested for some vector-borne diseases since a vector may efficiently search for its hosts, maintaining contacts even at low host populations. An analysis of a generic frequency dependent host parasite model where leads to a very similar pattern of host extinction, endemic and parasite extinction to the spatial model (Figure 3), however it is easy to show that the epidemiological R_0 is maximized by evolution in this case [**7**]. Evolution to extinction will therefore occur dependent only on the cost structures that occur which contrasts with evolution to a point before extinction in the spatial models.

3. Models between the completely local and the mean-field

The completely local interactions of the models described above and indeed the majority of spatial models in the literature are an extreme characterization of spatial structure. Interactions only occur between nearest neighbours. This abstraction is useful as when we want to ask "what is the effect of space?" it makes sense to use a model that is highly spatial. However, since the mean-field is also an extreme assumption since the population is completely mixed, it is dangerous to simply compare these two extreme abstractions. Most systems are in reality some where between these two extremes. Modelling intermediate population structures, however, often leads to models of much higher complexity than either of the local or mean-field models.

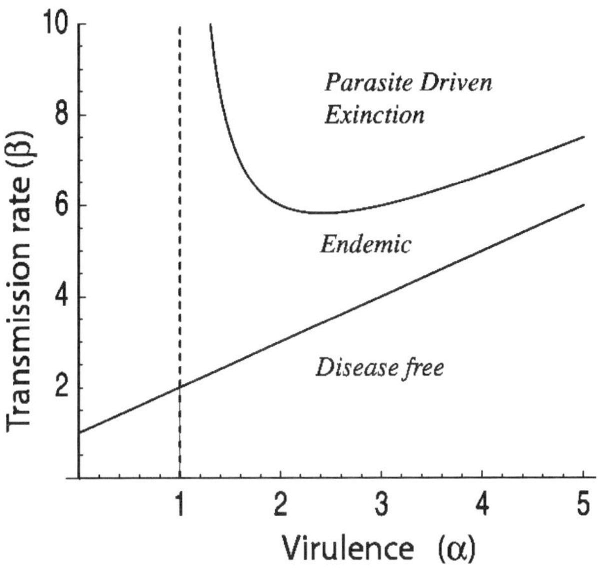

FIGURE 3. Characteristic regions in transmission (β) and virulence (α) parameter space in a simple host-parasite model with frequency dependent transmission where the infected individuals are able to reproduce. In a similar way to the spatial model, there is a parasite driven extinction (PDE) region. Although the ecological dynamics are very similar, the evolutionary dynamics are completely different.

One simple approach is to allow a proportion of the interactions to be local and the remaining ones to be at random throughout the population. This process allows population structures that are intermediate between the local and mean-field approximations. Effects such as the evolution of milder parasites that are subscribed to the affects of spatial structure can then be assessed for their robustness to varying levels of mixing. Boots and Sasaki [**6**] took this approach, in examining how the ESS transmission rate and virulence was affected by different proportions of local and global infection. Again we represented space by considering a regular network of sites, each of which contains either a single host individual or is empty. Susceptible host individuals again reproduce at rate r into neighbouring empty sites but this time infection occurs from the contact of an infected and susceptible locally at neighbouring sites and globally at sites chosen at random (P denotes the proportion of global infection where $0 \leq P \leq 1$). Hosts do not move between sites and infected individuals have a higher death rate ($d + \alpha$) than that of susceptibles (d). There are a number of strains (i) of the parasite that differ in their transmission rate (β_i) (the probability of causing infection) and have correlated changes in the increased death rate (α_i) that infection causes in their hosts (virulence). Evolution occurs through small mutations (in any single time step mutation to the strain with the next highest or next lowest transmission rate can occur). We initially assume that infected individuals do not reproduce.

The model is equivalent to the Haraguchi and Sasaki model, but while reproduction is local, transmission can vary from completely local to completely global. We can however determine the evolutionarily stable (ES) virulence for the parasite by invasion analysis from pair density dynamics as before. This time we replace:

$$\beta(1-\theta)q_{I/Sx} \Rightarrow \beta\left\{P\rho_I + (1-P)(1-\theta)q_{I/Sx}\right\}$$

where $x = S$ or 0, and

$$\beta\left\{\theta + (1-\theta)q_{I/SI}\right\} \Rightarrow \beta\left\{P\rho_I + (1-P)\left[\theta + (1-\theta)q_{I/SI}\right]\right\}$$

where $\rho_I = p_{II} + p_{IS} + p_{I0}$ is the global density of the infecteds and $\theta = 1/z$. From the pair density dynamics the global density of infecteds follows exactly as

$$\frac{d\rho_I}{dt} = \left[\beta_I\left\{P\rho_S + (1-P)q_{S/I}\right\} - (d+\alpha_I)\right]\rho_I. \tag{3.1}$$

If a mutant parasite strain J invades a population at endemic equilibrium with strain I, the growth rate of the mutant when rare is approximated as

$$\lambda(J|I) = \frac{1}{\rho_J}\frac{d\rho_J}{dt} = \beta_J\left\{P\hat{\rho}_S + (1-P)\hat{q}^0_{S/J}\right\} - (d+\alpha_J) \tag{3.2}$$

where $\hat{q}^0_{S/J}$ is the local density of susceptible individuals in the neighbourhood of the mutant parasite at a quasi-equilibrium (any conditional density including $q_{S/J}$ in the nearest neighbourhood of a rare mutant parasite J changes much faster than that of its global density x_J, and hence it can be approximated by the quasi-equilibrium value — the equilibrium value for a fixed global density [**20**]. Since

$$P\hat{\rho}_S + (1-P)\hat{q}_{S/I} = \frac{d+\alpha_I}{\beta_I} \tag{3.3}$$

holds at the endemic equilibrium of the resident strain I, the sign of

$$\lambda(J|I) = \beta_J\left[\left(\frac{d+\alpha_I}{\beta_I} - \frac{d+\alpha_J}{\beta_J}\right) + (1-P)(\hat{q}^0_{S/J} - \hat{q}_{S/I})\right] \tag{3.4}$$

determines the invasibility of the mutant.

From (3.4) it is easy to see that when transmission is completely global ($P = 1$) evolution maximises the basic reproductive rate $R_0 = (d+\alpha)/\beta$. The ESS is therefore the same as when we completely ignore spatial structure. Otherwise ($P < 1$) the ESS of the parasite is different from that in the completely mixing population and there is no simple quantity such as R_0 that is maximised. Local transmission ($P < 1$) favours the pathogen that increases the local density of its nearest neighbours that are susceptible ($q_{S/I}$). This in turn implies that a parasite with a lower basic reproductive rate can overcome one with a higher basic reproductive rate.

Therefore if infection is global ($P = 1$), the highest attainable transmission rate is predicted for the pathogen whether host reproduction is local or global (classical mean-field approximation) [**21**]. Simulation shows that when a proportion of the infection is local and a proportion global, we get the evolution of intermediate transmission rates between the two extremes. Figure 4 shows how an increased proportion of global infection leads to the evolution of higher virulence when there is a linear trade-off between transmission rate and virulence. These results were consistent for different trade-offs between transmission and virulence. Another related model [**6**] showed that varying the reproduction from local to global also affected the ES virulence, but that local infection with global reproduction did not

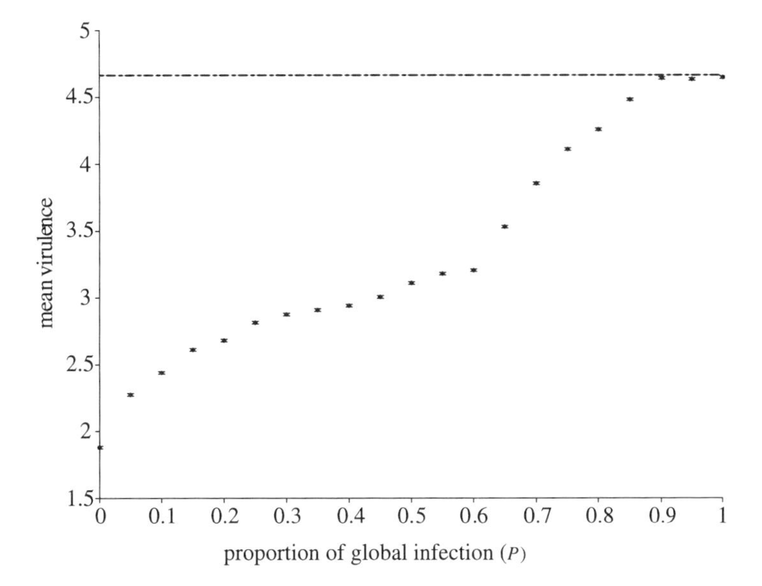

FIGURE 4. The ES virulence in an SI model where reproduction is local, but infection varies from completely local ($P = 0$) to completely global ($P = 1$). The maximum virulence is denoted by the dot/dash line, the mutation rate is 0.015, the reproductive rate (r) is 1.0, the natural deth rate (d) is 0.01 and the lattice size 150 by 150. The transmission rate (b) and the virulence is linked such that $\beta = 3\alpha^2$.

lead to the mean-field result of maximum transmission and virulence since it is the local nature of the infection process that constrains transmission.

The importance of this work is that it demonstrates that the effect of the spatial structure on the evolution of the parasite is not degenerate in the sense that it is only found for extreme spatial structure. This makes the application of the result much more robust as most systems are not completely local. There are other methods of creating models between local and global, including models of "small world" phenomena that have been examined in a number of contexts by models of static networks with a high degree of clustering and characteristically short path lengths between sites [**35**]. Boots and Sasaki's model differs from those of Watts and Strogatz [**35**] in that the connections between the different sites are dynamic. Since we are studying systems in evolutionary time, it may however be more realistic to adopt the dynamical approach. In addition, Ellner [**10**] developed multiscale pair approximations where different interactions occur over different scales and showed that it was often accurate enough to approximate all but the shortest-range interactions by a global interaction. The characterization of all interactions into either completely local or random as in [**6**] can also be argued to be a reasonable characterization for a number of natural systems where individuals tend to stay either remain locally or disperse. Other systems are better described by a description where individuals vary in the distance that they disperse. These systems can perhaps be more realisitically modelled in continuous space [**2, 3, 4, 5, 26, 27, 28**] where the location of each individual is tracked, although the resulting moment equations are the less analytically tractable partial- or integro differential equations.

4. Added realism: Biological and structural

The models discussed so far have been simple, both from a biological point of view and also in terms of the population structure that is assumed. Other models have examined some more complex biological scenarios and some more complex population structures. For example, the some of the implications of immune individuals in spatial populations has been examined recently. Space was again represented on a regular lattice of sites that can either be empty or occupied by an individual host. The state of each lattice is either empty (0), occupied by a susceptible host (S), occupied by the host infected by the strain j (I_j), ($j = 0, 1, \ldots, n-1$), or occupied by a recovered and immune host (R). Each site changes its state with the transition probabilities:

$$\begin{aligned}
&0 \to S \text{ with probability } (1-P)r[n(S)+n(R)]z + Pr[x(S)+x(R)],\\
&S \to I_j \text{ with probability } (1-P)\beta n(I_j)/z + P\beta x(I_j),\\
&S \to 0 \text{ with probability } d,\\
&I_j \to 0 \text{ with probability } d + \nu\kappa,\\
&I_j \to R \text{ with probability } \nu(1-\kappa),\\
&R \to 0 \text{ with probability } d,
\end{aligned}$$

where $n(S)$ and $n(R)$ respectively are the numbers of sites occupied by susceptible and recovered host in the nearest neighbor of the focal (empty) site while $x(S)$ and $x(R)$ are the global frequencies. Other parameters were:

κ: the case mortality (the probability that the infection ends by the mortality of host)
ν: the rate at which infection ends (either by host death or recovery)
r: the host reproductive rate
β: the transmission rate
P: the fraction of global transmission and reproduction
d: the disease free death rate

where ν and κ may vary with strains.

In addition it is assumed that the mutation occurs between adjacent pathogen strains

$$I_j \to I_{j\pm 1} \text{ with probability } u$$

(with $I_0 \to I_1$ and $I_{101} \to I_{100}$ at the limit). Note that in the classical SIR formulation the recovery rate is $\nu(1-\kappa)$ and the additional host mortality by infection, by $\nu\kappa$. We assumed a tradeoff between the mean infectious period and the case mortality of $\nu = \nu_0 + (\nu_1 - \nu_0)\kappa$ (where ν_0 and ν_1 are minimum and maximum recovery/removal rate).

By including spatial structure and acquired immunity in our model, we find the existence of bi-stability between avirulent and highly virulent strains, over a wide range of population structures (Figure 5). The bistability occurs with biologically reasonable parameter estimates when we have an accelerating trade-off (as virulence increases, recovery becomes as might be expected increasingly more difficult) between virulence and the recovery rate. The basic reproductive ratio of

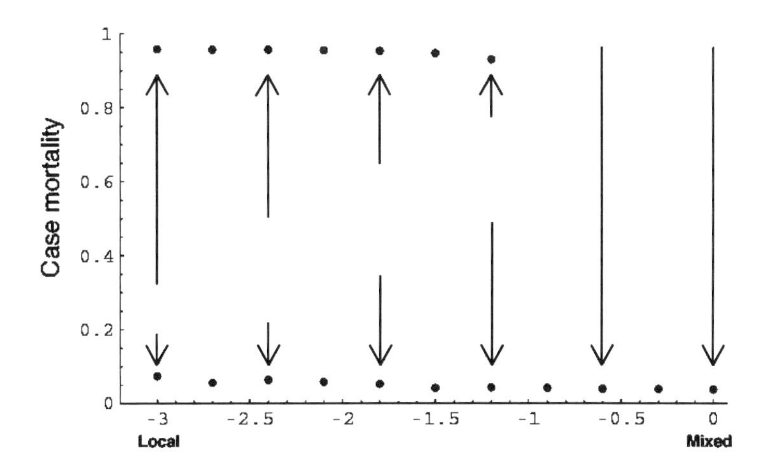

FIGURE 5. An example of the evolutionary dynamics in a spatial SIR model. Here we find evolutionary bi-stability in pathogen virulence expressed as case-mortality over a range of different population structures. The points represent locally evolutionarily stable case mortalities, while the degree of spatial structure is represented as the log of the proportion of global interactions. The arrows indicate the direction of evolution of the pathogen following local mutation from a particular initial virulence. There are 101 pathogen strains with the case mortalities equally divided between the harmless ($\alpha = 0$) and the lethal ($\alpha = 1$). The trade-off between the mean infectious period $1/\gamma$ and the case mortality α is $\gamma = 0.5 + 0.5\alpha$. The mutation rate in the case mortality between adjacent pathogen strains is $u = 0.5$. The mean case mortality at the evolutionary equilibrium is plotted as a function of the fraction PGlobal of global transmission/reproduction. The host reproductive rate $r = 5$, the natural mortality $d = 0.05$, the transmission rate $\beta = 5$.

the pathogen strain is independent of the case mortality α, such that if there is no trade off $R_0 = \beta/(d + \gamma)$, but with the trade-off assumed in Figure 5, R_0 attains the maximum ($R_{0|\alpha=0} = 9.09$) by minimizing the case mortality (the lowest R_0 is attained by the lethal pathogen, $R_{0|\alpha=1} = 4.76$). This corresponds to the simulation results for when all the interactions are global (PGlobal≈ 1), in which the minimum case mortality is globally evolutionarily stable. Once spatial structure is included, such that some of the interactions are local, very strong case mortality can also be locally evolutionarily stable.

The bi-stability arises because the avirulent strain is favored in a dense viscous population, while the virulent strain is favored in a more open one. Higher virulence may pay a pathogen strain in a sparse population as killing the host prevents immune individuals blocking the spread of the strain. Once a highly virulent strain is established in a sparse population, a less virulent strain will tend to block mostly itself by producing immune individuals and therefore not replace the resident. In contrast in a dense population, many individuals are already immune due to recovery from avirulent parasite strains. A more virulent strain invading such a population would kill itself, but still be blocked by immune individuals created by other strains, and would not therefore invade. Before any particular population structure was established, an initially virulent strain will tend to create a sparse population due to the mortality they cause and therefore favor even more virulent strains. Conversely, relatively avirulent strains will tend to maintain viscous populations thereby selecting for less virulence. Since both states are locally evolutionarily stable, the pathogen evolves to either low or high virulence, depending on the initial virulence.

Another example of additional biological realism is a model where parasites infect via free-living infective stages [**17**]. The idea that parasites with long-lived infective stages may evolve higher virulence has received considerable attention (see Bonhoeffer, Gandon and Day et al., this volume). This idea is called 'the curse of the pharaoh' because of the hypothesis that the death of Lord Carnavon was caused by very long-lived propagules of a highly virulent infectious disease. When examined in a spatial context, if virulence evolves independently of transmission, long-lived infective stages can select for higher virulence. There is always the evolution of a finite transmission rate, which becomes higher when the infective stages are shorter lived. When a trade-off occurs between transmission and virulence, they show that there is no evidence for the curse of the pharaoh. Indeed, higher transmission and therefore virulence may be selected for by shorter rather than long-lived infective stages.

The pattern of contacts in many systems and in particular human populations is characterised by social networks rather than by simple spatial relationships. At a very fundamental level, networks of social contacts differ from spatial networks in that the connections between individuals are often highly irregular, with a few core individuals have very many contacts. In a social network the number of contacts varies between individuals leading to a wealth of possible population structures. The implications of this wide diversity of different networks to the evolutionary dynamics of parasites is as yet little known, although [**31**] have compared evolution on highly aggregated local irregular networks, with many social cliques, and global irregular networks dominated by random long-distance connections and few cliques using simulations.

5. Mathematical analysis by approximation: Limitations and directions

As described above pair approximations are very useful in predicting the equilibrium states of the full spatial simulations. It is also possible to write down the approximate invasion dynamics of rare mutants and therefore calculate the value of a pair approximated ESS if it exists. Indeed since we have the approximate invasion criteria, we can use the techniques of adaptive dynamics [**12, 13, 24, 25**] to predict evolutionary outcomes. Here we produce analytical pair-wise invasibility plots (PIP) through approximating the invasion criteria of the Haraguchi and Sasaki model (see [**15**]) and compare it with the outcomes found in a full evolutionary simulation and an equivalent PIP drawn by direct simulation of the invasions. In this PIP the spatial simulations were carried out in the same way as Haraguchi and Sasaki [**15**] but with only one resident strain initially. Initial conditions are randomly assigned and after the simulation reaches an equilibrium, 10% of the randomly selected resident strain mutate. Then the simulation continues until either one of the strains goes extinct. This procedure is repeated for all combinations of transmission rates.

Approximated PIPs can be obtained analytically by the following procedures. The idea has been described in the Section 3 of this chapter and in [**34**]. We first extract the 'fast variable' dynamics from the linearlized mutant pair density dynamics, then obtain the equilibrium local densities in the neighborhood of the mutant ($q_{0/J}$, $q_{S/J}$, $q_{I/J}$ and $q_{J/J}$), and finally evaluate the slow variable dynamics (for mutant global densities) to see whether or not the mutant can increase when rare.

Linearized pair density dynamics for the mutants when they are rare are

$$
\begin{aligned}
\dot{P}_{J0} =& -(1+\alpha)P_{J0} - r(1-\theta)q_{S/0}P_{J0} + \beta_J(1-\theta)q_{J/S}P_{S0} \\
& + P_{JS} + (1+\alpha)(P_{IJ} + P_{JJ}) \\
\dot{P}_{JS} =& - P_{JS} - (1+\alpha)P_{JS} - \left[\beta_I(1-\theta)q_{I/S} + \beta_J\theta\right] P_{JS} \\
& + r(1-\theta)q_{S/0}P_{J0} + \beta_J(1-\theta)q_{J/S}P_{SS} \\
\dot{P}_{IJ} =& -2(1+\alpha)P_{IJ} + \beta_I(1-\theta)q_{I/S}P_{JS} + \beta_J(1-\theta)q_{J/S}P_{IS} \\
\dot{P}_{JJ} =& -2(1+\alpha)P_{JJ} + 2\beta_J\theta P_{JS}.
\end{aligned}
\tag{5.1}
$$

Note that variables without J are at their equilibrium values of the wildtype-endemic population (Equations (2.7)–(2.12)). Using the relationship, $P_{Jx} = \rho_J q_{x/J}$ (where $x = \{0, S, I, J\}$), we have

$$
\dot{q}_{x/J} = \frac{1}{\rho_J}\dot{P}_{Jx} - \frac{1}{\rho_J}\dot{\rho}_J q_{x/J},
\tag{5.2}
$$

where $\dot{\rho}_J$ is defined in (2.13).

We apply this relationship to obtain the dynamics for the local densities of a mutant-infected host. With noting $\rho_x q_{J/x} = \rho_J q_{x/J}$ we have

$$
\begin{aligned}
\dot{q}_{0/J} =& -(1+\alpha)q_{0/J} - r(1-\theta)q_{S/0}q_{0/J} + \beta_J(1-\theta)q_{0/S}q_{S/J} \\
& + q_{S/J} + (1+\alpha)(q_{I/J} + q_{J/J}) - \left[\beta_J q_{S/J} - (1+\alpha)\right] q_{0/J} \\
\dot{q}_{S/J} =& -(2+\alpha)q_{S/J} + \beta_J\left[(1-\theta)q_{S/S} + \{\theta + (1-\theta)q_{J/S}\}\right] q_{S/J} \\
& + r(1-\theta)q_{S/0}q_{0/J} - \beta_J(1-\theta)q_{I/S}q_{S/J} - \left[\beta_J q_{S/J} - (1+\alpha)\right] q_{S/J} \\
\dot{q}_{I/J} =& -2(1+\alpha)q_{I/J} + (\beta_J + \beta_I)(1-\theta)q_{I/S}q_{S/J} \\
& - \left[\beta_J q_{S/J} - (1+\alpha)\right] q_{I/J} \\
\dot{q}_{J/J} =& -2(1+\alpha)q_{J/J} + 2\beta_J\theta q_{S/J} - \left[\beta_J q_{S/J} - (1+\alpha)\right] q_{J/J},
\end{aligned}
\tag{5.3}
$$

where the local densities other than $q_{x/J}$ (where $x = \{0, S, I\}$) in the right hand sides are evaluated at the stable equilibrium in the resident population. The stable equilbiurm value $\hat{q}^0_{x/J}$ of $q_{x/J}$ of this dynamics is then substituted into (2.13), and its sign determines the invasibility of the mutant.

Figure 6a shows a PIP produced by simulation with the results of 20 replicates. In this figure, the frequencies that the mutants failed to invade in 20 independent runs is shown in a gray scale (it's black if the mutant failed to invade 20 times out of 20 replicates; white, if it could always invade). We can see that the PIP predicts a continuously stable, and evolutionarily stable transmission rate around $\beta = 25$,

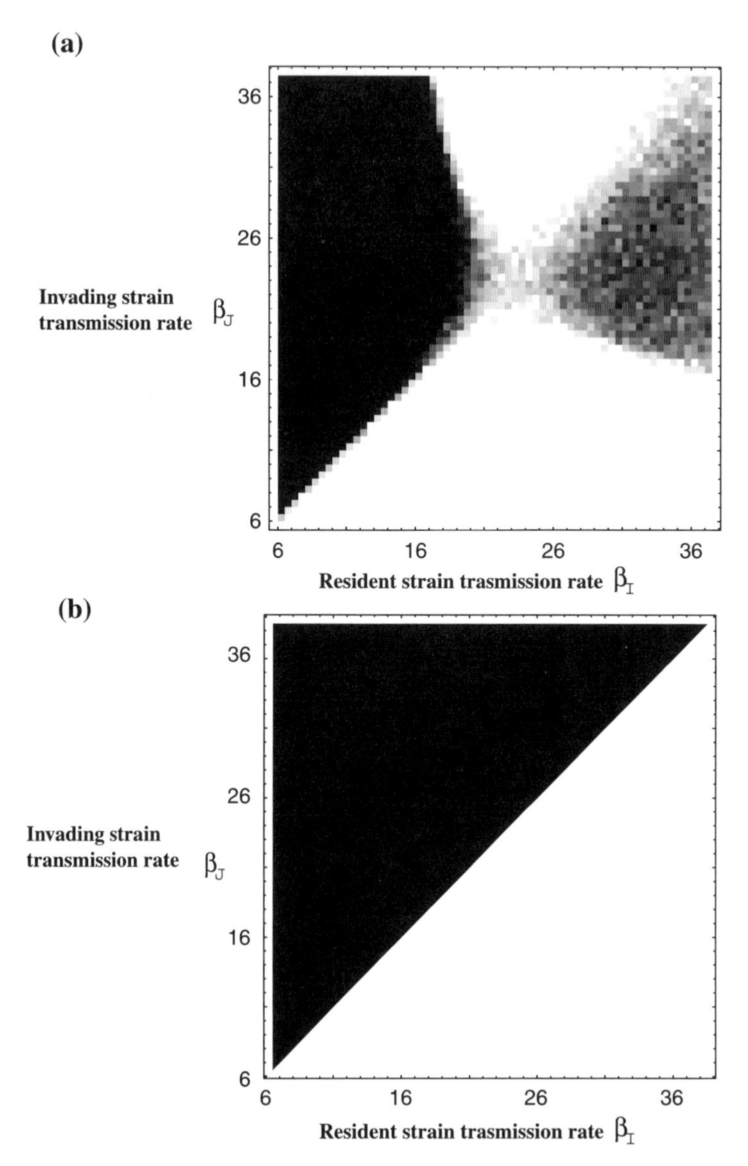

FIGURE 6. Spatial pairwise invasibility plots (PIPs) of the model [**15**] described in the text. Figure **(a)** illustrates a PIP obtained by simulation of the system that predicts the ESS transmission rate well. In contrast, the PIP in **(b)** which is obtained by the approximation the analysis fails. In both figures, white indicates the failure of an invasion of the mutant strain. In (a), the results of 20 replicates are shown in gray scale. If the mutant strain fails 20 times to invade among 20 replicates, it is shown at its lightest. Lattice size is 100×100. Parameters are the same as Figure 2a ($\alpha = 1$, $r = 8$) in [**15**]. They predict that the ESS transmission rate is around $\beta = 25$, which is also predicted in the invasion simulation PIP. Near the ESS transmission rate, since small difference of fitnesses between two strains are masked by demographic stochastisity, a mutant strain which has a smaller density than the resident is more likely to be excluded from the population.

which is the ESS transmission rate obtained by Haraguchi and Sasaki [**15**]. Around the ESS transmission rate, the gray scale tends to be darker. This blur is due to the random loss of mutants by demographic stochasticity, which is more exaggerated near the ESS where the mutant's invasition fitness is only slightly larger or smaller than that of the resident.

Figure 6b illustrates the results of the analytical PIP produced by pair approximation. This figure simply says that if the mutant strain has a higher transmission rate than the resident, it can invade. As such the PIP obtained by pair approximation does not predict the outcome of the full evolutionary simulation described in [**15**]. We cannot specify the reason why the pair approximation fails to predict ESS transmission rate. One possibility is that the local density around hosts infected by invading mutant parasites is not accurately evaluated. Another is that small deviations in the equilibrium spatial correlations of the resident population may cause realtively large deviations in the invisibility criteria. However this remains an open question.

6. Summary

In summary, there is little doubt that spatial structure in host populations can have dramatic effects on the evolution of parasites. Transmission can be constrained without trade-offs and there is the possibility of multiple stable states due to interactions with self-generated population structures. Models intermediate between local and the mean-field have demonstrated that these effects are not only the result of extreme local interactions and may therefore be widely applicable. Advances in this field have mostly been driven by the use of computer simulation, but moment closure approximations offer an opportunity to develop our analytical understanding of these processes. However, advances in these approximation techniques are required at least for models with extreme local interactions.

References

[1] R. M. Anderson, Populations and Infectious-Diseases – Ecology or Epidemiology – the 8th Tansley Lecture, *J. Animal Ecology* **60**(1) (1991), 1–50.

[2] B. Bolker and S. W. Pacala, Using moment equations to understand stochastically driven spatial pattern formation in ecological systems, *Theor. Pop. Biol.* **52**(3) (1997), 179–197.

[3] B. M. Bolker, Analytic models for the patchy spread of plant disease, *Bulletin Math. Biol.* **61**(5) (1999), 849–874.

[4] B. M. Bolker, Combining endogenous and exogenous spatial variability in analytical population models, *Theor. Pop. Biol.* **64**(3) (2003), 255–270.

[5] B. M. Bolker and S. W. Pacala, Spatial moment equations for plant competition: Understanding spatial strategies and the advantages of short dispersal, *Am. Nat.* **153**(6) (1999), 575–602.

[6] M. Boots and A. Sasaki, Small worlds' and the evolution of virulence: infection occurs locally and at a distance, *Proc. R. Soc. Lond. B* **266**(1432) (1999), 1933–1938.

[7] M. Boots and A. Sasaki, Parasite evolution and extinctions, *Ecology Letters* **6**(3) (2003), 176–182.

[8] H. J. Bremermann and H. R. Thieme, A Competitive-Exclusion Principle for Pathogen Virulence, *J. Math. Biol.* **27**(2) (1989), 179–190.

[9] D. Claessen and A. M. deRoos, Evolution of virulence in a host-pathogen system with local pathogen transmission, *Oikos* **74**(3) (1995), 401–413.

[10] S. P. Ellner, Pair approximation for lattice models with multiple interaction scales, *J. Theor. Biol.* **210**(4) (2001), 435–447.

[11] S. A. Frank, A kin selection model for the evolution of virulence, *Proc. R. Soc. Lond. B* **250** (1992), 195–197.

[12] S. A. H. Geritz, E. Kisdi, et al., Evolutionarily singular strategies and the adaptive growth and branching of the evolutionary tree, *Evol. Ecol.* **12**(1) (1998), 35–57.
[13] S. A. H. Geritz, J. A. J. Metz, et al., The dynamics of adaptation and evolutionary branching, *IIASA Working Paper WP-96-77* (1996).
[14] W. M. Getz and J. Pickering, Epidemic Models – Thresholds and Population Regulation, *Am. Nat.* **121**(6) (1983), 892–898.
[15] Y. Haraguchi and A. Sasaki, The evolution of parasite virulence and transmission rate in a spatially structured population, *J. Theor. Biol.* **203**(2) (2000), 85–96.
[16] M. Katori and N. Konno, Upper bounds for survival probability of the contact process, *J. Statistical Physics* **63** (1991), 115–130.
[17] M. Kamo and M. Boots, The curse of the pharaoh in space: free-living infectious stages and the evolution of virulence in spatially explicit populations, *J. Theor. Biol.* **231**(3) (2004), 435–441.
[18] R. E. Lenski and R.M. May, The Evolution of Virulence in Parasites and Pathogens - Reconciliation between 2 Competing Hypotheses, *J. Theor. Biol.* **169**(3) (1994), 253–265.
[19] M. J. Mackinnon and A. F. Read, Genetic relationships between parasite virulence and transmission in the rodent malaria Plasmodium chabaudi, *Evolution* **52**(3) (1999), 689–703.
[20] H. Matsuda, N. Ogita, et al., Statistical-Mechanics of Population - the Lattice Lotka-Volterra Model, *Progress of Theoretical Physics* **88**(6) (1992), 1035–1049.
[21] R. M. May and R. M. Anderson, Epidemiology and genetics in the coevolution of parasites and hosts, *Proc. R. Soc. Lond. B* **219** (1983), 281–313.
[22] R. M. May and R. M. Anderson, Epidemiology and genetics in the coevolution of parasites and hosts, *Proc. R. Soc. Lond. A* **390**(1798) (1983), 219–219.
[23] R. M. May and M. A. Nowak, Superinfection, metapopulation dynamics, and the evolution of diversity, *J. Theor. Biol.* **170**(1) (1994), 95–114.
[24] J. A. J. Metz, S. A. H. Geritz, et al., Adaptive dynamics: A geometrical study of the consequences of nearly faithful reproduction, in (S.J. van Strien and S. M. V. Lunel, eds.), *Dynamical Systems and Their Applications*, Elsevier, North Holland, 1995, 147–194.
[25] J. A. J. Metz, R. M. Nisbet, et al., How should we define fitness for general ecological scenarios, *Trends Ecol. Evol.* **7**(6) (1992), 198–202.
[26] D. J. Murrell, U. Dieckmann, et al., On moment closures for population dynamics in continuous space, *J. Theor. Biol.* **229**(3) (2004), 421–432.
[27] D. J. Murrell, D. W. Purves, et al., Uniting pattern and process in plant ecology, *Trends Ecol. Evol.* **16**(10) (2001), 529–530.
[28] D. J. Murrell, J. M. J. Travis, et al., The evolution of dispersal distance in spatially-structured populations, *Oikos* **97**(2) (2002), 229–236.
[29] M. A. Nowak and R. M. May, Superinfection and the evolution of parasite virulence, *Proc. R. Soc. Lond. B* **255**(1342) (1994), 81–89.
[30] D. A. Rand, M. Keeling, et al., Invasion, stability and evolution to criticality in spatially extended, artificial host-pathogen ecologies, *Proc. R. Soc. Lond. B* **259**(1354) (1995), 55–63.
[31] J. M. Read and M. J. Keeling, Disease evolution on networks: the role of contact structure, *Proc. R. Soc. Lond. B* **270**(1516) (2003), 699–708.
[32] C. J. Rhodes and R. M. Anderson, Dynamics in a lattice epidemic model, *Physics Letters A* **210**(3) (1996), 183–188.
[33] A. Sasaki and Y. Iwasa, Optimal-growth schedule of pathogens within a host – switching between lytic and latent cycles, *Theor. Pop. Biol.* **39**(2) (1991), 201–239.
[34] K. Sato, H. Matsuda, et al., Pathogen invasion and host extinction in lattice structured populations, *J. Math. Biol.* **32**(3) (1994), 251–268.
[35] D. J. Watts and S. H. Strogatz, Collective dynamics of 'small-world' networks, *Nature* **393** (1998), 440–442.

Department of Animal and Plant Sciences, The University of Sheffield, Alfred Denny Building, Western Bank, Sheffield S10 2TN, UK
E-mail address: m.boots@sheffield.ac.uk

Advanced Industrial Science and Technology Research Center for Chemical Risk Management, Tsukuba Onogawa 16-1 305-8569, JAPAN
E-mail address: masashi-kamo@aist.go.jp

Department of Biology, Faculty of Science, Kyushu University Graduate Schools, Fukukoka 812-8581, JAPAN
E-mail address: asasascb@mbox.nc.kyushu-u.ac.jp

DIMACS Series in Discrete Mathematics
and Theoretical Computer Science
Volume **71**, 2006

Insights from Price's equation into evolutionary epidemiology

Troy Day and Sylvain Gandon

ABSTRACT. We present an alternative theoretical framework for modeling the evolutionary and epidemiological dynamics of host-parasite interactions that is based on using the instantaneous rate of change of infected hosts as a measure of pathogen fitness, rather than the more commonly used quantity, R0. This alternative approach leads to a number of re-interpretations of predictions derived from previous theory, and it thereby provides a more thorough perspective on how various factors affect pathogen evolution. It also provides a relatively straightforward approach for modeling the dynamics of evolutionary change in pathogen populations when it cannot be assumed that the epidemiological dynamics occur on a time scale that is fast relative to that of the evolutionary dynamics.

1. Introduction

The basic reproduction number, denoted by R_0, is one of the most important quantities in epidemiological theory [**11, 23**]. It is defined as the expected number of new infections generated by an infected individual in an otherwise wholly susceptible population [**2, 12, 23**]. Part of the reason why R_0 plays such a central role in this body of theory undoubtedly stems from its relatively simple and intuitively sensible interpretation as a measure of pathogen reproduction. If R_0 is less than unity then we expect the pathogen to die out since each infected individual fails to generate at least one other infection during the lifetime of the infection.

Given that R_0 is a measure of pathogen reproductive success, it is not surprising that this quantity has also come to form the basis of most evolutionary considerations of host-pathogen interactions [**1, 18**]. For example, mathematical models for numerous epidemiological settings have been used to demonstrate that natural selection is often expected to favour the pathogen strain that results in the largest value of R_0 [**6, 18**]. In more complex epidemiological settings such optimization criteria typically cannot be derived and instead a game-theoretic approach is taken [**5**]. In this context a measure of the fitness of a rare mutant pathogen strain is used to characterize the evolutionarily stable strain (i.e., the strain that, if present within the population in sufficient numbers, cannot be displaced by any mutant strain that

Key words and phrases. Virulence, Infectious disease, Evolution, Quasispecies, Disease.

arises). Typically R_0 again plays a central role as the measure of mutant fitness in such invasion analyses [**10, 18, 30**].

In this chapter we consider an alternative approach for developing theory in evolutionary epidemiology. Rather than using the total number of new infections generated by an infected individual (i.e., R_0) as a measure of pathogen fitness we use the instantaneous rate of change of the number of infected hosts instead (see also [**3, 18**]). This shifts the focus from a consideration of pathogen reproductive success per generation to pathogen reproductive success per unit time. One very useful result of this change in focus is that we can then model the time dynamics of evolutionary change in the pathogen population simultaneously with the epidemiological dynamics, rather than simply characterizing the evolutionary equilibria that are expected. Even more importantly, however, this seemingly slight change in perspective can lead to a very different interpretation of what drives pathogen evolution than has been obtained from theory based on R_0. Our contention is thus that this alternative perspective provides a useful complement to theory based on R_0 and that it can sometimes yield fresh insights into old questions regarding the evolution of host-pathogen interactions.

Our approach is closely related to that of Day and Proulx [**10**] who developed theory based on the assumption that there is a continuum of pathogen strains. In that publication Day and Proulx likened their approach to quantitative genetic models. Here, however, we assume that there are a finite number of discrete pathogen strains in the population. This provides a much simpler and more general theoretical framework and it allows us to readily extend the approach to models for the evolution of multiple pathogen traits, as well as models for pathogens that can infect multiple host types. This discrete strain approach also provides a simple framework in which models of virulence evolution as well as models of antigenic site evolution and host immunity can be combined [**22**].

2. Price's equation

Price's equation has been used extensively in evolutionary biology to model the dynamics of allele frequencies [**7, 17, 28, 29**]. Although the original version of this equation was presented in discrete-time, here we provide a simple and very general continuous-time derivation of Price's equation that also accounts for mutation from one type to another. Similar derivations can be conducted in discrete-time. We then extend these results to allow for multiple habitats.

2.1. Mathematics of selection and mutation. Let's put all epidemiological considerations aside for the moment, and consider a population of asexual individuals of which there are n distinct strains. For example, these might be bacterial strains, asexual *Daphnia* clones, or strains of any other asexual species. Our goal here is derive an equation for the rate of change of the frequency of each strain under the action of natural selection and mutation, in the form of Price's equation. The rate of change of the number of individuals of strain i is assumed to be governed by the following equation:

$$\dot{N}_i = r_i N_i - \mu N_i + \mu \sum_{j=1}^{n} m_{ji} N_j, \tag{2.1}$$

where r_i is the per capita rate of change of strain i (sometimes referred to as its fitness), μ is the mutation rate for all strains, and m_{ji} is the probability that, if a mutation occurs in an individual of strain j, it mutates to an individual of strain i. Although equations (2.1) are perfectly fine descriptions of the evolutionary dynamics of this system, evolutionary biologists are often more interested in the frequency of different strains rather than their absolute numbers since evolutionary change is defined as a change in the frequency of different alleles. Equation (2.1) can be readily used to derive an equation for the rate of change of the frequency of strain i, defined as $q_i = N_i/N_T$ where $N_T = \sum_{i=1}^{n} N_i$. We have

$$\begin{aligned} \dot{q}_i &= \frac{\dot{N}_i}{N_T} - q_i \frac{\dot{N}_T}{N_T} \\ &= q_i \left(r_i - \bar{r}\right) - \mu q_i + \mu \sum_{j=1}^{n} m_{ji} q_j \end{aligned} \tag{2.2}$$

where $\bar{r} = \sum_{j=1}^{n} r_j q_j$ is the average rate of change (i.e., average fitness) of all strains in the population.

Equation (2.2) is the fundamental equation for the rate of change of strain frequencies. Now suppose that we are interested in the evolution of a particular trait, x. An equation for the rate of evolutionary change in the average value of x can be derived as

$$\dot{\bar{x}} = \sum_{i=1}^{n} x_i \dot{q}_i, \tag{2.3}$$

which, using equation (2.2) gives

$$\begin{aligned} \dot{\bar{x}} &= \left(\sum_{i=1}^{n} q_i x_i r_i - \bar{r}\bar{x}\right) - \mu \left(\bar{x} - \sum_{i,j} x_i m_{ji} q_j\right) \\ &= \operatorname{cov}\left(x, r\right) - \mu \left(\bar{x} - \bar{x}_m\right). \end{aligned} \tag{2.4}$$

In equations (2.4) $\operatorname{cov}(x, r)$ is the covariance between x and r across all strains, and $\bar{x}_m = \sum_{i,j} x_i m_{ji} q_j$, which is the average trait value of all of mutations that arise. Equation (2.4) is Price's equation with mutation, and it (or its extension: see next section) is the central equation of interest in this chapter.

Price's equation (2.4) has a simple and informative interpretation. The average trait value in the population changes as a result of two different processes. First, the average trait value changes in a direction given by the sign of the covariance between the trait and fitness; if strains with large values of x also have a large fitness, r, then this covariance will be positive and natural selection drives the average of x to higher values. Second, the average trait value changes in a direction governed by any mutational bias that might occur (the second term in (2.4)). Specifically, if the average trait value of mutants that arise is larger than that of the population as a whole at any given time, then the second term in (2.4) will be positive, leading to an evolutionary increase in $\bar{x}$ through mutational bias.

It is only when the average value of the mutants that arise is the same as that of the population that mutation has no directional effect on evolution. Although this is a theoretical possibility, in reality some degree of mutational bias is always expected. For example, even when the mutation rates among the different strains are all equal (i.e., $m_{ji} = 1/n$) we have $\bar{x}_m = \sum_i \frac{1}{n} x_i$, which is the average trait value if all strains were at equal frequency. Consequently, even when mutation is unbiased among the different strains, it will nevertheless impart a directional effect on the mean value of any given trait because selection on this trait will usually not favour equal frequencies of all strains (and hence $\bar{x} \neq \bar{x}_m$).

2.2. An extension to multiple habitats. Before connecting the above results to evolutionary epidemiology, we first generalize equation (2.4) to allow for multiple habitats. For example, a bacterial population of interest might exist in different habitats connected by migration. Similarly, a *Daphnia* population might inhabit different parts of a lake connected by migration.

To simplify the presentation we consider only two habitats, labeled A and B. The extension to an arbitrary number of habitats will become obvious from this case. The analogs of equation (2.1) for each of the two habitats are

$$\dot{N}_i^A = r_i^{AA} N_i^A - \mu N_i^A + \mu \sum_{j=1}^{n} m_{ji} N_j^A + r_i^{BA} N_i^B \tag{2.5}$$

$$\dot{N}_i^B = r_i^{BB} N_i^{BB} - \mu N_i^B + \mu \sum_{j=1}^{n} m_{ji} N_j^B + r_i^{AB} N_i^A. \tag{2.6}$$

In equations (2.5) and (2.6) r_i^{AA} is the per capita rate of production of offspring of strain i by such an individual in habitat A, and that end up in habitat A. On the other hand, r_i^{BA} is the per capita rate of production of offspring of strain i by such an individual in habitat B, that end up in habitat A. Analogous interpretations apply to r_i^{BB} and r_i^{AB}.

From equations (2.5) and (2.6) we can derive the analogues of equation (2.2) as

$$\begin{aligned}\dot{q}_i^A =& \frac{\dot{N}_i^A}{N_T^A} - q_i^A \frac{\dot{N}_T^A}{N_T^A} \\ =& \frac{r_i^{AA} N_i^A - \mu N_i^A + \mu \sum_{j=1}^{n} m_{ji} N_j^A + r_i^{BA} N_i^B}{N_T^A} \\ &- q_i^A \frac{\sum_i \left(r_i^{AA} N_i^A - \mu N_i^A + \mu \sum_{j=1}^{n} m_{ji} N_j^A + r_i^{BA} N_i^B \right)}{N_T^A}\end{aligned}$$

which yields

$$\begin{aligned}\dot{q}_i^A =& r_i^{AA} q_i^A - \mu q_i^A + \mu \sum_{j=1}^{n} m_{ji} q_j^A + \frac{N_T^B}{N_T^A} r_i^{BA} q_i^B - q_i^A \left(\bar{r}^{AA} + \frac{N_T^B}{N_T^A} \bar{r}^{BA} \right) \\ =& q_i^A \left(r_i^{AA} - \bar{r}^{AA} \right) - \mu q_i^A + \mu \sum_{j=1}^{n} m_{ji} q_j^A + \frac{N_T^B}{N_T^A} \left(r_i^{BA} q_i^B - q_i^A \bar{r}^{BA} \right) \\ =& q_i^A \left(r_i^{AA} - \bar{r}^{AA} \right) - \mu q_i^A + \mu \sum_{j=1}^{n} m_{ji} q_j^A + \frac{N_T^B}{N_T^A} q_i^B \left(r_i^{BA} - \bar{r}^{BA} \right) \\ &+ \frac{N_T^B}{N_T^A} \bar{r}^{BA} \left(q_i^B - q_i^A \right)\end{aligned} \tag{2.7}$$

with an analogous equation for habitat B (not shown). Note that there are two different probability distributions used in the averages calculated in (2.7). For example, the average $\bar{r}^{BA}$ is calculated over the distribution q_i^B whereas the average $\bar{r}^{AA}$ is calculated over the distribution q_i^A.

Finally, we can calculate the equation for the dynamics of the average value of trait x, specific to each habitat as $\dot{\bar{x}}^A = \sum_{i=1}^n x_i \dot{q}_i^A$ and $\dot{\bar{x}}^B = \sum_{i=1}^n x_i \dot{q}_i^B$. This gives

$$\begin{aligned}\dot{\bar{x}}^A =& \sum_i x_i q_i^A \left(r_i^{AA} - \bar{r}^{AA}\right) - \sum_i x_i \left(\mu q_i^A - \mu \sum_{j=1}^n m_{ji} q_j^A\right) \\ &+ \frac{N_T^B}{N_T^A} \sum_i x_i q_i^B \left(r_i^{BA} - \bar{r}^{BA}\right) + \frac{N_T^B}{N_T^A} \sum_i x_i \bar{r}^{BA} \left(q_i^B - q_i^A\right),\end{aligned} \tag{2.8}$$

or

$$\dot{\bar{x}}^A = \operatorname*{cov}_A\left(x, r^{AA}\right) - \mu\left(\bar{x}^A - \bar{x}_m^A\right) + \frac{N_T^B}{N_T^A} \operatorname*{cov}_B\left(x, r^{BA}\right) + \frac{N_T^B}{N_T^A} \bar{r}^{BA}\left(\bar{x}^B - \bar{x}^A\right). \tag{2.9}$$

Analogously, we have

$$\dot{\bar{x}}^B = \operatorname*{cov}_B\left(x, r^{BB}\right) - \mu\left(\bar{x}^B - \bar{x}_m^B\right) + \frac{N_T^A}{N_T^B} \operatorname*{cov}_A\left(x, r^{AB}\right) + \frac{N_T^A}{N_T^B} \bar{r}^{AB}\left(\bar{x}^A - \bar{x}^B\right). \tag{2.10}$$

Equations (2.9) and (2.10) are the multiple habitat versions of Price's equation (2.4), and they also have a useful interpretation. Let's focus on (2.9) (equation (2.10) can be interpreted analogously). The average trait value in habitat A changes as a result of four processes, corresponding to the four terms in (2.9). The first two terms are analogous to those of equation (2.4) and represent natural selection and mutational bias specific to habitat A. The third and fourth terms represent the evolutionary change in habitat A that results from migration of individuals from habitat B to habitat A.

Beginning with the fourth term, migration into habitat A causes evolutionary change in $\bar{x}^A$ if the mean trait value differs in the two habitats. The factor $\bar{r}^{BA}$ represents the average per capita rate of such migration, and this is weighted by the relative population sizes of the two habitats, N_T^B/N_T^A, to obtain the absolute effect of migration on the mean trait value. Furthermore, these migrants will cause an evolutionary increase in $\bar{x}^A$ if they have an average trait value larger than that of $\bar{x}^A$ and vice versa. This accounts for the factor $(\bar{x}^B - \bar{x}^A)$ in the fourth term.

Turning to the third term in equation (2.9), migration can also cause an evolutionary effect on $\bar{x}^A$ even if the average trait values in the two habitats are the same (in which case the fourth term disappears). In particular, if those individuals that migrate tend to have higher than average trait values, then this will drive $\bar{x}^A$ toward higher values and vice versa. This is represented by the covariance term $\operatorname*{cov}_A(x, r^{AB})$. Again this is weighted by the relative population sizes of the two habitats, N_T^B/N_T^A, to obtain the absolute effect of migration on the mean trait value.

3. Applying Price's equation to epidemiological models

The above equations are quite general, and track the dynamics of any collection of asexually reproducing entities. Of primary interest here is using this formalism to model the evolutionary dynamics of pathogen populations. To do so, we view the pathogen population from the perspective of infected hosts, and interpret q_i

as the frequency of all infected hosts that harbour a pathogen of strain i. This implicitly assumes that a host contains at most a single pathogen strain at any given time. Furthermore, mutation from strain j to strain i in the above formalism then corresponds to a host infected with strain j "becoming" a host infected with strain i.

Such transitions between infection types are assumed to take place as a result of two processes. First, a mutant pathogen strain must arise within an already infected host. Second, it is assumed that competition between these two pathogen strains then results in competitive exclusion. Thus, as is common in many models of evolutionary epidemiology, we assume that a polymorphism is never maintained within a host [**4, 27**]. In fact, to simplify matters further we assume that the competitive exclusion is effectively instantaneous (i.e., we assume superinfection [**27**]). Clearly these assumptions neglect some features of the reality of host-parasite interactions, but they have been used successfully in previous theory to provide considerable insight into evolutionary epidemiology [**4, 21, 22, 25, 27**]. Given the above assumptions, we next need to specify the parameters m_{ji} and r_i.

The parameter m_{ji} is the probability that an infection with strain j undergoes a transition to an infection with strain i. This can be decomposed into the product to two factors: (i) the probability that a strain j pathogen mutates into a strain i pathogen within an infected host, and (ii) the probability that, given strain i has appeared in the host, it competitively excludes the pathogen of strain j. There is no *a priori* reason to expect a particular bias in any of the mutations that arise, and therefore we assume that factor (i) is simply $1/n$ (i.e., strain j gives rise to all other strains with equal probability). Factor (ii) will depend on any competitive asymmetry between strains within a host, and we denote this probability by ρ_{ji}.

The parameter r_i is the per capita rate of change of hosts infected with strain i. This will be determined by the epidemiological dynamics of the host population. To specify r_i we must therefore specify the epidemiological model that is assumed to describe the between-host dynamics of the pathogen. We consider three examples.

3.1. A Simple S-I-R Model. As a very simple example, consider a standard SIR description of the epidemiological dynamics. Specifically, if S, I, and R and the densities of susceptible, infected, and recovered and immune hosts, we suppose

$$\dot{S} = \theta - dS - \beta SI \tag{3.1}$$

$$\dot{I} = \beta SI - dI - vI - cI \tag{3.2}$$

$$\dot{R} = cI - dR, \tag{3.3}$$

where θ is a constant immigration rate of susceptible hosts, d is the per capita mortality rate of hosts in the absence of infection, β is the transmission rate of the infection from infected to susceptible hosts, v is the pathogen-induced host mortality rate (i.e., virulence [**8**]), and c is the per capita rate of clearance of the infection through host defense mechanisms.

Equations (3.1)–(3.3) implicitly assume that there is only one parasite strain circulating in the host population. To connect this to the earlier results for the evolutionary dynamics of the pathogen population we need to consider how this epidemiological model is extended to multiple strains. If I_i is the number of hosts infected with strain i, then (3.1)–(3.3) extend to

$$\dot{S} = \theta - dS - S\sum_i \beta_i I_i \tag{3.4}$$

$$\dot{I}_i = S\beta_i I_i - dI_i - v_i I_i - cI_i \tag{3.5}$$

$$\dot{R} = \sum_i cI_i - dR. \tag{3.6}$$

Equations (3.4)–(3.6) assume that once a host is infected, it becomes immune to all reinfection regardless of strain type. They also assume that the clearance rate, c, is the same for all strains and that immunity is cross specific.

From equation (3.5) we can now readily identify the per capita rate of change of each strain. Writing this equation as $\dot{I}_i = I_i(S\beta_i - d - v_i - c)$, we can see that $r_i = S\beta_i - d - v_i - c$. We can now apply equation (2.4) to obtain the following equation for the evolutionary dynamics of the two characteristics of the pathogen that are assumed to vary between strains (i.e., the virulence, v and the transmission rate, β). We obtain

$$\dot{\bar{v}} = S\sigma_{\beta v} - \sigma_{vv} - \mu\left(\bar{v} - \bar{v}_m\right) \tag{3.7}$$

$$\dot{\bar{\beta}} = S\sigma_{\beta\beta} - \sigma_{\beta v} - \mu\left(\bar{\beta} - \bar{\beta}_m\right). \tag{3.8}$$

Here σ_{xy} is the covariance between x and y across the pathogen strains that are circulating in the population. We can also use equations (3.4)–(3.6) to obtain a system of equations governing the total number of susceptible, infected, and recovered individuals. Defining $I = \sum_i I_i$, we can write the summation in (3.4) as $SI\sum_i \beta_i I_i/I$ or $SI\bar{\beta}$ where $\bar{\beta}$ is the average transmission rate. Performing similar calculations for (3.5), (3.6) yields

$$\dot{S} = \theta - dS - SI\bar{\beta} \tag{3.9}$$

$$\dot{I} = SI\bar{\beta} - dI - \bar{v}I - cI \tag{3.10}$$

$$\dot{R} = cI - dR, \tag{3.11}$$

where all overbars denote an expectation over the distribution of strains in the population.

System (3.7)–(3.11) describes the coupled evolutionary-epidemiological dynamics. Equations (3.7), (3.8) reveal how natural selection and within-host mutation are acting on v and β at each instant in time as a function of the current epidemiological state of the population. Equations (3.9)–(3.11) reveal how the epidemiological state of the population changes over time as a function of the current average value of virulence and transmission, $\bar{v}$ and $\bar{\beta}$. Day and Proulx [**10**] present a similar derivation in which there is a continuum of strain types possible, but interestingly those previous results were based on an assumption of small strain variance whereas the above results are exact and require no such assumption. The difference in the two derivations stems from the fact that Day and Proulx [**10**] tracked the evolution of a single trait only. This typically means that the trait enters the epidemiological model in a nonlinear fashion, and therefore requires a small variance assumption in order to simplify the model. Once multiple trait evolution is considered, however, these nonlinearities often disappear (as above) removing the need for this assumption about the variance.

The variances and covariances in system (3.7)–(3.11) will also change through time, and equations for these dynamics will typically depend on higher moments of the strain distribution. Nevertheless, even in the absence of an explicit model for these dynamics, system (3.7)–(3.11) can be used to gain some insights into pathogen evolution. It can be helpful to take a geometric approach to this question, by first writing equations (3.7), (3.8) in matrix notation. We obtain:

$$\begin{pmatrix} \dot{\bar{v}} \\ \dot{\bar{\beta}} \end{pmatrix} = \mathbf{G} \begin{pmatrix} -1 \\ S \end{pmatrix} - \mu \begin{pmatrix} \bar{v} - \bar{v}_m \\ \bar{\beta} - \bar{\beta}_m \end{pmatrix} \tag{3.12}$$

where $\mathbf{G}$ is termed the genetic (co)variance matrix and $\begin{pmatrix} -1 & S \end{pmatrix}^T$ is termed the selection gradient. In general the (co)variances in equations (3.12) are functions of the strain frequencies. It is relatively easy to obtain explicit mathematical expressions for these when there are few strains. For example, the two-strain case is particularly simple and it also provides a natural connection with more standard invasion analyses involving rare mutants. Even in more complex scenarios with many strains, however, expressions (3.12) provide a nice way to interpret the action of natural selection within this epidemiological setting.

The product of G with the selection gradient in equation (3.12) describes the way in which natural selection changes the average level of virulence and transmission in the pathogen population. Natural selection always favours reduced virulence with a strength of -1. On the other hand, natural selection always favours an increased transmission rate with a strength that is proportional to the density of susceptible hosts. At equilibrium the force of mutation must balance the force of natural selection, as mediated through the genetic covariance structure of the pathogen population (Figure 1).

The first interesting insight that Price's equation yields is that, quite generally we expect an intermediate equilibrium level of virulence and transmission regardless of the pattern covariance (i.e., regardless of whether there is a tradeoff between transmission and virulence). This is in contrast to most classical theoretical results that predict virulence will evolve to be zero and transmission rate as large as possible whenever there is no tradeoff (but see [**4**]). The difference in prediction arises from the inevitable effects of mutational bias (Figure 2). The exact location of the equilibrium will depend on the suite of strains that are possible for a given pathogen species as well as the mutation rate. Higher mutation rates will lead to higher equilibrium levels of virulence. This can be appreciated from Figure 2 by noting that the hollow arrow in Figure 2c will be larger in this case, thereby pulling the population closer to $\bar{x}_m$.

It is also interesting to tie these results to previous research in quasi-species [**26**]. In many situations we might expect that, out of all the strains that might exist, relatively few of these will have high fitness (i.e., have zero virulence and very high transmission). This is because there is likely a very specific exploitation strategy that a pathogen must have to gain a high transmissibility while also inducing very little virulence. As such these high fitness strains will be akin to the "master sequences" of quasi-species theory. The vast majority of strains in the population will contain deleterious mutations leading to lower fitness because of a

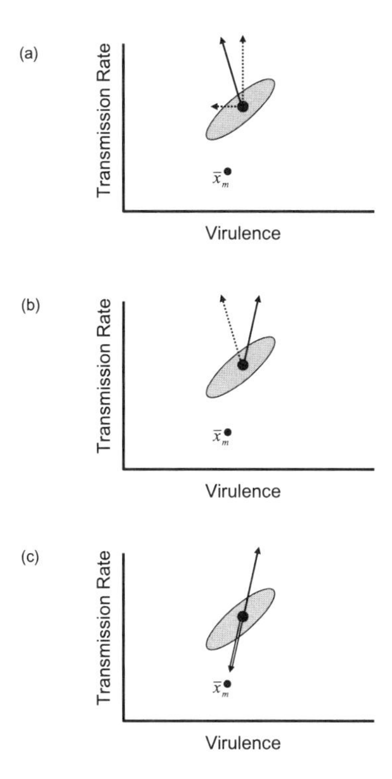

FIGURE 1. A schematic representation of a pathogen population at evolutionary and epidemiological equilibrium. The dot next to $\bar{x}_m$ denotes the mean transmission rate and virulence of all mutations that arise. The shaded ellipse represents the contour within which 95% of the pathogen genotypes lie, and the dot in its center is the mean transmission rate and virulence of the population. The fact that the major axis of the 95% ellipse has a positive slope implies a positive genetic covariance between transmission rate and virulence. (a) Dashed arrows represent the direction of selection on transmission rate and virulence, and the solid arrow is the net direction of selection. (b) Dashed arrow is the net direction of selection from panel (a) and the solid arrow is the direction of evolutionary change that results from this selection when mediated through the positive genetic covariance between transmission rate and virulence. (c) Solid arrow is the direction of evolution change from selection from panel (b) and the hollow arrow is the force of mutational bias that exactly balances this at equilibrium.

lower transmission rate and/or a higher virulence. Interestingly, as has been well-documented in quasi-species theory, there can be a threshold balance between the strength of selection (which, here, depends on the density of susceptible hosts) and mutation [**26**]. When mutation becomes too great (or selection becomes too weak – e.g., which can happen here if S is very small) then an "error threshold" is crossed. This leads to the extinction of the high fitness strain (i.e., the one with low virulence

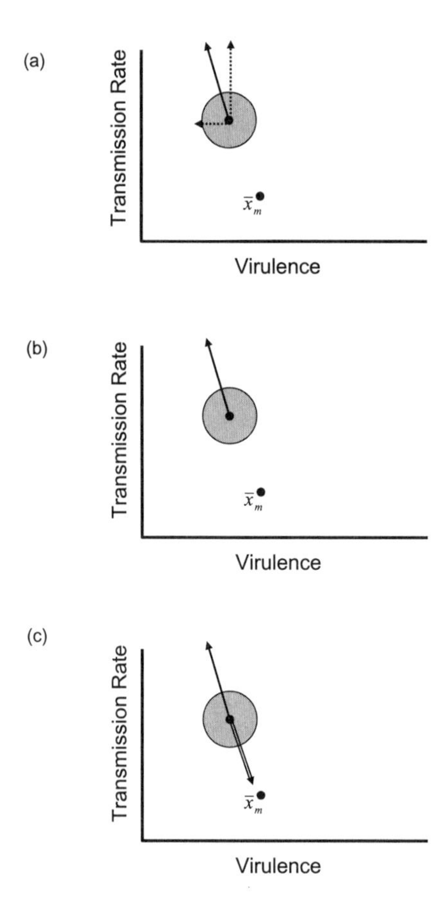

FIGURE 2. A schematic representation of a pathogen population at evolutionary and epidemiological equilibrium as in Figure 1 but with no covariance between transmission rate and virulence. The dot next to denotes the mean transmission rate and virulence of all mutations that arise. The shaded circle represents the contour within which 95% of the pathogen genotypes lie, and the dot in its center is the mean transmission rate and virulence of the population. (a) Dashed arrows represent the direction of selection on transmission rate and virulence, and the solid arrow is the net direction of selection. (b) Solid arrow is the direction of evolutionary change is the same as in panel (a) because there is no covariance between transmission rate and virulence. (c) Solid arrow is the direction of evolution change from selection from panel (b) and the hollow arrow is the force of mutational bias that exactly balances this at equilibrium.

and high transmissibility). Again this contrasts with classical results which predict that this strain should prevail (Figure 3). For highly mutable pathogens such as RNA viruses this might provide an explanation for the existence of intermediate levels of virulence even in the absence of tradeoffs between pathogen characters.

Aside from the inevitable effects of mutations bias, Price's equation yields other interesting insights and interpretations regarding how natural selection shapes pathogen evolution. Let's ignore the effects of mutation, in which case equations (3.7), (3.8) become

$$\dot{\bar{v}} = S\sigma_{\beta v} - \sigma_{vv} \tag{3.13}$$

$$\dot{\bar{\beta}} = S\sigma_{\beta\beta} - \sigma_{\beta v}. \tag{3.14}$$

One of the central predictions from the classical theory on virulence evolution is that a heightened background mortality rate of the host will select for increased pathogen virulence (assuming that a tradeoff between transmission rate and virulence exists [**18**]). Conceptually, this is explained as resulting from the decreased lifespan of an infected host (see [**10**] for a comparison of the approach presented here with more classical game-theoretic models of this question). This reduced lifespan of infections reduces the future transmission potential of the pathogen at any given time during an infection, and therefore it selects for an increased emphasis on current transmission. This is realized through the evolution of higher virulence. The same reasoning has been used to predict the evolution of higher virulence in response to any factor that shortens the lifespan of an infected host, including an increased clearance rate as well as the existence of secondary infections (see section 3.2).

Equations (3.13)–(3.14) reveal that there is something amiss with this interpretation. Host background mortality and clearance do not appear in the equations for the evolution of virulence and transmission, and therefore they have no direct effect on evolution. In fact this should be obvious from the outset. Background mortality and clearance are both assumed to act in a fashion that is independent of strain type, and therefore by definition they can play no direct role in the change in frequency of strain types. Both of these factors can have an indirect role in evolution, however, through their influence on the genetic parameters in equations (3.13)–(3.14) and/or their influence on the density of susceptible hosts.

Typically we might expect the influence of clearance and host mortality on the genetic variances and covariances to be relatively minor. They can play a large role, however, in determining the density of susceptible hosts. In standard epidemiological models such as equations (3.9)–(3.11) heightened host mortality or clearance rates lead to a higher density of susceptible hosts. It is this increased benefit of transmission that causes the evolution of higher virulence. This clearly illustrates that virulence evolves to be higher solely because of its positive covariance with transmission. Higher transmission rates are selected for because of the increased density of susceptible hosts, and virulence gets dragged along as a correlated response to selection. Interestingly, in this case the transient evolutionary dynamics occurring from an increased mortality rate can actually be opposite to those occurring from an increased clearance rate despite them reaching the same equilibrium values. This can occur because of differences in the transient dynamics of S that occur from these two different manipulations [**10**].

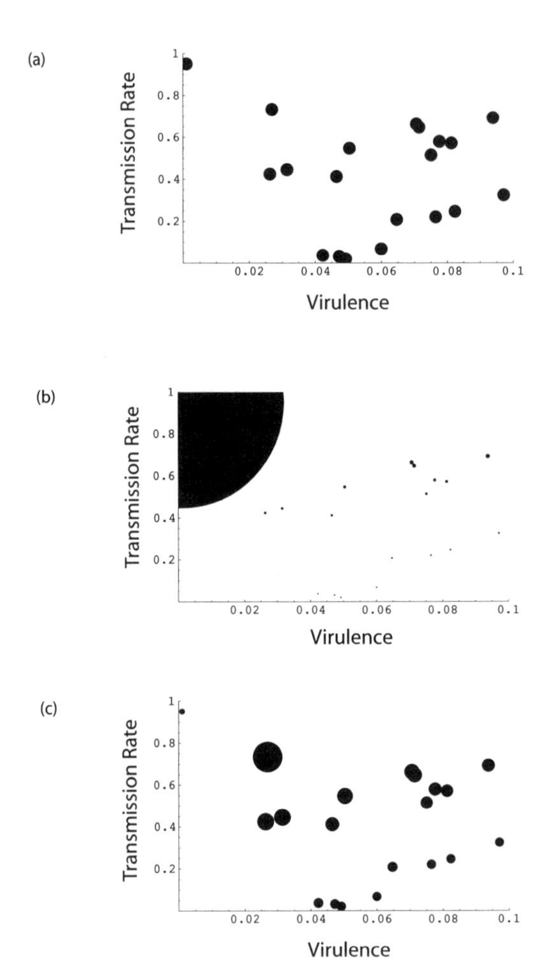

FIGURE 3. An example of how an intermediate level of virulence and transmission rate can evolve in the absence of tradeoffs as a result of crossing an error threshold. One pathogen strain was assigned high fitness parameters (i.e., high transmission rate and low virulence) and 19 others were assigned low fitness values of these parameters at random. These are represented by the dots plotted in each of the panels. Model (3.4)–(3.6) of the text was then used to simulate the evolutionary and epidemiological outcome under different conditions. Parameter values were $\theta = 10$, $\mu = .13$, and $d = 0$ or $d = 1$. Size of dots depict relative equilibrium frequency in the population. (a) all strains are introduced at equal frequency at time 0. (b) When $d = 1$. The equilibrium density of susceptible hosts is high and thus the fitness difference between the high fitness strain and the deleterious mutants outweighs mutational loss, causing the high fitness strain to prevail. (c) When $d = 0$. The equilibrium density of susceptible hosts is low and thus the fitness difference between the high fitness strain and the deleterious mutants is not enough to outweigh mutational loss. The error threshold is crossed and the high fitness strain goes extinct, leading to an intermediate average value of virulence and transmission rate.

Perhaps even more significantly, an increased mortality rate of infected hosts does not necessarily lead to the evolution of higher virulence under all epidemiological schemes. For example, if we increased host mortality rate but artificially maintained a constant density of susceptible hosts experimentally, then no evolutionary change in virulence or transmission rate should occur. Thus, rather than viewing host mortality and clearance as factors directly affecting pathogen evolution, we should view them as factors that will affect pathogen evolution only through their influence on the density of susceptible hosts (e.g., Figure 3). Previous theory has focused predominately on cases whether there is a positive relationship between mortality or clearance and the density of susceptible hosts but this need not always be the case. It is also worth noting that the perspective used here to elucidate a qualitative understanding of how various factors affect pathogen evolution does not require that we even attempt to calculate the evolutionarily stable level of virulence. Rather, a general understanding of these issues can be obtained from inspection of the equations governing the evolutionary dynamics.

Finally, by treating the evolution of transmission rate and virulence as distinct traits with some potential genetic covariance, the present approach opens the door to making concrete predictions regarding the magnitude of evolutionary change in response to various manipulations. The lack of these sorts of predictions has led to criticisms of previous theory on virulence evolution [**13**]. As an example, Ebert and Mangin [**14**] manipulated background host mortality and then quantified evolutionary changes in transmission rate and virulence. As already mentioned, this manipulation will cause evolution only through its effects on the density of susceptible hosts. Consequently, equations (3.13)–(3.14) predict that the manipulation should cause a stronger evolutionary response in transmission rate than in virulence. The reason is that the response in transmission rate is mediated directly through the genetic variance in transmission rate whereas the response in virulence occurs only indirectly through its genetic covariance with transmission. This difference in evolutionary response does appear to have occurred in the results of Ebert and Mangin [**14**] (see their Figure 1).

3.2. An S-I-R Model with Secondary Infections. As a second example, we consider the same model as in section 3.1, but we now allow for secondary infections to occur. Specifically, we allow for hosts that are already infected with a particular strain, to acquire secondary infections as a result of contact with other infected hosts. When such an event occurs, we again suppose that one of the two strains competitively excludes the other instantaneously (i.e., superinfection). This has previously been shown to result in the evolution of higher levels of virulence in many models [**21, 22, 27, 25**].

We begin with model (3.1)–(3.3), and extend this to multiple strains as

$$\dot{S} = \theta - dS - S\sum_i \beta_i I_i \tag{3.15}$$

$$\dot{I}_i = S\beta_i I_i - dI_i - v_i I_i - cI_i + I_i \sum_j I_j \left(\beta_i \rho_{ji} - \beta_j \rho_{ij}\right) \tag{3.16}$$

$$\dot{R} = \sum_i cI_i - dR. \tag{3.17}$$

Equations (3.15)–(3.17) are identical to equations (3.4)–(3.6) except for the inclusion of secondary infection. This is represented in equation (3.16) by terms reflecting the additional way in which infections of type i can be gained and lost. Contacts between hosts infected with strain i and hosts infected with strain j occur at a rate $I_i I_j$ (according to the mass action assumption used implicitly in equations (3.1)–(3.3)). Upon such contact, strain i pathogens are transmitted to the host infected with strain j at a rate β_i, and strain j pathogens are transmitted to the host infected with strain i at a rate β_j. Within-host competition then takes place, and strain i competitively excludes strain j in the j-type host with probability ρ_{ji}, whereas strain j competitively excludes strain i and the i-type host with probability ρ_{ij}. Thus the total rate of change of hosts infected with strain i due to contacts with hosts of strain j is $I_i I_j (\beta_i \rho_{ji} - \beta_j \rho_{ij})$. Summing this over all the possible strain types in the population then gives the new term in equation (3.16).

To apply equation (2.4) we need to identify the per capita rate of change of hosts infected with strain i. Using I to denote the total number of infected hosts, we can write equation (3.16) as

$$\dot{I}_i = I_i \left(S\beta_i - d - v_i - c + I \sum_j q_j \left(\beta_i \rho_{ji} - \beta_j \rho_{ij} \right) \right)$$

or

$$\dot{I}_i = I_i \left(S\beta_i - d - v_i - c + I\beta_i \bar{\rho}_{\bullet i} - I\overline{\beta \rho_{i\bullet}} \right),$$

where $\beta_i \bar{\rho}_{\bullet i}$ is the average rate at which strain i infections displace infections of other strains through secondary infection, and $\overline{\beta \rho_{i\bullet}}$ is the average rate at which they themselves are displaced. This illustrates that the per capita rate of change under this model is

$$r_i = S\beta_i - d - v_i - c + I\beta_i \bar{\rho}_{\bullet i} - I\overline{\beta \rho_{i\bullet}} .$$

If we further simplify matters by assuming that all strains have equivalent within-host competitive abilities (i.e., $\rho_{ji} \equiv \rho$), we have

$$r_i = S\beta_i - d - v_i - c + I\beta_i \rho - I\bar{\beta}\rho .$$

Applying equation (2.4) then gives

$$\dot{\bar{v}} = (S + \rho I)\, \sigma_{\beta v} - \sigma_{vv} - \mu \left(\bar{v} - \bar{v}_m \right) \tag{3.18}$$

$$\dot{\bar{\beta}} = (S + \rho I)\, \sigma_{\beta\beta} - \sigma_{\beta v} - \mu \left(\bar{\beta} - \bar{\beta}_m \right) . \tag{3.19}$$

Note the similarity between these equations and equations (3.7), (3.8). Secondary infection has a very simple effect on the evolutionary dynamics; it changes S in equations (3.7), (3.8) to $S + \rho I$ in equations (3.18), (3.19). Under secondary infection, hosts that are already infected now represent an additional 'resource' that can be used for transmission, and ρ is a factor that weights the susceptibility of already infected hosts to secondary infections, relative to the susceptibility of hosts with no infection. We can also use equations (3.15)–(3.17) to obtain a system of equations governing the total number of susceptible, infected, and recovered individuals, yielding exactly the same system as before (i.e., equations (3.9)–(3.11)). Secondary infection has no direct effect on the epidemiological dynamics.

Again Price's equation yields some interesting new insights in the case of secondary infection. As mentioned in section 3.1, secondary infection has previously been suggested to cause the evolution of higher virulence because it reduces the lifespan of an infected host in much the same way that increased mortality or clearance does. The suggestion is that this reduces the future transmission potential of the pathogen at any given time during an infection and therefore it selects for higher virulence. Examination of equations (3.18)–(3.19) reveals that, as with the case of mortality and clearance, this interpretation is incorrect. Secondary infection selects for higher virulence solely because it increases the benefit of transmission, and because transmission is positively genetically correlated with virulence.

Another way to appreciate this difference in interpretation is to imagine conducting an experiment with a host-pathogen system that does not normally have secondary infection (e.g., some phage-bacteria systems [**24**]), but in which you can experimentally induce secondary infection. Suppose you maintained a control population with no secondary infection, and an experimental population in which you caused secondary infections with randomly chosen pathogen strains. The loss rate of infected hosts in the experimental population through secondary infection would be elevated, and therefore previous interpretations would lead you to expect the evolution of higher virulence. This will not actually occur, however, and equations (3.18)–(3.19) reveal why. Secondary infection selects for higher virulence solely because it increases the benefit of transmission, and because transmission is positively genetically correlated with virulence. The experimental design suggested here has removed this benefit while maintaining the level of secondary infection. Specifically, strains causing secondary infection are chosen randomly and therefore the covariance between transmission rate and virulence is zero for all secondary infections. In this case equations (3.18), (3.19) reduce to equations (3.7), (3.8) revealing that secondary infection will have no evolutionary consequences.

3.3. The Curse of the Pharaoh. Our final example examines the so-called "Curse of the Pharaoh" hypothesis [**3, 16, 20, 31**]. This hypothesis applies to pathogens that have free-existing environmental stages (e.g., spores) such as *Bacillus anthrax* and many of the nucleopolyhedrosis viruses of insects. The hypothesis postulates that long-lasting environmental stages (which we will refer to here as spores) result in the evolution of higher virulence. In such cases the pathogen can kill its host quickly without compromising transmission to other hosts because of the existence of transmissible spores persisting in the environment. The hypothesis derives its name from the suggestion that virulent pathogens with long-lived spores might have been the cause of the mysterious death of Lord Carnarvon after opening the tomb of King Tutankhamen [**3, 20**].

As with the previous two examples, we begin by specifying an epidemiological model. Several authors have examined this hypothesis [**3, 9, 20**], and here we use a special case of the model by [**9**]:

$$\dot{S} = \Phi \tag{3.20}$$

$$\dot{I} = \gamma S F - (d + v) I \tag{3.21}$$

$$\dot{F} = \kappa I + v \omega I - \delta F \,. \tag{3.22}$$

The variables S, I, and F are the populations sizes of susceptible hosts, infected hosts, and freely-existing spores. We leave the dynamics of the susceptible class unspecified. The parameter γ is the transmission rate of spores from the environment to susceptible hosts, d is the natural host death rate, v is the pathogen-induced death rate (i.e., the virulence), κ is the rate at which infected hosts shed spores into the environment, ω is the number of spores shed into the environment upon pathogen-induced death of the host, and δ is the per capita loss rate of spores from the environment. Model (3.20)–(3.22) assumes that infections are generated only through contact between susceptible hosts and environmental spores, and that this has a negligible effect on the number of spores in the environment.

Model (3.20)–(3.22) embodies two different pathogen habitats: (i) the host habitat, and (ii) the environmental habitat. As a result, we will need to use the multiple-habitat version of Price's equation to model evolutionary change. Doing so requires that we first extend model (3.20)–(3.22) to multiple strains. Focusing only on equations (3.21), (3.22) gives

$$\dot{I}_i = \gamma S F_i - (d + v_i) I_i \tag{3.23}$$

$$\dot{F}_i = \kappa_i I_i + v_i \omega I_i - \delta F_i \ . \tag{3.24}$$

In equations (3.23), (3.24) we have assumed that the rate of transmission of spores to susceptible hosts, γ, is independent of strain type, as is the number of spores produced upon pathogen-induced host death, ω. Model (3.23)–(3.24) does not allow for superinfection, but it can be extended to do so relatively easily. This requires only that equation (3.23) be changed to

$$\dot{I}_i = \gamma S F_i - (d + v_i) I_i + F_i \sum_j I_j \gamma \rho_{ji} - I_i \sum_j F_j \gamma \rho_{ij} \ . \tag{3.25}$$

The second-to-last term in equation (3.25) represents the influx of strain i infections through secondary infection. It is the rate at which hosts infected with strain j come into contact with spores of strain i (i.e., $F_i I_j$) multiplied by the probability of transmission, γ, and the probability that strain i competitively excludes strain j, ρ_{ji} (summed over all strains j). Similarly, the last term in equation (3.25) represents the loss of strain i infections through secondary infection. It is the rate at which hosts infected with strain i come into contact with spores of strain j (i.e., $F_j I_i$) multiplied by the probability of transmission, γ, and the probability that strain j competitively excludes strain i, ρ_{ij} (summed over all strains j).

To apply Price's equation (2.9)–(2.10) we now need to identify r_i^{II} and r_i^{FI}. These represent the per capita rate of production of strain i infected hosts by strain i infected hosts, and by strain i spores respectively. Similarly, we need to identify, r_i^{IF} and r_i^{FF}, which represent the per capita rate of production of strain i spores by strain i infected hosts and by strain i spores respectively. From equations (3.23)–(3.24) we can see that $r_i^{II} = -(d + v_i)$, $r_i^{FI} = \gamma S$, $r_i^{IF} = \omega_i + v_i \omega$, and $r_i^{FF} = \delta\delta$. If there are secondary infections then r_i^{IF} and r_i^{FF} remain unchanged (because equation (3.24) remains unchanged) while r_i^{II} and r_i^{FI} become $r_i^{II} = -(d + v_i) - F\gamma\bar{\rho}_{i\bullet}$ and $r_i^{FI} = \gamma S + \bar{\rho}_{\bullet i}\gamma I$, where $\bar{\rho}_{i\bullet}$ is the average probability that strain i infections gets competitively excluded during secondary infection, and $\bar{\rho}_{\bullet i}$ is the average probability that strain i secondary infections competitively exclude other strains in secondary infections.

For simplicity we will assume that ρ_{ji} is a constant (and equal to ρ) and we will ignore mutational effects on virulence and focus only on the effects of natural selection. In this case applying equations (2.9)–(2.10) to virulence, v, gives

$$\dot{\bar{v}}^F = \frac{I}{F}\left(\sigma_{\kappa v}^I + \omega \sigma_{vv}^I\right) + \frac{I}{F}\left(\bar{\kappa}^I + \omega \bar{v}^I\right)\left(\bar{v}^I - \bar{v}^F\right) \tag{3.26}$$

and

$$\dot{\bar{v}}^I = -\sigma_{vv}^I + \frac{F}{I}\gamma S\left(\bar{v}^F - \bar{v}^I\right), \tag{3.27}$$

where I and F to denote the total population size of all infected hosts and spores respectively. In the case of secondary infection equation (3.26) remains unchanged but equation (3.27) becomes

$$\dot{\bar{v}}^I = -\sigma_{vv}^I + \frac{F}{I}\gamma\left(S + \rho I\right)\left(\bar{v}^F - \bar{v}^I\right). \tag{3.28}$$

Even though the pathogen spores do not express virulence, each spore can still be characterized by its strain type, which indicates the level of virulence it would cause in a host. Equation (3.26) then gives the dynamics of the average level of virulence in this spore population. Finally, the equations governing the total number of infected hosts and spores are

$$\dot{I} = \gamma SF - (d + \bar{v}^I)I \tag{3.29}$$

$$\dot{F} = \bar{\kappa}^I I + \bar{v}^I \omega I - \delta F. \tag{3.30}$$

Equations (3.26)–(3.28) provide an interesting perspective on virulence evolution in spore-producing pathogens. We are primarily interested in the average level of virulence that is expressed when pathogens infect the host, $\bar{v}^I$, and thus equations (3.27) or (3.28) are of most interest. The first term in equation (3.27) or (3.28) (i.e., $-\sigma_{vv}$) shows that virulence is always selected against in the host population regardless of whether there is secondary infection or not. The reason is simply that low virulence strains will persist for longer in the host population than high virulence strains because the latter will kill their hosts quickly and then be shed into the spore population. Indeed the only reason why virulent strains are able to persist in the host population is that they "migrate" into hosts from the spore population.

The average level of virulence in the spore population (even though it is unexpressed) is higher than that in the host population ($\bar{v}^F > \bar{v}^I$). This can be seen directly from equation (3.26), which reveals that $\bar{v}^F$ will always increase above the value of $\bar{v}^I$. Biologically, this occurs because it is the most virulent strains that tend to be shed into the spore population. Thus the second term in equation (3.27) or (3.28) will be positive, and will eventually counterbalance the selection against virulence in the host population. Furthermore, we can see that the strength of this effect of "migration" is determined by the magnitude of the flow of spores into the host population (i.e., the force of infection). This will be higher when there is secondary infection because spores then move into the host population through both susceptible and infected hosts. This is reflected by the fact that the only effect of superinfection is to change S in equation (3.27) to $S + \rho I$ in equation (3.28). This selects for higher virulence in a fashion analogous to that in the model of section 3.2.

We can now ask how the lifespan of spores is expected to affect the evolution of virulence. The Curse of the Pharaoh hypothesis states that an increase in δ (which results in a decreased spore life-span) should lead to the evolution of lower virulence. The parameter δ does not appear anywhere in equations (3.26)–(3.28) and therefore spore lifespan does not have a direct effect on the evolution of virulence. It can nevertheless have an indirect effect, however, because it will affect the values of

S, I, and F. At equilibrium, these epidemiological variables cancel out of the evolutionary equation (3.26) and therefore changes in their values (as a result of changes in δ) will have an evolutionary effect on the average level of virulence only through equation (3.27) or (3.28). We treat each of these in turn.

In the case of no secondary infection equation (3.27) applies and thus we need to know how the value of the quantity FS/I changes as δ is increased. Assuming that the population is at an epidemiological equilibrium, equation (3.29) reveals that the relationship $FS/I = (d + \bar{v}^I)/\gamma$ must always hold. Consequently, even though the equilibrium values of all of the variables S, I, and F change as δ is increased, the ratio FS/I remains constant. As has been noted in previous epidemiological models [**3**], the prediction is therefore that spore lifespan has no effect on the evolution of virulence provided we assume that an epidemiological equilibrium is reached.

In the case of secondary infections equation (3.28) applies and we then need to know how the quantity $F(S + \rho I)/I$ changes. We can re-write this as $FS/I + F\rho$. From the above analysis we know that the first of these two terms remains constant as δ increases, and we expect the second to decrease (the size of the spore population will decrease if its death rate increases). Consequently, there is a lower migration rate of spores into the host population and thus the effect of this migration on maintaining virulence in the host population is diminished; the average level of virulence $\bar{v}^I$ decreases. This is exactly in accord with the Curse of the Pharaoh.

The logic of why the hypothesis holds under superinfection is as follows. The average level of virulence in spore populations is always higher than that in host populations for reasons outlined above. Longer spore life spans also lead to a higher spore population size. So long as secondary infections occur, this larger spore population then results in a greater flux of virulent spore strains into the host population, yielding a higher equilibrium level of virulence. Interestingly, secondary infection has previously been noted to result in predictions consistent with the Curse of the Pharaoh but for different reasons [**20**]. In these previous results changes in spore lifespan result in changes in the degree of relatedness among coinfecting pathogens, and it is this effect of relatedness that leads to evolutionary changes in virulence.

Finally, we consider what the model of this section can tell us about virulence evolution in pathogens such as the nucleopolyhedrosis viruses of insects, that release spores only upon pathogen-induced host death. In this case $\kappa = 0$, and we are left with the epidemiological-evolutionary system

$$\dot{I} = \gamma SF - (d + \bar{v}^I)I \tag{3.31}$$

$$\dot{F} = \bar{v}^I \omega I - \delta F \tag{3.32}$$

$$\dot{\bar{v}}^F = \frac{I}{F}\omega \left(\sigma_{vv}^I + \bar{v}^I \left(\bar{v}^I - \bar{v}^F\right)\right) \tag{3.33}$$

$$\dot{\bar{v}}^I = -\sigma_{vv}^I + \frac{F}{I}\gamma S \left(\bar{v}^F - \bar{v}^I\right). \tag{3.34}$$

We can now use these equations to deduce how the average level of virulence in the host, $\bar{v}^I$, is expected to evolve. Assuming that the epidemiological dynamics are fast relative to evolutionary change, equation (3.31) tells use that $FS\gamma/I = d + \bar{v}^I$.

Thus, equations (3.33), (3.34) become

$$\dot{\bar{v}}^F = \frac{I}{F}\omega\left(\sigma_{vv}^I + \bar{v}^I\left(\bar{v}^I - \bar{v}^F\right)\right) \tag{3.35}$$

$$\dot{\bar{v}}^I = -\sigma_{vv}^I + \left(d + \bar{v}^I\right)\left(\bar{v}^F - \bar{v}^I\right). \tag{3.36}$$

As discussed earlier, we always expect $\bar{v}^F > \bar{v}^I$.

Can equations (3.35), (3.36) reach an equilibrium if there is always some genetic variation in the population? If so equation (3.35) requires that $\bar{v}^F - \bar{v}^I = \sigma_{vv}^I/\bar{v}^I$. Substituting this into equation (3.36) and re-arranging shows that, in this case, $\dot{\bar{v}}^I \propto d$. This means that virulence in the host will evolve to be large whenever the level of virulence in the spore population is at equilibrium. Thus there is no joint equilibrium of the two. Once $\bar{v}^I$ increases, $\bar{v}^F$ will evolve to higher values as well leading to yet further evolutionary increases in $\bar{v}^I$.

The above considerations reveal that, for pathogens that release spores only upon pathogen-induced host death, virulence is expected to evolve to be as large as possible. Eventually, however, the inevitable forces of mutational bias outlined earlier will come into play, halting evolutionary change. Moreover, it is possible that other factors not included in the above model would also halt evolution towards extreme virulence. For example, if there were a tradeoff between the speed at which the host is killed and the number of spores produced, this too could result in the evolution of intermediate levels of virulence [**15**].

4. Summary

In this chapter we have presented an alternative theoretical framework for modeling the evolutionary and epidemiological dynamics of host-parasite interactions. The approach is based on using the instantaneous rate of change of infected hosts as a measure of pathogen fitness rather than the more commonly used quantity, R_0. Our alternative approach leads to a number of re-interpretations of predictions derived from previous theory, and it thereby provides a more thorough perspective on how various factors affect pathogen evolution. It also provides a relatively straightforward approach for modeling the dynamics of evolutionary change in pathogen populations when it cannot be assumed that the epidemiological dynamics occur on a time scale that is fast relative to that of the evolutionary dynamics (see also [**10**]).

The approach used here is also more amenable to integrating the somewhat disparate bodies of theory that have developed in the study of the evolutionary ecology of host-parasite interactions. For example, there is a large body of theory devoted to understanding the evolution of the harm that pathogens induce on their hosts (i.e., virulence as defined here). There is also a large and relatively independent body of theory that is focused on predicting the evolutionary dynamics of antigenic matching and avoidance between host and pathogen (e.g., gene-for-gene and matching-allele models [**19**]). The approach based on Price's equation that we have developed here offers one framework in which these two bodies of theory might be integrated. Similarly, we have illustrated how the ideas of quasispecies theory can be integrated into theory on the evolution of virulence using this framework as well.

There are several potentially important areas for future development, but there is one in particular that is especially important. All of the theory developed and discussed here has assumed that the transmission rate and virulence of difference pathogen strains are determined by their genotypes alone. In reality the extent to which a pathogen strain causes mortality and is transmitted is a function of both its genotype and that of its host. Developing models that elucidate the additional complexities of such coevolutionary dynamics will be an important challenge for future research.

Acknowledgements

This research was conducted during a visit by T.D. to Montpellier, France that was supported the Centre National de la Recherche Scientifique (CNRS). We thank Jean-Baptiste André and an anonymous referee for comments.

References

[1] R. M. Anderson May and R. M. May, Coevolution of hosts and parasites, *Parasitology* **85** (1982), 411–426.
[2] R. M. Anderson May and R. M. May, *Infectious diseases of humans: dynamics and control*, Oxford University Press, Oxford, 1991.
[3] S. Bonhoeffer, R. E. Lenski, et al., The curse of the pharaoh: The evolution of virulence in pathogens with long living propagules, *Proc. R. Soc. Lond. B* **263**(1371) (1996), 715–721.
[4] S. Bonhoeffer and M. A. Nowak, Mutation and the Evolution of Virulence, *Proc. R. Soc. Lond. B* **258**(1352) (1994), 133–140.
[5] H. J. Bremermann and J. Pickering, A game-theoretical model of parasite virulence, *J. Theor. Biol.* **100** (1983), 411–426.
[6] H. J. Bremermann and H. R. Thieme, A competitive exclusion principle for pathogen virulence, *J. Theor. Biol.* **27** (1989), 179–190.
[7] M. Bulmer, *Theoretical evolutionary ecology*, Sinauer Associates, Sunderland, 1994.
[8] T. Day, On the evolution of virulence and the relationship between various measures of mortality, *Proc. R. Soc. Lond. B* **269**(1498) (2002), 1317–1323.
[9] T. Day, Virulence evolution via host exploitation and toxin production in spore-producing pathogens, *Ecology Letters* **5**(4) (2002), 471–476.
[10] T. Day and S. R. Proulx, A general theory for the evolutionary dynamics of virulence, *Am. Nat.* **163** (2004), E40–E63.
[11] O. Diekmann and J. A. P. Heesterbeek, *Mathematical epidemiology of infectious disease*, Wiley, New York, 2000.
[12] O. Diekmann, J. A. P. Heesterbeek, et al., On the definition and computation of the basic reproduction ratio R0 in models for infectious diseases in heterogeneous populations, *J. Math. Biol.* **28** (1990), 365–382.
[13] D. Ebert and J. J. Bull, Challenging the trade-off model for the evolution of virulence: is virulence management feasible? *Trends in Microbiology* **11**(1) (2003), 15–20.
[14] D. Ebert and K. L. Mangin, The influence of host demography on the evolution of virulence of a microsporidian gut parasite, *Evolution* **51**(6) (1997), 1828–1837.
[15] D. Ebert and W. W. Weisser, Optimal killing for obligate killers: The evolution of life histories and virulence of semelparous parasites, *Proc. R. Soc. Lond. B* **264**(1384) (1997), 985–991.
[16] P. W. Ewald, *Evolution of infectious disease*, Oxford University Press, Oxford, 1994.
[17] S. A. Frank, George Price's Contributions to Evolutionary Genetics, *J. Theor. Biol.* **175**(3) (1995), 373–388.
[18] S. A. Frank, Models of parasite virulence, *Quarterly Review of Biology* **71** (1996), 37–78.
[19] S. A. Frank, Polymorphism of attack and defense, *Trends Ecol. Evol.* **15**(4) (2000), 167–171.
[20] S. Gandon, The curse of the pharaoh hypothesis, *Proc. R. Soc. Lond. B* **265**(1405) (1998), 1545–1552.
[21] S. Gandon, V. A. A. Jansen, et al., Host life history and the evolution of parasite virulence, *Evolution* **55**(5) (2001), 1056–1062.
[22] S. Gandon, M. van Baalen, et al., The evolution of parasite virulence, superinfection, and host resistance, *Am. Nat.* **159**(6) (2002), 658–669.
[23] H. W. Hethcote, The mathematics of infectious diseases, *Siam Rev.* **42**(4) (2000), 599–653.

[24] S. L. Messenger, I. J. Molineux, et al., Virulence evolution in a virus obeys a trade-off, *Proc. R. Soc. Lond. B* **266**(1417) (1999), 397–404.
[25] J. Mosquera and F. R. Adler, Evolution of virulence: a unified framework for coinfection and superinfection, *J. Theor. Biol.* **195**(3) (1998), 293–313.
[26] M. A. Nowak, What is a quasispecies? *Trends in Ecology and Evolution* **7** (1992), 118–121.
[27] M. A. Nowak and R. M. May, Superinfection and the Evolution of Parasite Virulence, *Proc. R. Soc. Lond. B* **255**(1342) (1994), 81–89.
[28] G. Price, Selection and covariance, *Nature* **227** (1970), 520–521.
[29] G. R. Price, Extension of covariance selection mathematics, *Annals of Human Genetics* **35** (1972), 485–490.
[30] M. vanBaalen and M. W. Sabelis, The dynamics of multiple infection and the evolution of virulence, *Am. Nat.* **146**(6) (1995), 881–910.
[31] B. A. Walther and P. W. Ewald, Pathogen survival in the external environment and the evolution of virulence, *Biological Reviews* **79** (2004), 1–21.

Departments of Mathematics/Statistics & Biology, Queen's University, Kingston, Ontario, K7L 3N6 CANADA
E-mail address: tday@mast.queensu.ca

Génétique et Evolution des Maladies Infectieuses, UMR CNRS/IRD 2724, Institut de Recherche pour le Développement, 911 Avenue Agropolis, 3, 4394 Montpellier Cedex 5, France
E-mail address: gandon@mpl.ird.fr

DIMACS Series in Discrete Mathematics
and Theoretical Computer Science
Volume **71**, 2006

Within-host pathogen dynamics: Some ecological and evolutionary consequences of transients, dispersal mode, and within-host spatial heterogeneity

Robert D. Holt and Michael Barfield

ABSTRACT. The ecology and evolution of infectious disease occur at multiple spatial scales. In this paper, we explore some consequences of transient dynamics of pathogens within individual hosts. If infected hosts die quickly, relative to internal equilibration in pathogen dynamics, within-host transients may influence between-host transmission and spread. We develop a formulation for characterizing the overall growth rate of an infectious disease, which includes both within-host dynamics and between-host transmission, when the disease is sufficiently rare that the supply of available hosts can be viewed as a constant. This formulation is analogous to the familiar Euler equation in age-structured demography. We suggest that the pathogen growth rate estimated this way may be a better measure of pathogen fitness than is R_0. We point out that even simple models of within-host pathogen dynamics can have phases in which numbers overshoot the final equilibrium, and that such phases may influence pathogen evolution. We touch on the potential importance of within-host spatial heterogeneities in pathogen dynamics, and suggest that an interesting question for future work is understanding the interplay of spatial structure and transient dynamics in the within-host infection process.

1. Introduction

The ecology of infectious disease plays out in arenas at vastly different spatial scales. Traditional epidemiology focuses on between-host infection dynamics, either between individual hosts within well-mixed host populations, or among spatially segregated populations [**1**]. In recent years, there has been increasing attention given to an important arena of infection embedded within the host population scale, namely that of within-host infection dynamics (e.g., [**2, 3, 7, 12, 13, 15, 16, 17, 20**]). Following successful infection by a virus, bacterium, or fungus, a population of the pathogen is established within an individual host, which in effect is a "patch" being colonized by that pathogen. As in any colonization, there is a phase of population growth before an equilibrium (if any) becomes established in that host. This transient phase has two broad implications for disease dynamics, which will provide the interwoven themes for this chapter.

We would like to thank the University of Florida Foundation and NIH (grant 5 R01 GM60792-04) for support.

First, the transient dynamics of the within-host infection process may contribute significantly to the spread of the infection in the host population as whole. General ecological theory has long been dominated by a focus on the asymptotic behavior of systems (e.g., characterizing equilibria) as the determinant of natural population and community dynamics. There is a growing recognition that an understanding of transient dynamics may be crucial for understanding ecological systems [**9**]. Epidemiological models that assume host individuals are either susceptible or infected (or immune, etc.) in effect assume that the transient phase of within-host establishment dynamics can be ignored. One of our goals will be to relax this assumption and explore some consequences of explicitly tracking the size of the within-host pathogen population during each bout of infection.

Second, it may be a gross oversimplification to simply track total pathogen numbers within a host. Considerable heterogeneity in pathogen numbers and even genotype exists among tissues within individual hosts (e.g. [**23**]). Spatial heterogeneity and patterns of connectivity can in turn have a strong influence on the kind of transient dynamics observed in within-host infection [**18**]. For instance, the initial site of infection will often be distinct from the sites within which the greatest population growth is expected to occur. This implies that there will be a lag observed in the growth rate of the pathogen, following initial infection, reflecting the internal spread of the pathogen population into tissues other than that provided by the original site of entry.

In the first part of this paper, we discuss some simple aspects of transient within-host dynamics, and how this bears on between-host transmission dynamics. We then turn to some more complex issues involving transient conditions, first for a spatially unstructured host, and then briefly for one with multiple spatial compartments, and touch on how transient dynamics influence between-host transmission dynamics and pathogen evolution.

Figure 1 displays four patterns of within-host dynamics, for four different scenarios (three of which correspond to special cases of model (3.1) discussed below; note that in Figure 1B, C, and D, abundances are shown on a log scale). In Figure 1A, we portray within-host dynamics implicit in the usual SI model (e.g., [**1**]), in which individuals are simply divided into infected and non-infected classes (the latter may be further subdivided, for instance into susceptible vs. immune), and all infected individuals are assumed equivalent (relative to their capacity to infect other hosts). In effect, this assumes that following infection, the population of pathogens in a host immediately reaches some kind of carrying capacity, where the within-host pathogen population stays until the host either dies or recovers.

In Figure 1B, we depict another scenario, in which all the dynamics within the host are transient. Here, we assume that the initial infective propagule is tiny, relative to the size of the pathogen population that could be produced inside the host. The pathogen simply grows exponentially, until the host dies, either because of the effect of the pathogen itself on the host, or for other reasons. In either case, the death of the host eliminates the local population of the pathogen entirely. There is a sense in which the entire within-host dynamic in this case is a transient, since there is no local equilibrium reached before the host is removed. (This picture could also apply to a scenario in which the host mounts an effective, abrupt immune defense; below, we mainly consider cases involving host death.)

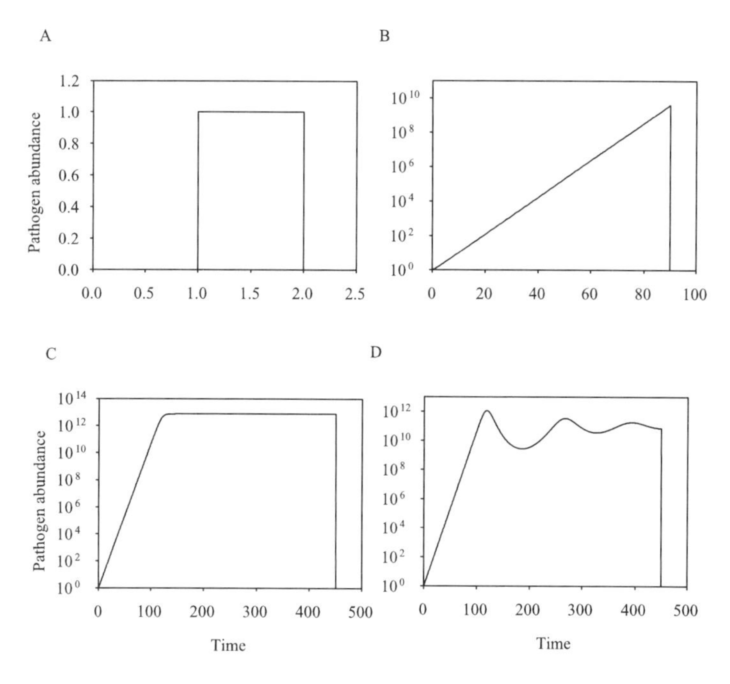

FIGURE 1. Patterns of within-host pathogen abundance. A. The standard SI model effectively assumes that abundance is fixed during time of infection. B. The pathogen increases exponentially throughout life of the host (note logarithmic scale on last 3 panels). C. Pathogen increases exponentially at first, but then saturates, in a logistic-like fashion. This can approximate the pattern in A, if the time of exponential increase is short compared to the time at equilibrium. D. More complex dynamics, with a large overshoot. The last three panels were all generated with the virus model (3.1) in the main text with parameters $\beta = 5 \times 10^{-14}$, $\mu' = 2$, $\nu = 250$, $\mu^* = 0.7$, $\lambda = 2.45 \times 10^9$. Panels B and C have $\mu = 0.7$, D uses $\mu = 0.01$ (the parameters are drawn from [**21**]).

Figure 1C is intermediate between A and B. Here, again there is a phase of exponential growth, but the pathogen approaches an internal equilibrium, before host death. If the host lives sufficiently long after infection, then the pattern of Figure 1A (the usual assumption of SI and SIR models) may be a reasonable approximation to the actual within-host dynamics.

Finally, in Figure 1D, there is a more complex pathogen dynamic within the host. In the example shown, there is an initial rapid exponential growth. Susceptible host cells are used up, and slowly replenished by the host; this generates a predator-prey-like dynamic leading to a series of damped oscillations, in which the maximal pathogen population achieved greatly exceeds the final, equilibrial population. Comparable transient dynamics can emerge in a wide range of models (e.g., incorporating within-host spatial structure, or immune responses).

In all cases, the within-host dynamic of the pathogen occurs within a time-frame set by host death. In general, the magnitude of this time-frame should itself depend upon the pathogen's own dynamics. For instance, host death rates might increase with pathogen load, or pathogen growth rate. If the host dies prior to the establishment of a within-host equilibrium, then analyzing transients becomes crucial to understanding the overall host-pathogen system.

2. Shedding or Bursting

To link models for within-host dynamics to the dynamics of the pathogen in the entire host population, we have to relate within-host pathogen population size to transmission between this focal, infected host individual, and other susceptible hosts. There are several ways one can imagine such linkage to occur mechanistically. As idealized extremes, we will contrast "shedding" and "bursting."

"Shedding" occurs when pathogens are emitted from an infected host throughout the course of infection. In this case, the degree of infectivity of the infected host in contact with another, susceptible host individual might scale with the instantaneous population size of the pathogen in the host. One issue that arises here is that shedding pathogens can act as a loss term for the microbial population within the host, which then should grow more slowly because of the "drain" of shed pathogens to the external environment. If disease transmission occurs via an environmental pool of infective propagules, then allocation of within-host pathogen growth to emission of such propagules should reduce the rate of continued growth within the focal host itself. (More broadly, the shedding may impact only a portion of the pathogen population, when there is internal spatial structuring and only certain tissues are sites of emission; see below.)

"Bursting" would instead apply to a system in which no pathogens are emitted until the host dies. This might pertain to some disease systems in which the host has to be consumed for pathogen transmission to occur (e.g., prions in bovine spongiform encephalopathy), and closely matches the life cycles of many baculoviruses infecting insects. Such a pathogen abstractly resembles an insect parasitoid, in that host death is required for successful reproduction. The degree of infection of other host individuals should then scale with the final pathogen population size within a host, at the time of that host's death.

We assume below that the host is a free-living organism. However, it is worth noting that the same equations can also describe dynamics within a multicellular host individual, with the reinterpretation that the "host" in the models would then be a single cell that is either healthy or infected by a parasite, for example, a virus. Some types of virus tend to be continuously released from an infected cell until it dies, which is equivalent to the shedding model below (assumed for instance by [**17**]). Other types of virus tend to remain within the infected cell until there are a sufficient number of them to cause the cell to break open, releasing the virions to attack other host cells (corresponding to the burst model; see [**11**]). The chapter by Kelly [**10**] shows that this distinction does not alter the evolutionary generation time of the virus. However, it could affect the overall reproductive rate of different clones, and so the ideas presented below could have implications for the evolution of virulence within viral populations inhabiting single hosts (we thank John Kelly for this insight).

In our initial model, we assume that the pathogen is a microparasite (a virus, bacterium, or fungus) which infects a host starting with a single infective particle (we will ignore stochastic factors), establishing a local population. Clearly, ignoring stochastic factors is not reasonable for the initial phases of infections starting from a single particle. However, the basic equations below apply if it is assumed that the initial infection is by an inoculum consisting of a fixed number of infectious particles (by simply rescaling the function that relates pathogen density to host mortality, and the constant that relates free pathogens to new infections). Also, if infection is by a single particle, then given the initial growth rate in the host, the probability of loss due to stochastic factors could be calculated and incorporated into the new infection probability parameter.

As noted above, the pathogen can spread from the infected host individual in one of two ways. It can be shed from the host continuously at a per-capita rate s, until the host dies, at which time all remaining pathogens in the host are lost. Alternatively, all the pathogens may be retained in the host until the host dies, at which time all pathogens are released (in a burst). In either case, the host mortality (m) may be a function of the pathogen level (denoted V) in the host. (An alternative mechanistic assumption might be that mortality scales with the rate of pathogen growth, dV/dt, a quantity which should be related to the instantaneous rate of conversion of host resources into pathogen resources.) In some cases, for simplicity we will assume a linear relationship between pathogen load and host mortality, so that $m(V) = a + bV$, where a is the intrinsic mortality rate and b gives the increase in mortality due to each additional pathogen. If the pathogen level is time-varying, the host mortality rate also varies over time. The probability that an individual host survives from the time of initial infection at time 0 to time t is given by

$$P_s = \exp\left\{-\int_0^t m[V(\tau)]\, d\tau\right\}.$$

We assume that each individual pathogen, once released (either by shedding or in a burst), has a probability c of infecting another host. If it does not successfully infect another host, following release, the pathogen is assumed to die. In general, the quantity c will be a function of the density of susceptible hosts. For now, we will assume that c stays constant.

2.1. Shedding. With pathogen shedding, the shedding rate s is the fraction of the host pathogens released per unit time, so the rate at which pathogens are released is sV. Since a fraction c of these are assumed to infect another host, the rate of new infections is csV. As a deliberate simplification of within-host pathogen dynamics, in some cases we assume that the pathogen tends to grow exponentially (as in Figure 1B), up to the point of host death, with intrinsic growth rate r. With this assumption, the rate of growth of pathogens in the focal individual host is reduced by the shedding rate s, so that starting from a single pathogen at time 0, the pathogen level (as long as the host is alive) is given by

$$V(t) = \exp\{(r - s)t\}.$$

Given density dependence in the pathogen (e.g., because host resources are used up), or defensive responses, more complex forms of $V(t)$ are likely.

The expected value of the rate of new infections from a host at time t is the product of the probability that the host is alive at time t and the rate of new infections assuming the host is alive at t, which is

$$csV(t) \exp\left\{-\int_0^t m[V(\tau)]\, d\tau\right\}. \tag{2.1}$$

The expected value of the total number of secondary infections is the integral of this rate over all time:

$$\int_0^\infty csV(t)\exp\left\{-\int_0^t m[V(\tau)]\,d\tau\right\}\,dt.$$

If the number of pathogens is constant at $\hat{V}$ (as in Figure 1A), the total number of pathogens released that infect other hosts, generally known as the basic reproduction number or "rate" (R_0) is simply $cs\hat{V}/m(\hat{V})$. For any time-varying V, to make further progress, we need to make an assumption about how changes in pathogen load translate into changes in host mortality.

For instance, as a limiting case we can assume that uninfected hosts only die due to infection. If the mortality rate m is proportional to V, so that $m(V) = bV$, then the above expression becomes

$$\frac{-cs}{b}\int_0^\infty -bV(t)\exp\left\{-\int_0^t bV(\tau)\,d\tau\right\}\,dt.$$

Note that the integrand is the derivative of the exponential term (which is the probability of host survival), so the integral is that term, evaluated at the limits. At the lower limit, this is 1, and at the upper limit, this is the probability of survival for an infinite time after infection, which we assume is 0 [this is true for any $V(t)$ that does not eventually go extinct within the living host, such as those shown in Figure 1B–D]. Taking into account the appropriate signs at the two limits, the total expected number of secondary infections produced per primary infection is simply $R_0 = cs/b$.

Note that this result does not depend upon the details of within-host dynamics. A pathogen population within a host can reach a high level, at which point it will be shedding many pathogens into the environment, but such a pathogen will also tend to kill off its host quickly. Conversely, if the pathogen stays at a low level, it sheds at a low rate, but its host lives much longer. The two effects cancel out, given our linear assumption relating host mortality to pathogen load.

If the pathogen grows exponentially, the condition that host survival eventually goes to 0 limits s to be no greater than r, the intrinsic rate of within-host pathogen growth. (If this is not true, the pathogen goes extinct within the host because it is shed more rapidly than it can grow; the realized rate of growth is $r - s$). Comparing different pathogen strains, the one with the greatest R_0 is the one with the greatest value of the ratio cs/b, so the maximal value is approximately cr/b.

However, as other authors have noted (e.g., [**6**]), R_0 is not typically the best measure of pathogen fitness. One reason for this is that in considering infection of novel hosts, one must consider the tertiary infections spawned by each secondary infection, the quarternary infections then generated, and so on (other reasons have to do with the potential for co-infection, which we are assuming does not occur). Another measure of fitness might thus be the total growth rate of the pathogen population, summed over all infected hosts, a fitness measure which is called the Malthusian parameter [**5**]. In the next few paragraphs we derive an expression for the Malthusian parameter of a pathogen, taking into account both within-host dynamics and between-host transmission.

We can derive a measure for the total growth rate of the pathogen in the host population, if we assume that the number of susceptible hosts does not change over the time scale in question (e.g., because the number is large, or replenishment occurs rapidly). We assume that after a sufficient time has passed since an initial infection, the total amount of pathogen in the population will grow exponentially at a rate of r', so that the total pathogen at time T from an initial infection at time 0 for sufficiently large T has the form $A\exp\{r'T\}$ (the quantity A reflects initial transients). To estimate r', we use expression (2.1) above, which gives the expected rate of secondary infections produced at time t from the initial infected host. Each of these secondary infections should likewise result in an exponentially growing pathogen population, of the form $A\exp\{r'(T-t)\}$ (since we assume that the susceptible host density remains constant, the only difference between secondary and initial infection is the start time). The total pathogen population at time T resulting from secondary infections in the interval $(t, t+dt)$ from an individual infected at time 0 is therefore

$$csV(t)\exp\left\{-\int_0^t m[V(\tau)]\,d\tau\right\} A\exp\left\{r'(T-t)\right\}\,dt. \tag{2.2}$$

The total pathogen population at time T resulting from the initial host is this quantity integrated over all time (or at least until the time at which the rate of new infections from this host becomes negligible), which we set equal to $A\exp\{r'T\}$, since this is the assumed ultimate form of the total pathogen population from the initial host. After dividing both sides by $A\exp\{r'T\}$, we obtain

$$\int_0^\infty csV(t)\exp\left\{-\int_0^t m[V(\tau)]\,d\tau\right\}\exp\{-r't\}\,dt = 1. \tag{2.3}$$

Given assumptions about the functional forms of within-host pathogen dynamics, and the mortality term, this equation can be solved numerically for r'. [In the Appendix, we provide some numerical results verifying that this expression for asymptotic growth rate is correct, assuming within-host pathogen dynamics are exponential up to the point of host death. In general, $r' < r$.]

Note that our derivation of this implicit expression for total pathogen population growth (which incorporates both within-host dynamics, and between-host transmission) is analogous to the familiar derivation of Euler's equation in age-structured demography (e.g., [**19**]; see [**4**] and [**14**] for other applications of the Euler's equation in pathogen population dynamics). In place of age-specific mortality, we have a mortality term that depends upon pathogen load; in place of age-specific fecundity, we have an effective rate of transmission per infected host, which also depends upon pathogen load (as well as shedding rate).

If the pathogen level is fixed at $\hat{V}$, then the mortality is fixed at $m(\hat{V})$, and (2.3) can be solved for $r' = cs\hat{V} - m(\hat{V})$. Recall that in this case $R_0 = cs\hat{V}/m(\hat{V})$, so $r' = m(\hat{V})[R_0 - 1]$. For a given $R_0 > 1$, higher mortality gives a higher growth rate, since the same number of pathogens is released, but they are released sooner.

If the host mortality rate in the absence of the pathogen is a, we can write the mortality rate as $m(V) = a + f(V)$. Then equation (2.3) above becomes

$$\int_0^\infty csV(t)\exp\left\{-\int_0^t f[V(\tau)]\,d\tau\right\}\exp\left\{-(r'+a)t\right\}\,dt = 1.$$

Since a appears only in the term in which it is added to r', we can solve the equation for $a = 0$ to get r'_0, and then find r' using $r'_0 = r' + a$, or $r' = r'_0 - a$. Therefore, the background host mortality rate a directly reduces r'.

As a limiting case, if we assume that host mortality is independent of the pathogen population within the host [$f(V) = 0$], and that there is exponential within-host growth, then $V(t) = \exp\{(r - s)t\}$ and the above equation can be solved, giving $r' = r - s(1 - c) - a$. In this case, the maximum r' is obtained by the minimal shedding of the pathogen. With constant mortality rate, all hosts that are alive have the same expected mortality, and the same pathogen growth rate. Therefore, unless $c = 1$, there is a penalty associated with being shed (the possibility of not finding a new host) but no benefit. In this case, the expected number of secondary infections from an infected host is $cs/(a + s - r)$ (unless the denominator is negative or 0, in which case it is infinite, since in that case the pathogen population size grows faster than the probability of host death). Assume that $r < a$ (e.g., a "slow virus"), and that we are comparing pathogen clones which may vary in their shedding rate. It is interesting to note that in this limiting case, the pathogen that generates the *highest* number of secondary infections per primary infection (over the lifetime of that infection, viz. R_0) is the one that has the lowest growth rate (as measured by its Malthusian parameter), in the host population as a whole.

The effect of varying the other parameters can be found by numerically solving the above equation. One question of interest may be how the rate of increase depends upon shedding rate. We earlier saw that R_0 increases linearly with s (with linear mortality). If R_0 is used as a measure of pathogen fitness, then more shedding is better. However, the result of numerical explorations shows that often r' increases with increasing s for low values of s, but eventually peaks, and then decreases with further increase in s (Figure 2A). The reason is basically that a pathogen remaining in the host can multiply, but faces the risk that the host will die, after which all pathogens in the host also die. However, a pathogen being shed also faces a risk of death. If there is too low a shedding rate, the pathogen population rises quickly and increases the host mortality rate, and thus many pathogens are lost at host death. Too high a shedding rate and the host lives longer, but the pathogen level increases more slowly within each host.

Thus, an intermediate rate of shedding may lead to the maximal overall rate of increase of the pathogen in the host population. Note that this arises because of the interwoven dynamics of within-host pathogen growth, and between-host infection, and not because of any assumption about tradeoffs among parameters in the model. If pathogen clones differ in their shedding rate, but are otherwise identical, the one with an intermediate rate of shedding will have the greatest growth rate in the host population.

Surprisingly, increasing the shedding rate can lead to an increase in the average time to release of pathogens (with exponential pathogen growth). With low shedding, the pathogen population rises quickly and kills the host early, so that the average time of shedding is short. With high shedding, the pathogen population grows more slowly and so the host lives longer. Since the pathogen population is growing (if $r > s$), the number of pathogens shed per unit time increases as long as the host is alive. Therefore, the average time of pathogen release tends to increase with higher shedding. Somewhat counter-intuitively, with a decrease in the probability of infecting a new host (c), the overall growth rate r' decreases, as expected, but the shedding rate at which r' is maximized can *increase*, because increased shedding means each pathogen population within a host on average spends more

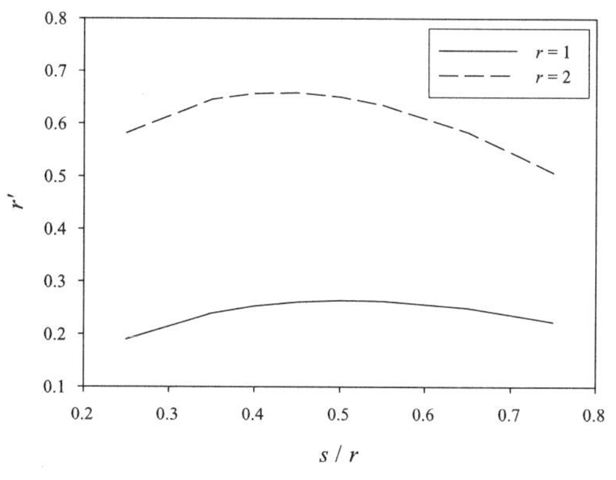

A. $a = 0$, $b = 0.01$, $c = 0.1$.

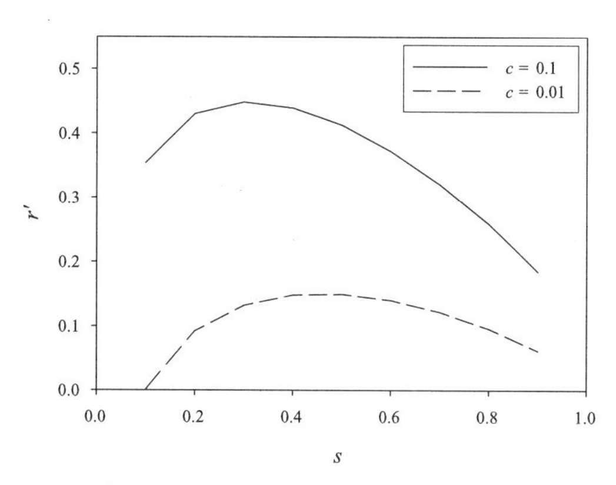

B. $a = 0$, $b = 0.001$ and intrinsic growth rate $r = 1$.

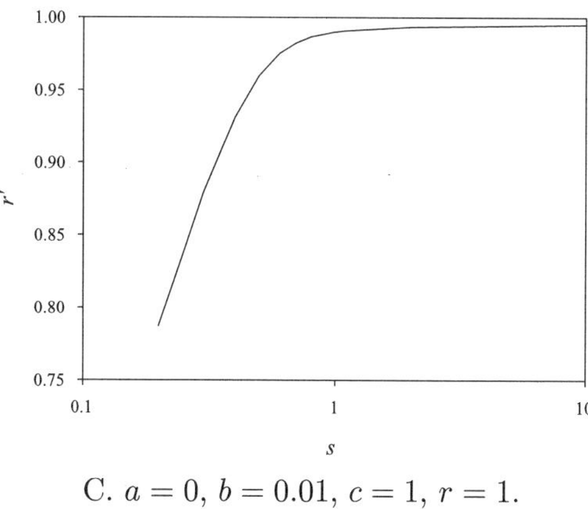

C. $a = 0$, $b = 0.01$, $c = 1$, $r = 1$.

FIGURE 2. Pathogen population growth rate for shedding model with linear mortality rate. Note there is an intermediate optimum in A and B, but not C.

time in the protective environment of the host, before killing its host. In effect, increased shedding provides a mechanism for decreased virulence, so this result is consistent with familiar models for the evolution of virulence. Figure 2B shows a numerical example of this effect, assuming a linear relationship between pathogen load and host mortality.

For these results on pathogen demography to be translated directly into assertions about the likely direction of pathogen evolution requires additional assumptions. For instance, if there are distinct pathogen clones, and each host has a single such clone (i.e., there is no admixture of pathogen clones within individual hosts), then relative growth rates can translate into relative Malthusian fitnesses.

Although there is often an optimum shedding rate, if $c = 1$ (i.e., each shed pathogen infects a new host), then r' monotonically increases with increasing s (see Figure 2C for an example). In this case, each shed pathogen finds a new host, so there is no longer any pathogen mortality associated with being shed. Therefore, the pathogen does better if shedding is higher, since this minimizes the chance of being in the host when it dies. The growth rate r' with $c = 1$ and high shedding approaches the within-host pathogen growth rate r.

Finally, the quantity c is likely to change over time, not least because as the infection spreads in the host population, the density of susceptibles will decline. Incorporating the demographic impact of infection on host numbers is required for a full ESS analysis of evolution in the pathogen.

An approximate expression for r' is the logarithm of R_0 divided by the average time from host infection to secondary infection. This expression is

$$\frac{R_0 \ln[R_0]}{\int_0^\infty tcsV(t)\exp\left\{-\int_0^t m[V(\tau)]\,d\tau\right\}dt}$$

where

$$R_0 = \int_0^\infty csV(t)\exp\left\{-\int_0^t m[V(\tau)]\,d\tau\right\}dt.$$

This expression would be exact if all pathogens were released at the average release time. The closer the release times cluster around the mean, the better this approximation becomes. It is not good for constant virus loads, since in this case release times have an exponential distribution, which has a high variance. The approximation is much better with exponential pathogen increase and pathogen-dependent mortality.

2.2. Bursting. The other model of pathogen spread has all pathogens retained until the death of the host, at which time a fraction c of the released virions infect a new host. With bursting, the rate at which new infections are generated is the product of the death rate, the pathogen number at the time of death, and the probability that a pathogen infects a new host (c). The death rate at time t is the probability that the host has survived to time t, multiplied by the mortality rate. Therefore, the new infection rate with bursting is

$$cV(t)m[V(t)]\exp\left\{-\int_0^t m[V(\tau)]\,d\tau\right\}.$$

If we assume exponential pathogen growth within the host, $V(t) = \exp\{rt\}$ with bursting, since there is no shedding to reduce the within-host growth rate. The equation for population growth rate of the pathogen can be found in a similar manner to that described above for shedding:

$$\int_0^\infty cV(t)m[V(t)]\exp\left\{-\int_0^t m[V(\tau)]\,d\tau\right\}\exp\{-r't\}\,dt = 1. \tag{2.4}$$

Note that in contrast to expression (2.3), the "fecundity" analogue now includes host mortality. With bursting, and within-host exponential growth, the overall pathogen population growth rate is an increasing function of the within-host pathogen growth rate r and infection rate c, and is equal to r when $c = 1$. For $c < 1$, it is a decreasing function of the mortality parameters, since pathogen growth proceeds at its maximum rate as long as the host is alive, and host death leads to a loss of some pathogens. In a more realistic scenario in which pathogen numbers saturate or decrease at some point in the infection, host mortality is likely to be beneficial to the pathogen at or before the time the pathogen level approaches its maximum.

With exponential pathogen growth, if mortality is independent of pathogen level ($m = a$), then the above equation can be solved for $r' = r - (1 - c)a$. If the host instead dies when the pathogen reaches a threshold level V_T, then the total pathogen population grows at rate r for a time $\ln(V_T)/r$ [the time it takes pathogen in a host to increase from the initial propagule size (scaled to unity) to the threshold level], after which host death occurs. Only a fraction c of the released pathogens finds a new host. Therefore, over one cycle (and all hosts will be synchronized if we start from one host), the pathogen population increases by cV_T in time $\ln(V_T)/r$. The equivalent r' is the logarithm of the increase divided by the time between infection and host death, leading to

$$r' = \frac{r\ln(cV_T)}{\ln(V_T)} = r\left(1 + \frac{\ln(c)}{\ln(V_T)}\right)$$

where the last entry in the bracketed term is negative, since $c < 1$.

We have thus far focused on the consequences of assuming that host death occurs prior to the equilibration of the pathogen within the host, but we have largely assumed a very simple model for within-host growth: exponential growth following infection. In the following sections, we will provide some explorations of other aspects of transient dynamics. Note that expressions (2.1) through (2.4) above do not depend upon explicit assumptions about within-host pathogen dynamics (e.g., exponential growth), provided all pathogens in the host are potentially available for shedding or release at host death. One way to generalize the model would be to allow the shedding rate itself to depend upon pathogen population size. Another route of generalization, which we touch on below, is to have a spatial structure for the pathogen within individual hosts, so that only a fraction of the total population is available for export to the external environment.

3. Within-host transient dynamics

3.1. Single-compartment, single-cell-type model. We now consider a particular class of pathogens - viruses. Because viruses must replicate within host cells, and are transmitted via free virion particles, one must monitor levels within the host of healthy cells, infected cells, and free virions. A simple dynamic model with the virus attacking only one cell type within one tissue or organ (e.g., [**11**]) is

$$\begin{aligned} \frac{dn}{dt} &= \lambda - \mu n - \beta q n, \\ \frac{dn^*}{dt} &= \beta q n - \mu^* n^*, \\ \frac{dq}{dt} &= \nu\mu^* n^* - \mu' q - \beta q n \end{aligned} \tag{3.1}$$

where n is the number of uninfected cells, n^* is the number of infected cells and q is the number of free virions. Uninfected cells are input at a constant rate λ, die at a per-capita rate of μ, and are infected at rate β. Infected cells die at rate μ^*, and release ν virions at death. Virions are cleared at rate μ'. The model is quite simple, yet illustrates some general points about transient dynamics.

3.2. Increase in virus when rare. If there is no infection, then there are no infected cells or free virions, and the number of uninfected cells should equilibrate at $\hat{n}_0 = \lambda/\mu$. Given that uninfected cells are at this equilibrium, the basic reproduction ratio (the average number of free virions produced within an infected host per free virion, or the number of infected cells produced per infected cell, when virus is rare) is

$$R_0 = \frac{\beta\lambda\nu}{\beta\lambda + \mu\mu'} = \frac{\beta\hat{n}_0\nu}{\beta\hat{n}_0 + \mu'}.$$

This is derived as follows. Virions either infect cells (at rate $\beta\hat{n}_0 = \beta\lambda/\mu$ if virus is rare), which on death release ν virions each, or are cleared (at rate μ'), producing no virions. It is assumed that both infection and clearance of virions are Poisson processes. If so, then the probability that a virion infects a cell before it is cleared is $(\beta\lambda/\mu)/(\beta\lambda/\mu + \mu') = \beta\lambda/(\beta\lambda + \mu\mu')$, which is the rate of infection divided by the sum of the two rates. The expected value of the number of virions produced by a free virion is the probability of infection multiplied by the number of free virions produced by an infection (ν), plus the probability of clearance multiplied by the number of free virions produced in this case (which is 0, so this second term drops out). Therefore, the expected value of the number of free virions produced by a free virion is $\beta\lambda\nu/(\beta\lambda + \mu\mu')$, which is the basic reproductive ratio, R_0. (A similar argument can be used projecting from infected cell to infected cell.)

For the virus to increase when rare, R_0 must exceed 1, which implies that $\beta(\nu - 1)\hat{n}_0 > \mu'$. If this condition is met, virus introduced at a low level will increase exponentially (after a transient period during which infected cells and virions reach their steady-state ratio, and until the virus reaches high enough levels to affect the uninfected cell number). The rate of exponential increase of the virions and infected cells is the dominant eigenvalue of the Jacobian of the infected cell and virion equations above, evaluated at the no-infection equilibrium. The Jacobian is

$$\begin{bmatrix} -\mu^* & \beta\hat{n}_0 \\ \nu\mu^* & -\beta\hat{n}_0 - \mu' \end{bmatrix}$$

and the dominant eigenvalue is

$$s_d = \sqrt{\left(\frac{\mu^* - \mu' - \beta\hat{n}_0}{2}\right)^2 + \beta\hat{n}_0\nu\mu^*} - \frac{\mu^* + \mu' + \beta\hat{n}_0}{2}. \tag{3.2}$$

(The non-dominant eigenvalue (s_n) is the same except for a negative sign before the first term.)

During the initial phase of the infection, infected cells and virions will reach a steady-state ratio which can be found from the eigenvector corresponding to the dominant eigenvalue, and both then increase exponentially at the same rate (until they start to have an effect on uninfected cell number). The ratio of virions to infected cells during this initial phase of increase is

$$\left[\sqrt{\left(\frac{\mu^* - \mu' - \beta\hat{n}_0}{2}\right)^2 + \beta\hat{n}_0\nu\mu^*} - \frac{\mu^* - \mu' - \beta\hat{n}_0}{2}\right]\frac{1}{\beta\hat{n}_0}.$$

Virions and infected cell will not, in general, have this ratio at the start of the infection. However, the initial transient response (the part of the transient response occurring when the virus is rare) for either virions or infected cells will have the form $A\exp\{s_d t\} + B\exp\{s_n t\}$. The first term is the exponential increase at the dominant eigenvalue. This is reached as soon as the second term, due to the non-dominant eigenvalue, becomes negligible. Therefore, the non-dominant eigenvalue (which is always negative for this system) determines how fast the virions and infected cells reach an exponential increase, with the ratio between them then given by the above expression. If uninfected cell number drops slowly enough, the virion and infected cell numbers could adjust their growth rates and ratio according to the above expressions with the actual uninfected cell numbers substituted for the no-virus uninfected cell numbers. However, as the uninfected cell number drops, the magnitude of the non-dominant eigenvalue drops, so that the speed with which the ratio of virions to infected cells adjusts drops. Often, the uninfected cell number drops too fast for this approximation to hold.

3.3. Equilibria. The above dynamical system has a stable equilibrium, provided the virus can increase when rare. In this case, the eventual equilibrium is

$$\begin{aligned}\hat{n} &= \frac{\mu'}{\beta(\nu-1)} \approx \frac{\mu'}{\beta\nu}\\ \hat{n}^* &= \frac{\lambda}{\mu^*} - \frac{\mu\mu'}{\beta\mu^*(\nu-1)} \approx \frac{\lambda}{\mu^*}\\ \hat{q} &= \frac{\lambda(\nu-1)}{\mu'} - \frac{\mu}{\beta} \approx \frac{\lambda\nu}{\mu'}.\end{aligned} \tag{3.3}$$

[The approximate equilibria are obtained by assuming that at the equilibrium the vast majority of uninfected cells become infected before they die, and that the vast majority of virions are cleared before they infect a cell; see [**11**].]

If there is a trade-off between burst size ν and infectivity β, so that as burst size increases infectivity decreases (an infected cell can produce either many virions that are poor at infecting new cells, or fewer virions that are each better at infection), then this trade-off may have different consequences for equilibrium virion numbers, infected cell numbers and initial growth rate. The equilibrium virion number is $\hat{q} = \lambda(\nu-1)/\mu' - \mu/\beta$. The effect of varying burst size and infectivity are in opposite directions, but the effect of burst size is likely to be larger. If $\beta\nu$ is constant $= d$, then virion equilibrium abundance increases with increasing ν as long as $\lambda/\mu' > \mu/d$. Substituting $\beta\nu$ for d and rearranging, this condition is $\beta\nu\lambda/\mu > \mu'$, while the condition for virus to increase when rare is $\beta(\nu-1)\lambda/\mu > \mu'$. Hence, as long as the virus can increase when rare, virion abundance will increase with increasing ν. The infected cell equilibrium is $\hat{n}^* = \lambda/\mu^* - \mu\mu'/(\beta\mu^*\nu - \beta\mu^*)$.

In this case, with the above trade-off, increasing β decreases the magnitude of the denominator of the second term, which increases the magnitude of this term, which decreases infected cells. So again, increasing ν would increase infected cell numbers. However, this would likely have a smaller effect than the effect on virions. The direction of the effect on the initial rate of increase is in the same direction.

3.4. Transient response - initial overshoot. The transient response of the system of equations given in (3.1) often contains significant oscillations (Figure 1D). Particularly important is the initial overshoot of the equilibrium. The tendency to overshoot depends on the relative values of the uninfected and infected cell death rates and virion clearance rate. If there is an overshoot, the magnitude is also mostly determined by these quantities, although the other parameters have some effect also on the magnitude of the overshoot (we assume in the results below that λ is proportional to μ, so the initial uninfected cell number is the same). If the uninfected cell death rate (μ) is lower than both the infected cell death rate (μ^*) and virion clearance rate (μ'), all three quantities (uninfected and infected cells, and virions) tend to overshoot their respective equilibria (uninfected cell numbers decrease as the virus increases, so when it overshoots its equilibrium, it becomes less than its equilibrial value). If $\mu^* > \mu, \mu'$, then only the infected cell number will usually overshoot. If both are true ($\mu^* > \mu' > \mu$), then all three quantities tend to overshoot. The infected cell death rate would normally be higher than the uninfected cell death rate. This generally results in overshoot of all quantities.

Figure 3 shows the maximum virion abundance divided by the equilibrium virion abundance (a measure of the overshoot) when the virus increases from low levels, for a virion clearance rate of 1 (time and other rates can be scaled by virion clearance rate, so this assumption does not lead to a loss in generality). The surface is at 1 (i.e., no overshoot) for uninfected cell death rate of 1 or more, since in this region we have $\mu \geq \mu'$. It is also 1 on and to the left of the $\mu = \mu^*$ diagonal, since in this region $\mu \geq \mu^*$. The non-unity values at the right and back of the Figure represent overshoot of the virion number, which increases as μ^*/μ increases, with the value of μ being the more important. The overshoot of uninfected cell number has the same form (the equilibrium over the minimum value, since uninfected cells actually undershoot), but slightly lower values where overshoot occurs (maximum difference of 10% in the right corner where the overshoot reaches the maximum values). Infected cell number overshoot is similar for low μ^* but much higher for high μ^*. It also is equal to 1 only to the left of the $\mu = \mu^*$ diagonal.

These patterns describe within-host dynamics. Note that if virions released from the infected host are the means of transmission of the infection to novel hosts, then we could link the above model (3.1) to our earlier treatment of pathogen population growth over the entire host population. To do that for the shedding scenario, we would need to explicitly include a loss term due to shedding (which could be added to the virion clearance rate). For instance, a clone with a higher μ^* has a higher initial exponential growth rate in its host, and if the host dies while the dynamics are in this transient phase, it should also have an overall higher growth rate in the host population as a whole (whether or not the mode of transmission is via virions, or infected cells). But if the pathogen reaches a dynamic equilibrium in the host, and the host is long-lived, this parameter becomes irrelevant to the number of virions present. And if between-host transmission is via infected cells, at equilibrium the abundances of those cells are inversely related to their death rates

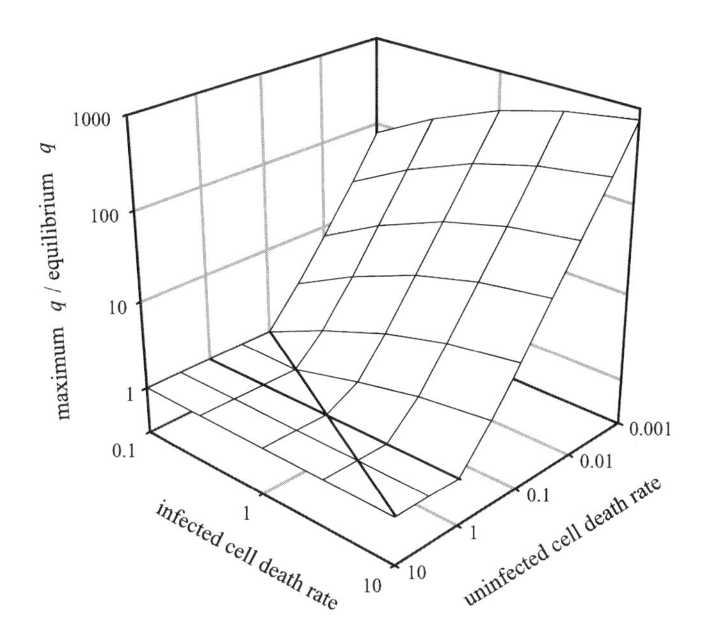

FIGURE 3. Magnitude of initial overshoot of virion abundance, relative to its equilibrium value, for $\beta = 0.0001$, $\nu = 50$, $\mu' = 1$, and $\lambda = 10000\mu$ (so the initial number of uninfected cells is 10000). There is no overshoot unless the infected cell death rate and virion clearance rate are both higher than the uninfected cell death rate.

[see (3.3)]. So whether or not evolution favors viruses which rapidly kill host cells, or disfavors such viruses, may depend on whether or not within-host dynamics is largely transient, or at equilibrium, over the lifetime of the infected host.

As a step towards this evolutionary question, one can couple within-host dynamics to between-host transmission using the machinery derived above, to generate the growth rate of the pathogen in the host population. Figure 4 shows two examples of pathogen population growth rate as a function of shedding rate calculated using equations (3.1) and some approximations based on them. The dotted lines were calculated numerically using the dynamics in equations (3.1), with the shedding rate added to the virion clearance rate parameter μ', (since shedding represents a loss of virions). This corresponds to Figure 1D, with the full intrahost dynamics incorporated into calculating r'. The solid lines were calculated assuming that the virion level was constant at the virion equilibrium level of equations (3.1) (with shedding rate added to μ'; note that this corresponds to the assumption of Figure 1A). The dashed lines were calculated assuming that the virions increased exponentially at the rates given by the dominant eigenvalues given in equation (3.2) (with shedding rate added to μ'; this matches the scenario of Figure 1B).

In both panels the host death rate is assumed to be directly proportional to the virion level. As we demonstrated above, this means that $R_0 = cs/b$, independent of the virion dynamics, as long as host death eventually occurs. We chose $c = 10b$ in both cases, so that $R_0 = 1$ at $s = 0.1$. Therefore, all curves have $r' = 0$ at this

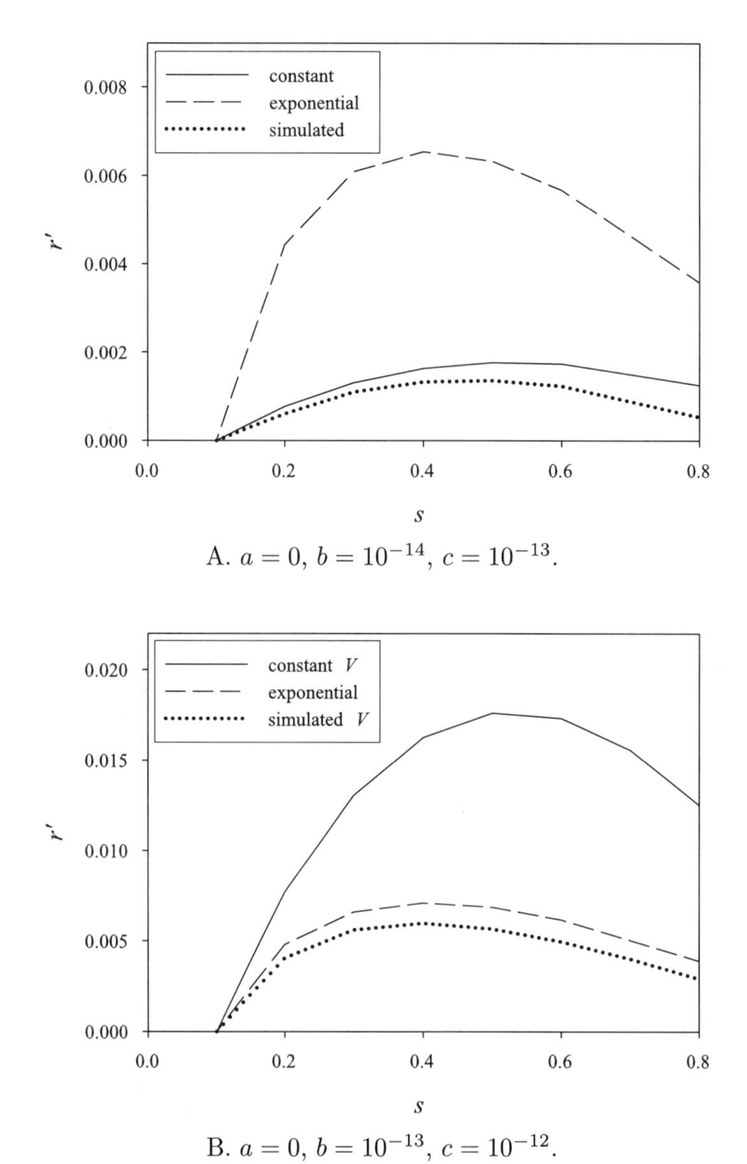

A. $a = 0$, $b = 10^{-14}$, $c = 10^{-13}$.

B. $a = 0$, $b = 10^{-13}$, $c = 10^{-12}$.

FIGURE 4. Pathogen population growth rate for shedding model with linear mortality rate and virus dynamics calculated using equations (3.1) (dotted curve), assuming virion level constant at equilibrium virion level from those equations (solid curve) or assuming a virion level exponentially increasing at the dominant eigenvalue in equation (3.2). The examples depicted show that at times one can accurately assume either a simple exponential pattern for within-host dynamics, or a rapid approach to a within-host equilibrium. Parameters for virus model as in Figure 1D.

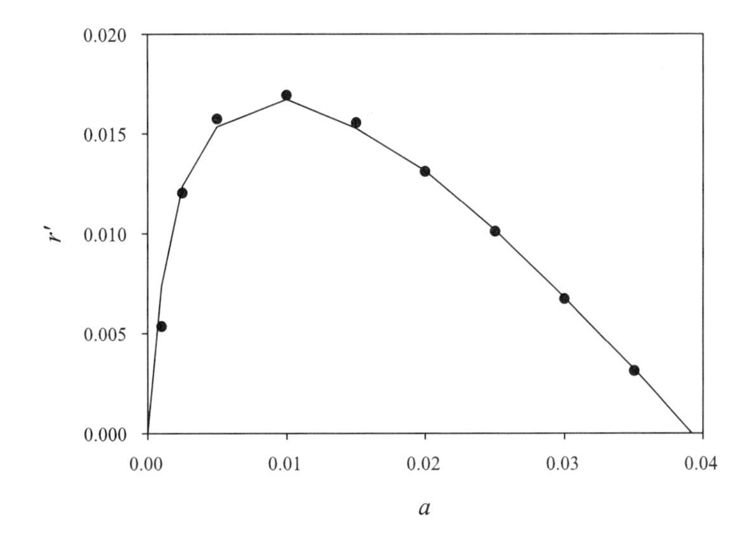

FIGURE 5. Pathogen population growth rate for a bursting model with virus-independent mortality and virus dynamics calculated using equations (3.1) (parameters for virus model as in Figure 1D), as a function of mortality rate (a). Other parameters are $b = 0$, $c = 10^{-10}$. Solid lines gives the calculated population growth rate. Solid circles represent an approximation based on an estimated probability of release at the peak abundance during the transient (see text); in this example, the approximation is quite accurate.

shedding rate. In Figure 4A, the death rate parameter b is set low enough that the host has a fair probability of surviving the initial transient, after which the virion level approaches the equilibrium. Therefore, the dotted line using the full dynamics of (3.1) is close to the results using the equilibrium virion levels. In Figure 4B, the death rate parameter is increased an order of magnitude, which makes it unlikely that a host could survive the initial transient. Therefore, in this case, the numerical results with the virion dynamics described by equations (3.1) are close to the results produced assuming exponential increase.

For the burst model, the existence of transients could potentially have a large impact on the number of virions present at the time of host death. If infected hosts are short-lived, relative to the transient phase of infection dynamics, then small differences in the expected time of host death could have large impacts on the likelihood of viral transmission. For instance, in Figure 1D, viral abundance plummets several orders of magnitude over just a few time steps, after the initial peak abundance. If the pathogen could determine the time of host death, so as to maximize its own fitness, it is clear that this death should occur no later than the initial peak in virion abundance. Early host death in this example both capitalizes on the internal transient overshoot, and permits rapid release of the pathogen.

Figure 5 shows the virus population growth rate as a function of host death rate, for the burst model with a fixed host death rate, as a function of that death rate (in this example, for illustrative purposes we assume that the pathogen is avirulent, as measured by effects on host mortality rates). For the example shown, the growth rate peaks at about 0.01, which gives a mean time of death of 100 time units. This is approximately the time that the virion level peaks in the virus model (the virion dynamics used in Figure 5 are shown in Figure 1D). If the death rate is higher, then individuals tend to die before the peak and so fewer virions are released. If the death rate is lower, then individuals tend to live longer and again release fewer virions (because the virus settles into an equilibrial within-host abundance less than observed during the peak of the transient phase). In the latter case, the virions also tend to be released later, which further depresses the growth rate.

The growth rate in this case depends largely on the host death rate at the time of the peak virion abundance, which is at time 112 in this example. Since the time of death has an exponential distribution with mean $1/a$, the probability density for death at this time is $a \exp\{-112a\}$. The expected number of secondary infections due to virions released at this time will be equal to the product of this value, the infection constant c, the peak virion density, and the effective duration of the peak (to convert the probability density above to a probability of death). If this peak release were the only release, then the population growth rate r' would be equal to the natural logarithm of this number of secondary infections divided by the time of the peak (112). The only quantity that has to be estimated is the effective duration of the peak. For illustration, this was approximated by dividing the integral of the virion density from the start of the infection to the time of the first relative minimum by the peak density. The solid circles in Figure 5 show the results of this calculation. The agreement is very good, indicating that variation in the probability of host mortality at the peak virion time can account for the shape of the curve.

This model thus suggests that there may be a nonlinear relationship between host mortality rates and the overall rate of spread of a pathogen in a host population, when the pathogen relies upon host death for infective propagules to be released, and when there are strong transient effects in within-host dynamics. In the example shown (corresponding to Figure 1D), an intermediate death rate of the host catches the pathogen population during a substantial overshoot of numbers, above the potential within-host equilibrium. Again, the existence of an optimal mortality rate for the host, with respect to maximizing the overall rate of growth of the pathogen, does not reflect a tradeoff in the usual sense, but rather emerges from the interplay of within-host and among-host dynamics.

3.5. Two or more compartments. Transient dynamics can also be important for more complex hosts with internal structure. Here, we briefly consider a more general virus model that includes multiple compartments and cell types (see also [**10**]). The variables now are the number of virions in each compartment $i (q_i)$, and the number of uninfected (n_{ji}) and infected (n^*_{ji}) cells of each type j and compartment i. We assume that cells and virions can move between compartments.

The system is described by the following equations.

$$\begin{aligned}
\frac{dq_i}{dt} &= \sum_j \mu^*_{ji}\nu_{ji}n^*_{ji} - \mu'_i q_i - \sum_j \beta_{ji}n_{ji}q_i + \sum_{k\neq i} m_{ki}q_k - \sum_{k\neq i} m_{ik}q_i,\\
\frac{dn_{ji}}{dt} &= \lambda_{ji} - \mu_{ji}n_{ji} - \beta_{ji}n_{ji}q_i + \sum_{k\neq i} M_{ki}n_{jk} - \sum_{k\neq i} M_{ik}n_{ji},\\
\frac{dn^*_{ji}}{dt} &= \beta_{ji}n_{ji}q_i - \mu^*_{ji}n^*_{ji} + \sum_{k\neq i} M^*_{ki}n^*_{jk} - \sum_{k\neq i} M^*_{ik}n^*_{ji}
\end{aligned}$$

which include the same parameters as the simpler model, with the subscripts j indicating the cell type, and i the compartment. The new parameters are migration rates: m for virions, M for uninfected cells and M^* for infected cells. The first subscript on the migration parameters denotes the source compartment, and the second the destination. Migration rates are assumed to be independent of cell type.

Analysis of this set of equations is difficult even with one cell type and two compartments. Some initial steps towards such an analysis are reported in [**18**], and elsewhere we intend to explore spatially structured hosts in more detail. Rather than explore the above model in full, here we merely touch on some limiting cases.

First, consider what happens when movement of virions and cells between compartments is slow, compared to dynamics within compartments. During the initial exponential growth phase, the overall growth rate of the infection is approximately just the growth rate expected in the compartment with the highest growth rate [**18**]. Thus, all else being equal, the compartment where the virus growth rate is highest will dominate the initial transient dynamics of the system. For instance, the compartment with the highest value for β or μ^* will also permit the most rapid initial growth rate of the pathogen. If the host dies while the pathogen is still in this transient phase of initial increase, then differences between pathogen clones in these habitat-specific parameters will translate into differences in Malthusian fitness. [Quantitatively describing these effects would require a more complex analogue of expressions (2.3) and (2.4), permitting for instance different shedding rates from different compartments, and different compartment-specific impacts of the pathogen on host mortality.] If the pathogen settles into a within-host equilibrium, then at low movement rates the density of virions and infected cells should approximate that given by equations (3.3). At equilibrium, the local density of virions does not depend on infected cell death rates, and it only weakly depends on infection rates of healthy host cells, so within-host heterogeneity in these parameters is not likely to influence overall transmission to new hosts.

At high movement rates, in this model if there is heterogeneity in parameters among compartments, the growth rate is depressed below the maximal possible. Moreover, one finds that the potential for dramatic overshoots in the transient phase as the infection moves towards a within-host equilibrium can be reduced (for details see [**18**]). As with the single-compartment case discussed above, a full treatment would need to couple these within-host transient dynamics with among-host transmission. We intend to address this theme in a future contribution.

4. Conclusions

In conclusion, we have argued that a consideration of transient dynamics of within-host infection may be important in understanding the epidemiology of infectious diseases. We have presented a formulation for characterizing the overall growth rate of an infectious disease, which includes both within-host dynamics and between-host transmission, when the disease is sufficiently rare that the supply of available hosts can be viewed as a constant. We have shown that even simple models of within-host pathogen dynamics can have phases in which numbers overshoot the final equilibrium, and suggest that such phases may be of selective importance in pathogen evolution. The distinction between shedding and bursting in pathogen transmission is relevant to pathogen dynamics within individual hosts, as well as to pathogen dynamics in host populations as a whole. Finally, we have noted that many hosts have substantial internal spatial heterogeneities, and that such heterogeneities can influence the magnitude and character of transient dynamics. There is a growing appreciation in general ecology of the importance of transient dynamics and spatial processes [**8, 9, 22**]. In like manner, we suggest that the interplay of transient dynamics and internal spatial structure should provide a fruitful avenue for future empirical and theoretical studies of infectious disease dynamics.

5. Appendix

We wrote a program to simulate the shedding process, so as to assess expression (2.3) in the main text. The growth of the pathogen numbers within each host was assumed to be deterministic (assumed to start at 1 at infection and to increase exponentially at a rate of $r - s$). Host mortality and new host infections were assumed to be Poisson processes which occurred randomly at the rates determined by the equations in the text. Host mortality was assumed to be a linear function of pathogen number. Small, fixed time steps were used. For each time step, the pathogen population of each host was determined using $V(T) = \exp\{(r - s)(T - t)\}$, where T is the time at the start of the time step and t is the time at which the host became infected (the only information we needed to keep about each host). The pathogen level of each host was used to determine its mortality rate, from which the probability of mortality was determined for that time step (the product of mortality rate and time step). A random number generator was used to determine whether death occurred. If so, the host was deleted. If not, the rate of production of secondary infections from that host (and corresponding probability) was determined, and another random number generator used to determine if a secondary infection occurred (the time step was small enough so that there was a very small probability of more than one secondary infection in a step). If so, a new infected host was created with the current time as the time of infection. Initially, there was a single host infected with a single pathogen, and the total pathogen population size was determined at each time step. This process was repeated 400 times and the results averaged at each time. The population size initially increased at a rate close to r (since the virus numbers at the beginning are dominated by the first host). Later, the growth rate dropped, and eventually reached exponential growth at the predicted rate. Figure 6 shows an example. Note that after an initial phase, the lines on the log plot are parallel (the solid line is a simulation run, and the dashed line is the exponential increase predicted, using the rate of increase calculated from equation (2.3), and starting from time 0). In this example, the

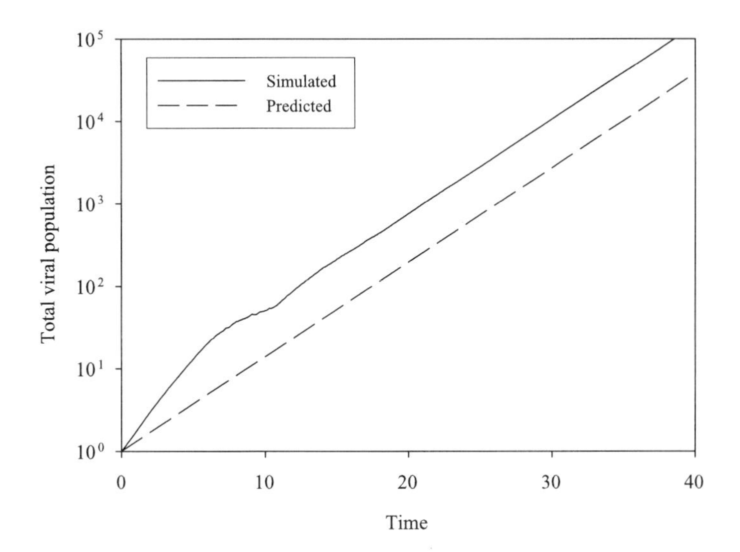

FIGURE 6. Simulated shedding scenario with parameters $a = 0$, $b = 0.01$, $c = 0.1$, $s = 0.5$ and $r = 1$. The solid line is the simulation results. The dashed line is $\exp\{r't\}$ with the value of r' calculated from equation (2.3) in the main text. The assumption used in deriving (2.3) is that eventually $V(t) = A\exp\{r't\}$. Since the lines are parallel (on a logarithmic scale) after an initial period, the assumption of eventual exponential growth is clearly valid for this example.

prediction from equation (2.3) does describe quite well the asymptotic growth rate of the pathogen population.

Acknowledgements

We thank Erin Taylor for editorial assistance, and John Kelly for very useful suggestions.

References

[1] R. M. Anderson, and R.M. May, *Infectious Diseases of Humans: Dynamics and Control*, Oxford University Press, Oxford, 1991.

[2] R. Antia, B. R. Levin, and R. M. May, 1994. Within-host population dynamics and the evolution and maintenance of microparasite virulence, *Am. Nat.* **144** (1994), 457-472.

[3] D. S. Callaway, and A. S. Perelson, HIV-1 infection and low steady state viral loads, *Bull. Math. Biol.* **64** (2002), 29-64.

[4] T. Day, Virulence evolution and the timing of disease life-history events, *Trends in Ecology and Evolution* **18** (2003), 113-118.

[5] G. De Jong, The fitness of fitness concepts and the description of natural selection, *Quarterly Review of Biology* **69** (1994), 3-29.

[6] U. Dieckmann, Adaptive dynamics of pathogen-host interaction, in (U. Dieckmann, J. A. J. Netz, M. W. Sabelis and K. Sigmund, eds.), *Adaptive Dynamics of Infectious Diseases*, Cambridge University Press, Cambridge, 2002, 39–59.

[7] O. Diekmann, J. A. P. Heesterbeck, and J. A. J. Metz, On the definition and the computation of the basic reproduction ratio R0 in models for infectious diseases in heterogeneous populations, *J. Math. Biol.* **28** (1990), 365-382.

[8] I. Hanski, and O. Gaggiotti, eds., *Ecology, Genetics, and Evolution of Metapopulations*, Academic Press, San Diego, 2004.

[9] A. Hastings, Transients: the key to long-term ecological understanding? *Trends in Ecology and Evolution* **19** (2004), 39-45.

[10] J. K. Kelly, Evolutionary and dynamic models of infection with internal host structure, this volume.

[11] J. K. Kelly, S. Williamson, M. E. Orive, M. S. Smith, and R. D. Holt, Linking dynamical and population genetic models of persistent viral infection. *Am. Nat.* **162** (2003), 14-28.

[12] T. B. Kepler, and A. S. Perelson, Drug concentration heterogeneity facilitates the evolution of drug resistance, *Proc. Nat. Acad. Sci. USA* **95** (1998), 11514-11519.

[13] D. E. Kirschner, R. Mehr, and A. S. Perelson, Role of the thymus in pediatric HIV-1 infection, *Journal of Acquired Immune Deficiency Syndromes and Human Retrovirology* **18** (1998), 95-109.

[14] B. R. Levin, J. J. Bull, and F. M. Stewart, The intrinsic rate of increase of HIV/AIDS: Epidemiological and evolutionary implications, *Mathematical Biosciences* **132** (1996), 69-96.

[15] M. Lipsitch, and B. Levin, The within-host population dynamics of antibacterial chemotherapy: conditions for the evolution of resistance, *Ciba Foundation Symposium* **207** (1997), 112-130.

[16] P. W. Nelson, and A. S. Perelson, Mathematical analysis of delay differential equation models of HIV-1 infection, *Math. Biosci.* **179** (2002), 73-94.

[17] M. A. Nowak, and R. M. May, *Virus Dynamics: Mathematical Principles of Immunology and Virology*, Oxford University Press, Oxford, 2000.

[18] M. E. Orive, M. N. Stearns, J. K. Kelly, M. Barfield, M. S. Smith, and R. D. Holt, Viral infection in internally structured hosts. I. Conditions for persistent infection, *J. Theor. Biol.* **232** (2004), 453-466.

[19] J. H. Pollard, *Mathematical Models for the Growth of Human Populations*, Cambridge University Press, Cambridge, 1973.

[20] V. H. Smith, and R. D. Holt, Resource competition and within-host disease dynamics, *Trends in Ecology and Evolution* **11** (1996), 386-389.

[21] S. J. Snedecor, Comparison of three kinetic models of HIV-1 infection: Implications for optimization of treatment, *J. Theor. Biol.* **221** (2003), 519-541.

[22] M. G. Turner, Landscape ecology: the effect of pattern on process, *Ann. Rev. Ecol. Syst.* **20** (1989), 171-197.

[23] T. H. Wang, Y. K. Donaldson, R. P. Brettle, J. E. Bell, and P. Simmons, Identification of shared populations of human immunodeficiency virus type I infecting microglia and tissue macrophages outside the central nervous system, *J. Virol.* **75** (2001), 11686-11699.

Department of Zoology, University of Florida, 223 Bartram Hall, PO Box 118525, Gainesville, FL 32611.

E-mail address: `rdholt@zoo.ufl.edu`

E-mail address: `mjb01@ufl.edu`

DIMACS Series in Discrete Mathematics
and Theoretical Computer Science
Volume **71**, 2006

Evolutionary and dynamic models of infection with internal host structure

John K. Kelly

ABSTRACT. A large body of mathematical theory has been developed to characterize persistent viral infections within vertebrate hosts. Most of the theory can be classified as either "dynamical models" that predict the population dynamic interaction between virus and host cells or "population genetic models" that predict gene sequence evolution of the pathogen. These two bodies of theory can be linked by considering the demography of the viral population. Gene sequence evolution is usually modeled as a mutation-limited process in which the rate of evolution is proportional to the mutation rate per replication cycle and the number of replication cycles (pathogen generations) per unit time. The latter is clearly dependent on dynamical parameters such as the clearance rate of free virus or the death rate of infected cells. Here, I review analytical methods that explicitly link dynamical and population genetic theories. These methods are extended to consider the evolutionary consequences of internal host structure, the tendency for a virus to infect multiple different compartments (e.g. tissue types). Infection of multiple compartments, coupled with virus migration, may establish "sources" and "sinks" of viral production within the host. Paradoxically, the existence of reproductive sinks can simultaneously reduce the number of viruses within a host and accelerate the genetic evolution of the viral population.

Many viruses establish persistent infections, in which the viral population undergoes many generations of replication within a single infected host. As a consequence, these intra-host populations have their own ecology and evolution, often changing substantially in both numbers and composition over the course of infection. Lentiviruses are particularly notable in this regard, with infection usually lasting the entire lifespan of the host. The most well studied Lentiviruses are the Human Immunodeficiency Viruses (HIV-1 and HIV-2). However, this genus also contains a number of important animal diseases including the Visna-Maedi Virus of sheep, Bovine Immunodeficiency Virus (cattle), Feline Immunodeficiency Virus (cats), Caprine Arthritis Virus (goats), Equine Infectious Anemia Virus (horses), as well as the Simian Immunodeficiency Viruses that are most closely related to HIV.

Key words and phrases. Cell types, Compartments, Dynamics, HIV, SIV.

Author received support from NIH grant 1 R01 GM60792-01A1 and a KBRIN award for Biomedical research.

Lentiviruses can often infect multiple different cell types. For example, HIV infects CD4 T cells, macrophages, peripheral blood mononuclear cells, and dendritic cells of the monocyte lineage, among others [**1, 4, 17**]. Like other retroviruses, the life cycle of a lentivirus involves two distinct stages. The free-living virion stage has an RNA genome. Virions fuse with and infect cells. Once inside the cell, their RNA genome is reverse transcribed to produce a DNA copy, which may subsequently incorporate into the host cell genome. In this second stage, the viral genome persists as proviral DNA for the remainder of the cells lifetime and also in its descendents if that cell proliferates [**41**]. Transcription of the proviral DNA to produce the RNA of daughter virions occurs at varying rates depending on the characteristics of both cell and provirus. In sum, these factors combine to produce a complex demography within the viral population. The viral generation time consists of two intervals, each of which may be variable in length. The duration of the proviral stage is likely to depend on the type of cell that is infected. The duration of the virion stage will depend on a range of factors including the strength of immune surveillance and the abundance of uninfected target cells.

Empirical studies have shown that the abundance of the virions can fluctuate dramatically over the course of infection (see [**38**] and references therein). Comparable fluctuations occur in the numbers of target cells, and in the numbers of immune cells that defend the body from attack. This dynamic of viral and cell populations is a primary determinant of disease progression [**45, 49**]. Mathematical models for viral/cell dynamics have been adapted from population biology [**36, 38, 42, 54**]. These models are similar to those for predator-prey interactions or epidemics on a large spatial scale. Key parameters are the death rates of virions and infected cells, the rate that killed cells are replenished by the host, and the infection constant(s) that determine the rate of new infections given the respective abundances of virions and uninfected cells.

Many viral populations also evolve substantially over the course of a single infection. In part, this is a consequence of their high mutation rates [**29, 44**]. However, many genetic changes reflect adaptive evolution of the viral population in response to the dynamic nature of the host environment [**19, 36, 41, 49, 50, 53**]. Population genetic theory provides a natural framework for the study of gene sequence evolution (e.g. [**6, 7, 10, 12, 13, 22, 23, 32, 33, 46, 47, 48, 57**]). Key parameters in these models are the distribution of selection coefficients on new mutations, recombination rates, the effective population size, population structure statistics (migration rates and local population extinction rates), and the mean generation time of the virus.

In a previous paper [**24**], we began to develop methods that link evolutionary and dynamical models by considering the accumulation of replication cycles within the viral population. Insofar as gene sequence evolution is a mutation-limited process (see below), the rate of evolution should be proportional to the number of replication cycles (pathogen generations) per unit time. In this way, the evolutionary effect of a dynamical parameter, say the death rate of infected cells, can be evaluated by determining its effect on the pathogen generation time (really on the probability distribution associated with generation times given the demographic complexity of the viral population). Of course, this is not likely to be the only evolutionary effect associated with a dynamic parameter. However, given the importance of generation time to the rate of gene sequence evolution, this is a sensible place to start.

The rate of gene sequence evolution within a population, measured say by the number of nucleotide substitutions per unit time, is typically modeled as a mutation limited process [**15**, Chapter 2]. This is most obviously the case for neutral models of sequence evolution, in which the substitution rate is equal to the mutation rate [**26**]. Coalescent approaches, which are based on the neutral theory [**27**], have been extensively used in studies of viral sequence evolution [**46, 47**]. The rate of evolution is also mutation limited in a variety of models involving selection (e.g. [**39**]). With selection, the rate of evolution will also depend on the distribution of selection coefficients associated with new mutations (particularly the fraction that are advantageous vs deleterious) and the effective population size. However, whenever it is possible to associate new mutations with fixed probabilities, the rate of evolution should still be proportional the rate that new mutations are introduced into the population.

I will review two different approaches to predicting the distribution of replication cycle counts (or at least moments of this distribution) as a function of dynamic parameters. The prospective approach, which is developed looking forward in time, is a direct extension of the differential equations that describe cell and virus dynamics. The retrospective approach considers the ancestry of viruses sampled in the present. The "lineage history" of a virus can be treated as a Markovian stochastic process with transition probabilities determined by the dynamical parameters. A number of different results from Markov chain theory can be used to analyze the accumulation of replication cycles in a viral population. Previously, we derived the mean and variance of viral generation using Renewal Theory [**11**, Chapter 13], [**24**]. Here, I solve for the stationary probability distribution associated with the different demographic states present within the viral population. This yields the mean generation time directly.

The second aim of this chapter is to extend the theory to consider internal host structure. To an infecting virus, a vertebrate host is an enormous spatially-structured landscape. Different tissue types or "compartments" are analogous to different habitats for a terrestrial animal. There are compelling reasons to believe this structure impacts viral dynamics and evolution. For example, HIV has been shown to exhibit substantial genetic differentiation among virus populations within different organs [**5, 9, 43, 48, 52**, and references therein] and even different subsections of the same organ [**8, 12**]. While the cause of this differentiation has not been established, it could not persist without substantial compartmentalization of viral dynamics.

There are at least two distinct components to within host-structure. First, within any given tissue compartment, there may be multiple infectable cell types, as well as free virions. Second, as noted above, a host individual is comprised of a suite of local "habitats" coupled by fluid flow and cell movement. A number of authors have explored multiple compartment models, with particular attention to the consequences for anti-viral drug treatment [**3, 25, 28, 34, 35**]. Here, I will analyze a model with both cell and compartment structure. I will first apply the prospective analysis to series of increasingly complicated special cases, and then returning to these cases using the retrospective approach.

1. The basic model

To introduce ideas and notation, I first consider a model without structure. A simple dynamical model of infection predicts the number of virions (q), the number of uninfected cells (n), and the number of infected cells (n^*) as a function of time. Changes in these quantities are governed by the following system of differential equations:

$$\frac{d}{dt}q = \nu\mu^* n^* - \mu' q - \beta qn \tag{1.1}$$

$$\frac{d}{dt}n = \lambda - \mu n - \beta qn \tag{1.2}$$

$$\frac{d}{dt}n^* = \beta qn - \mu^* n^* \tag{1.3}$$

where ν is the "burst size", the number of virions released when an infected cell dies, μ is death rate of virions (clearance rate), β is the infection constant, λ is the input rate of uninfected cells, μ is the death rate of uninfected cells, and μ^* is the death rate of infected cells. This is essentially the basic model of Nowak and Bangham [**37**] (see also [**38**, and references therein]). If $\frac{\beta\lambda(\nu-1)}{\mu\mu'} > 1$, the infection persists and the system rapidly converges to the following equilibrium

$$\hat{q} = \frac{\lambda(\nu-1)}{\mu'} - \frac{\mu}{\beta}, \quad \hat{n} = \frac{\mu'}{\beta(\nu-1)}, \quad \hat{n}^* = \frac{\lambda}{\mu^*} - \frac{\mu\mu'}{\beta\mu^*(\nu-1)} \tag{1.4}$$

Nowak and May [**38**, Chapter 3] describe the basic dependencies of this equilibrium on the parameters. Briefly, the proportion of cells that are infected is positively related to β but inversely related to μ^*. The total number of viral genomes ($q+n^*$) increases with β, ν and λ, but decreases with μ, μ^*, and μ.

The viral population, at any point over the course of infection, can be considered as a collection of lineages. A new replication cycle is added to the ancestral lineage of virus with each successful cellular infection. We wish to determine the distribution of replication cycle counts across viruses (and provirus) that are present within the population at time t. We can extend (1.1), (1.2) and (1.3) for this purpose subdividing q and n^* into subgroups based on replication cycle count. Let $q_{[i]}(t)$ denote the number of virions with i replication cycles in their ancestry and $n^*_{[i]}(t)$ the number of infected cells in which the proviral DNA has i replication cycles in its ancestry.

There is no zero class for provirus ($n^*_{[0]}(t) = 0$) because all infected cells have at least one replication cycle in their ancestry (accounted with the RNA to DNA transition). These subdivided variables are constrained to sum to the dynamical equilibria of (1.4). The initial virion inoculum, $q_{[0]}(t)$, declines exponentially due to both successful infection of cells and clearance by the host immune system:

$$q_{[0]}(t) = q(0)\exp\{-(\mu' + \beta\hat{n})t\} = q(0)\exp\left\{-\mu'\left(1 + \frac{1}{\nu-1}\right)t\right\} \tag{1.5}$$

where the equilibrium uninfected cell number from (1.4) is used to get the last expression.

The following differential equations describe changes in $q_{[i]}(t)$ and $n^*_{[i]}(t)$ for $i > 0$:

$$\frac{d}{dt}q_{[i]}(t) = \nu\mu^* n^*_{[i]}(t) - \mu'\left(1 + \frac{1}{\nu-1}\right)q_{[i]}(t) \tag{1.6}$$

and

$$\frac{d}{dt}n^*_{[i]}(t) = \frac{\mu'}{\nu - 1}q_{[i-1]}(t) - \mu^* n^*_{[i]}(t). \tag{1.7}$$

The two terms in each differential equation represent the flow into and out of a given category. The change in $n^*_{[i]}(t)$ is determined by new infections from virions with $i-1$ replication cycles (the inflow) minus the death of infected cells with i replication cycles (the outflow). The change in $q_{[i]}(t)$ is determined by the influx of new virions from the death of infected cells in the i'th class, minus the clearance of virions from this category. Equations (1.5)–(1.7) are based on two important assumptions. First, I assumed that each cell is infected by a single virion. This neglects the possibility of "superinfection", where multiple viruses infect the same cell. Second, I assumed that the dynamical equilibrium of (1.4) is obtained immediately after primary infection (at $t = 0$). This insures that the coefficients of equations (1.6)–(1.7) are constant and permits the simple solution given below. Support for this approximation is provided by the rapid approach of n, n^*, and q to fixed values under a wide range of parameter values [**24**].

Equations (1.6)–(1.7) are linear differential equations, but the solutions are cumbersome, particularly as i increases. For this reason, it is useful to summarize the distribution of replication cycles in terms of moments. Let Z_x denote the mean replication cycle count for category x of the viral population. In this simple unstructured model, there are only two categories, virions (q) and provirus (n^*). The mean number of replication cycles for virions and provirus at time t:

$$Z_q(t) = \frac{1}{\hat{q}}\sum_{i=0}^{\infty} i q_{[i]}(t), \quad Z_{n^*}(t) = \frac{1}{\hat{n}^*}\sum_{i=0}^{\infty} i n^*_{[i]}(t). \tag{1.8}$$

Taking derivatives of (1.8) and substituting derivatives from (1.6) and (1.7),

$$\frac{d}{dt}Z_q(t) = \frac{\hat{n}^*}{\hat{q}}\nu\mu^* Z_{n^*}(t) - \mu'\left(1 + \frac{1}{\nu - 1}\right)Z_q(t) \tag{1.9}$$

and

$$\frac{d}{dt}Z_{n^*}(t) = -\mu^* Z_{n^*}(t) + \frac{\hat{q}}{\hat{n}^*}\frac{\mu'}{\nu - 1}\left(Z_q(t) + 1\right). \tag{1.10}$$

This is a simple 2×2 linear system with initial conditions $Z_q(0) = 0$ and $Z_{n^*}(0) = 1$. Noting that ν will generally be much greater than 1 and using (1.4) for $\hat{n}^*$ and $\hat{q}$, we obtain the following solution:

$$Z_{n^*}(t) \approx 1 + \frac{\mu'\mu^*}{(\mu' + \mu^*)^2}[e^{-t(\mu'+\mu^*)} - 1 + t(\mu' + \mu^*)] \tag{1.11}$$

and

$$Z_q(t) \approx \frac{\mu'\mu^*}{(\mu' + \mu^*)^2}[\frac{\mu'}{\mu^*}\left(1 - e^{-t(\mu'+\mu^*)}\right) + t(\mu' + \mu^*)] \tag{1.12}$$

Equations (1.11)–(1.12) are asymptotically linear functions of t with a common slope, $\frac{\mu'\mu^*}{(\mu'+\mu^*)}$. A non-linear increase occurs immediately following initial infection, but both functions rapidly converge to parallel lines with $Z_{n^*}(t) - Z_q(t) = \frac{\mu^*}{(\mu'+\mu^*)}$. The reciprocal of this slope, $\frac{(\mu'+\mu^*)}{\mu'\mu^*}$, is perhaps the most meaningful measure of "generation time" from an evolutionary perspective. The fact that this is the sum of the reciprocals of each death rate, $\frac{1}{\mu'} + \frac{1}{\mu^*}$, makes clear the fact that the smaller death rate will be the primary determinant of viral generation time. The average time a virion exists before infecting a cell is approximately $\frac{1}{\mu'}$ (with the assumption that burst size is much greater than 1), while the average time an infected cell lives is $\frac{1}{\mu^*}$.

It is noteworthy that the effective generation time depends on only two of the six parameters, μ' and μ^*. It is tempting to conclude that generation time is determined by μ and μ^* because viral reproduction directly linked to cell death in model of (1.1)–(1.3). Virions are released in a single burst when an infected cell is killed. Nowak and May [**38**] consider a slightly different model in which virions are produced continuously over the lifespan of an infected cell. In their formulation, (1.2) and (1.3) are the same but the differential equation for virion numbers is:

$$\frac{d}{dt}q = \kappa n^* - \mu' q,$$

where κ is the rate of virion production from infected cells. Interestingly, if we analyze this model following the procedures of (1.5)–(1.10), we obtain the same solution for $Z_{n^*}(t)$ and $Z_q(t)$, i.e. (1.11)–(1.12). Thus, the predicted generation time is not a consequence of burst virion production.

It is straightforward to determine the variance of replication cycle numbers using the same basic procedures (see [**24**]). In this paper, I will focus on distribution means. This is mainly for brevity of presentation and the various approaches that I describe can be extended to predict the higher moments of the distribution of replication cycles. One final comment before progressing to the structured model: an important prediction of the simple model is that the mean replication cycle counts increases in an approximately linear way with time. This will also prove true for the structured model and provides a simple means to determine the slope.

2. Structured model

Consider a model with an arbitrary number of target cells that can migrate among an arbitrary number of compartments. The dynamical variables are the number of virions in each compartment i (q_i), and the number of uninfected (n_{ji}) and infected (n^*_{ji}) cells of each type j within each compartment i. We assume that movement between compartments can occur via migration of virions, cells, or both virions and cells. Changes in these quantities are described by the following system of differential equations:

$$\frac{d}{dt}q_i = \sum_j \mu^*_{ji}\nu_{ji}n^*_{ji} - \mu'_i q_i - \sum_j \beta_{ji}n_{ji}q_i + \sum_{k\neq i} m_{ki}q_k - \sum_{k\neq i} m_{ik}q_i \tag{2.1}$$

$$\frac{d}{dt}n_{ji} = \lambda_{ji} - \mu_{ji}n_{ji} - \beta_{ji}n_{ji}q_i + \sum_{k\neq i} M_{ki}n_{jk} - \sum_{k\neq i} M_{ik}n_{ji} \tag{2.2}$$

$$\frac{d}{dt}n^*_{ji} = \beta_{ji}n_{ji}q_i - \mu^*_{ji}n^*_{ji} + \sum_{k\neq i} M^*_{ki}n^*_{jk} - \sum_{k\neq i} M^*_{ik}n^*_{ji} \tag{2.3}$$

where ν_{ji} gives the "burst size" (number of virions released when an infected cell dies), μ_i is the death rate of virions (clearance rate), β_{ji} is the infection constant, λ_{ji} is the input rate of uninfected cells, μ_{ji} is the death rate of uninfected cells, and μ^*_{ji} is the death rate of infected cells. The quantities m_{ik}, M_{ik}, and M^*_{ik} give the migration rates of virions, uninfected cells and infected cells from compartment i to compartment k. The summations over j are for cell types, while the summations over k are for compartments. With this model, we are assuming that cell migration rates (upper case Ms) are not cell-type specific. We would simply need to add an additional subscript to the M and M^* terms to include this further level of detail.

Orive et al. [**40**] provide a detailed analysis of this model in terms of the conditions that allow infection to persist. Briefly, migration between compartments can create "sources" and "sinks" within the virus population, where realized viral growth rate and abundance is lowered in some compartments compared to what would be observed in isolation. A boundary analysis of (2.1)–(2.3) suggests that migration between compartments makes the conditions for the initial increase of viral abundance (in terms of both q_i and n^*_{ji}) more stringent than in the absence of migration [**40**]. In other words, within-host spatial structure combined with viral movement decreases the likelihood of viral establishment. This result may help to explain the tissue specificity observed for many viruses. There are, however, a number of important exceptions to this result. Most obviously, migration is essential when virus initially invades a compartment that is unfavorable to population growth. To persist, it must then migrate to infect other parts of the host body.

If infection does persist, the numbers of virions and infected cells tend to converge on equilibrium values (although not so rapidly as in the unstructured model). We have not found simple analytical approximations for these equilibria that are accurate over the range of reasonable parameter values. We have found different approximations that are useful with different regions of parameter space. A summary of equilibrium results will be given elsewhere. For the purpose of the present paper, predicting viral generation time, it is sufficient just that the dynamical variables q_i and n^*_{ji} converge on fixed values.

There are quite a number of distinct categories of viral genome when we consider the accumulation of replication cycles in the system of equations (2.1)–(2.3). Let Z_{q_i} denote the mean replication cycle count for virions in compartment i and $Z_{n^*_{ji}}$ denote the mean for provirus within cells of type j within compartment i. The differential equations for these quantities are structurally similar to those for the simple model:

$$\begin{aligned}\frac{d}{dt}Z_{q_i}(t) =& \sum_j \left(\frac{n^*_{ji}}{q_i}\right) \mu^*_{ji}\nu_{ji}Z_{n^*_{ji}}(t) - \left[\mu'_i + \sum_j \beta_{ji}n_{ji} + \sum_{k\neq i} m_{ik}\right] Z_{q_i}(t) \\ &+ \sum_{k\neq i}\left(\frac{q_k}{q_i}\right) m_{ki}Z_{q_k}(t)\end{aligned} \tag{2.4}$$

and

$$\begin{aligned}\frac{d}{dt}Z_{n^*_{ji}}(t) =& \beta_{ji}n_{ji}\left(\frac{q_i}{n^*_{ji}}\right)[1 + Z_{q_i}(t)] - \left(\mu^*_{ji} + \sum_{k\neq i} M^*_{ik}\right) Z_{n^*_{ji}}(t) \\ &+ \sum_{k\neq i} M^*_{ki}\left(\frac{n^*_{jk}}{n^*_{ji}}\right) Z_{n^*_{jk}}(t)\end{aligned} \tag{2.5}$$

for each i and j. Equations (2.4)–(2.5) are a system of non-homogeneous linear differential equations. Once the q_i and n^*_{ij} reach dynamic equilibrium, the coefficients become constants and, in principle, it is possible to analyze this system by standard methods (e.g. [**2**], Ch. 7). Here, I note one important feature of these equations and then move on to special cases. The coefficients within each equation must sum to zero as a consequence of dynamical equilibrium (using (2.1) and (2.3)). This implies that

$$\mu_i' + \sum_j \beta_{ji} n_{ji} + \sum_{k \neq i} m_{ik} = \sum_j \left(\frac{n_{ji}^*}{q_i} \right) \mu_{ji}^* \nu_{ji} + \sum_{k \neq i} \left(\frac{q_k}{q_i} \right) m_{ki} \tag{2.6}$$

and

$$\beta_{ji} n_{ji} \left(\frac{q_i}{n_{ji}^*} \right) = \mu_{ji}^* + \sum_{k \neq i} M_{ik}^* - \sum_{k \neq i} M_{ki}^* \left(\frac{n_{jk}^*}{n_{ji}^*} \right). \tag{2.7}$$

These relations allow us to reduce the number of parameters in the differential equations for means. There are basically two different sorts of complication in the structured model, cellular population structure and compartment structure. We consider their effects separately in the following sections.

An arbitrary number of cell types in a single compartment. For this special case, we can drop the subscript for compartment and neglect the terms related to migration among compartments from equations (2.4)–(2.5):

$$\begin{aligned} \frac{d}{dt} Z_q(t) &= \sum_j \left(\frac{n_j^*}{q} \right) \mu_j^* \nu_j Z_{n_j^*}(t) - \left[\mu' + \sum_j \beta_j n_j \right] Z_q(t) \\ &= \sum_j \left(\frac{n_j^*}{q} \right) \mu_j^* \nu_j \left[Z_{n_j^*}(t) - Z_q(t) \right] \end{aligned} \tag{2.8}$$

where the rightmost expression uses the constraints on the coefficients (2.6). Likewise,

$$\begin{aligned} \frac{d}{dt} Z_{n_j^*}(t) &= \beta_j n_j \left(\frac{q}{n_j^*} \right) [1 + Z_q(t)] - \mu_j^* Z_{n_j^*}(t) \\ &= \mu_j^* [1 + Z_q(t) - Z_{n_j^*}(t)] \end{aligned} \tag{2.9}$$

where the rightmost expression uses (2.7).

This system behaves like the simple, unstructured model. After a brief transitory, each variable increases linearly with time and they share a common slope, Φ. We can use this to obtain a closed form expression for the slope. For large t, $Z_q(t) = \alpha + \Phi t$, and $Z_{n_j^*}(t) = Z_q(t) + \Delta_{n_j^*}$, where α, Φ, and the $\Delta_{n_j^*}$ are constants. Substituting,

$$\frac{d}{dt} Z_q(t) = \Phi = \sum_j \left(\frac{n_j^*}{q} \right) \mu_j^* \nu_j \Delta_{n_j^*} \tag{2.10}$$

and

$$\frac{d}{dt} Z_{n_j^*}(t) = \Phi = \mu_j^* [1 - \Delta_{n_j^*}] \tag{2.11}$$

which yields the following solution for the common slope:

$$\Phi = \frac{\sum_j (\frac{n_j^*}{q})\nu_j\mu_j^*}{1+\sum_j (\frac{n_j^*}{q})\nu_j} \approx \frac{\mu'}{1+\sum_j (\frac{n_j^*}{q})\nu_j}. \tag{2.12}$$

The evolutionary generation time is $1/\Phi$. Equation (2.12) indicates that this depends on a weighted average of μ_j^*, the death rates of infected cells of each type. The weights are the relative contributions of each cell type to the pool of virions, $n_j^*\nu_j$. The generation time also depends, implicitly, on the death rate of virions, μ, which influences the ratio of infected cells to virions, $\left(\frac{n_j^*}{q}\right)$. The primary conclusion that follows from (2.12) is that cell types with lower death rates can have a disproportionate effect in reducing the generation time [**24**]. Equation (2.12) is slightly more general than the corresponding equations in [**24**]. The former obtained a solution for Φ explicitly in terms of constants from the dynamical model by substituting an approximate solution for q and n_j^*.

A single cell type in two compartments. Here we can drop the subscript associated with cell type (j) from (2.1)–(2.3) and there are only six dynamical variables: the number of virions in each compartment (q_1 and q_2), the number of uninfected cells in each compartment (n_1 and n_2), and the number of infected cells in each compartment (n_1^* and n_2^*). As an example, for HIV, we might consider T cells in both lymph nodes (compartment 1) and blood (compartment 2). Asymptotically, this model behaves like the two previous cases. After a transitory, the mean replication cycle count increases linearly with time and the various categories share a common stope, Φ. For large t,

$$q_1\Phi = n_1^*\mu_1^*\nu_1\Delta_1 + q_2 m_{21}\Delta_q \tag{2.13}$$

$$q_2\Phi = n_2^*\mu_2^*\nu_2\Delta_2 - q_1 m_{12}\Delta_q \tag{2.14}$$

$$n_1^*\Phi = n_1^*(\mu_1^* + M_{12}^*)(1-\Delta_1) + n_2^* M_{21}^*(\Delta_2 + \Delta_q - 1) \tag{2.15}$$

$$n_2^*\Phi = n_2^*(\mu_2^* + M_{21}^*)(1-\Delta_2) + n_1^* M_{12}^*(\Delta_1 - \Delta_q - 1) \tag{2.16}$$

where $\Delta_1 = Z_{n_1^*}(t) - Z_{q_1}(t)$, $\Delta_2 = Z_{n_2^*}(t) - Z_{q_2}(t)$, $\Delta_q = Z_{q_2}(t) - Z_{q_1}(t)$. The common slope, Φ, is

$$\Phi = \frac{\sum_{i=1}^{2} n_i^*\mu_i^*\nu_i[q_i m_{ij}(M_{ji}^*\mu_i^* + M_{ij}^*\mu_j^* + \mu_i^*\mu_j^*) + \nu_j\mu_j^* M_{ij}^*(n_i^*(M_{ij}^* + \mu_i^*) - n_j^* M_{ji}^*)]}{\sum_{i=1}^{2} A_{ij} + B_{ij} + C_{ij}} \tag{2.17}$$

where

$$A_{ij} = n_i^* M_{ij}^*[q_j m_{ji}\nu_j\mu_j^* - q_j M_{ji}^*\nu_i\mu_i^* + n_i^*\nu_i\mu_i^*\nu_j\mu_j^*] \tag{2.18}$$

$$B_{ij} = q_i^2 m_{ij}[M_{ji}^*\mu_i^* + \mu_j^*(M_{ij}^* + \mu_i^*)] \tag{2.19}$$

$$C_{ij} = n_i^* q_i[M_{ij}^*\nu_j\mu_j^*(\mu_i^* + M_{ij}^*) + m_{ij}\nu_i\mu_i^*(\mu_j^* + M_{ji}^*)] \tag{2.20}$$

and $j \neq i$ ($j = 2$ if $i = 1$ and vice versa).

3. The Retrospective Approach

This section describes an alternative means to derive formula for the evolutionary generation time. The retrospective approach considers the ancestral lineage of a viral genome as a stochastic process. The history of this lineage is a series of intervals with the genome persisting either as provirus or virion. Since we are sampling a random viral genome from the current population, we can treat the number and duration of these intervals as a series of random variables. Under the assumptions of our models ((1.1)–(1.3) or (2.1)–(2.3)), the amount of time spent as either provirus or virion within a particular compartment and/or cell type is exponentially distributed. This satisfies the "memoryless" property of a Markov process. Moreover, the various genomic categories, e.g. provirus within cell type (j), can be considered the discrete states of a Markov Jump process [**14**, Chapter 5]. Time is measured retrospectively because we are only interested in ancestry of genomes that are present in the current population. Each extant viral genome must have a continuous lineage back to the founders of the infection.

In the simple unstructured model, it is convenient to distinguish three distinct states: free virions, infecting virions, and provirus (Figure 1(A)). I assume that the process of infecting a cell (crossing the cell membrane, reverse-transcription of the RNA genome to DNA, incorporation of the DNA in the host cell genome) requires a small amount of time, dt, which we take as the basic time interval (assumed to be small enough so that there is at most one transition per interval). Tracking time retrospectively (backwards), a viral lineage will spend a single time step of length dt in the infecting virion state and then automatically pass to the virion state. Once in the virion state, two transitions are possible. It may remain virion, which occurs with probability $P[V \to V]$, or it may pass to the proviral state, with probability $P[V \to PV] = 1 - P[V \to V]$. Each time interval as provirus is also associated with two possible transitions: the lineage remains provirus with probability $P[PV \to PV]$ or becomes an infecting virion with probability $P[PV \to IV] = 1 - P[PV \to PV]$.

For a population in dynamical equilibrium, these transition probabilities are constant and can be directly determined from the dynamical model. Referring back to equations (1.1)–(1.3),

$$P[IV \to V] = 1 \tag{3.1}$$

$$P[V \to PV] = (\mu^* \nu n^*/q)dt = (\mu' + \beta n)dt, \quad P[V \to V] = 1 - (\mu' + \beta n)dt \tag{3.2}$$

$$P[PV \to IV] = (\beta n q/n^*)dt = \mu^* dt, \quad P[PV \to PV] = 1 - \mu^* dt \tag{3.3}$$

A number of results from the theory of Markov processes can be used to derive results regarding the accumulation of replication cycles. In Kelly et al. [**24**], we derived the mean and variance of replication cycle counts from the mean and variance of "recurrence times." The recurrence time for a give state is the amount of time it takes for the process (in this case a viral lineage) to return to that state. Over a sufficiently long stretch of time, the number of visits to a particular state is normally distributed and the parameters are simple functions of the mean and variance of the recurrence time [**11**, pp. 320–321]. For either the simple model, or the structured model with an arbitrary number of cell types in a single compartment, the replication cycle count is equal to the number of visits to the infecting virion state within a lineage.

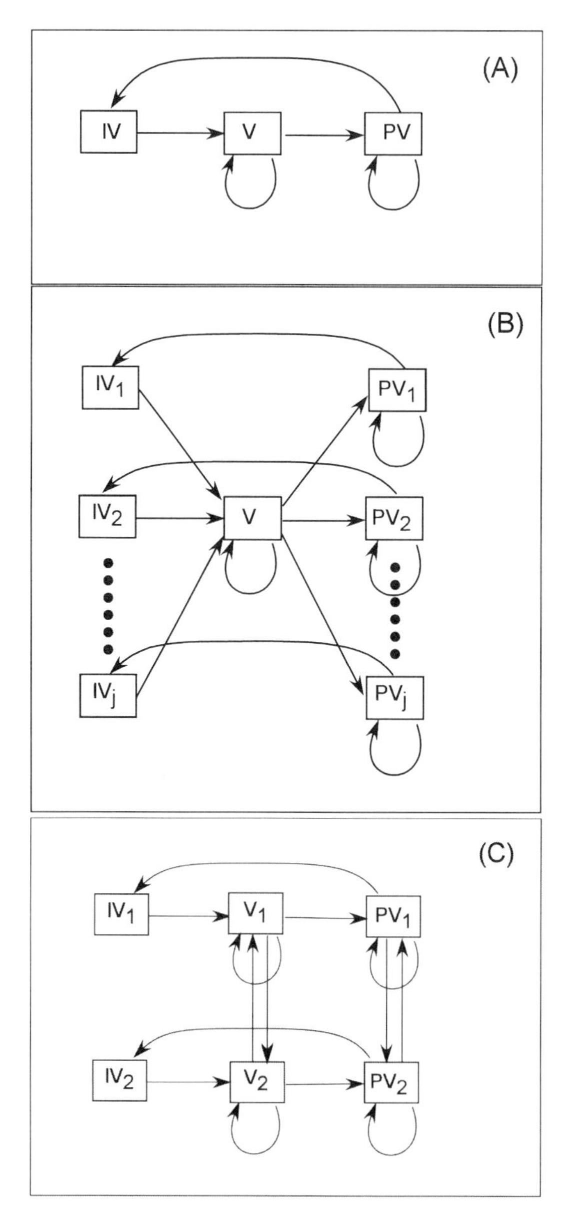

FIGURE 1. Loop diagrams describing the transitions among the possible states for a viral lineage for (A) the unstructured model, (B) the structured model with an arbitrary number of cell types within a single compartment, and (C) the structured model with an arbitrary number of cell types within a single compartment.

Here, we use a different Markov property based on the stationary distribution associated with states. This is the probability that a random genome will be in a particular state at any given time. The rate of increase in replication cycle count, Φ, is equal to the probability that a lineage resides in the infecting virion state divided by the time step dt. Since there are multiple infecting virion states in the structured model, corresponding to different cell types and compartments, Φ is determined by summing probabilities across categories. The justification for this procedure is simple: over the long run, the number of replication cycles will equal the probability that lineage is infecting a cell (per unit time) times the amount of time that has elapsed. This procedure assumes that a stationary distribution exists and that a viral lineage can visit all states. This "ergodic property" places some constraints on migration parameters.

Let $U[x]$ denote the stationary probability associated with viral genomes in state x. For the simple model, the stationary probabilities satisfy the following relations:

$$U[IV] = U[PV]P[PV \to IV] = U[PV]\mu^* dt \tag{3.4}$$

$$\begin{aligned} U[V] &= U[V]P[V \to V] + U[IV]P[IV \to V] \\ &= U[V](1 - (\mu' + \beta n)dt) + U[IV] \end{aligned} \tag{3.5}$$

$$\begin{aligned} U[PV] &= U[V]P[V \to PV] + U[PV]P[PV \to PV] \\ &= U[V](\mu' + \beta n)dt + U[PV](1 - \mu^* dt). \end{aligned} \tag{3.6}$$

These three equations are not linearly independent. So using any two of them plus the fact that the total probability $U[IV] + U[PV] + U[V] = 1$ we find that

$$U[PV] = \frac{\mu' + \beta n}{\mu^* + (1 + \mu^* dt)(\mu' + \beta n)} \approx \frac{\mu'}{\mu^* + \mu'} \tag{3.7}$$

where the rightmost term relies on the approximations that $\nu \gg 1$ and dt is small. Using (3.7),

$$U[IV] = \Phi dt \approx \left(\frac{\mu' \mu^*}{\mu^* + \mu'}\right) dt \tag{3.8}$$

which retrieves the result of the prospective analysis (equations (1.11)–(1.12)).

We need to expand the set of possible states when considering the first special case of the structured model, an arbitrary number of cell types in a single compartment (Figure 1(B)). Let IV_j denote a virion infecting cell type j and PV_j denote provirus within cell type j. Referring back to (2.1)–(2.3),

$$P[IV_j \to V] = 1 \tag{3.9}$$

$$P[V \to PV_j] = \frac{n_j^* \mu_j^* \nu_j}{q} dt, \quad P[V \to V] = 1 - \sum_j \frac{n_j^* \mu_j^* \nu_j}{q} dt \tag{3.10}$$

$$P[PV_j \to IV_j] = \mu_j^* dt, \quad P[PV_j \to PV_j] = 1 - \mu_j^* dt. \tag{3.11}$$

The stationary probabilities satisfy the following equations,

$$U[IV_j] = U[PV_j]\mu_j^* dt \tag{3.12}$$

$$U[V] = U[V]\left(1 - \sum_j \frac{n_j^* \mu_j^* \nu_j}{q} dt\right) + \sum_j U[IV_j] \tag{3.13}$$

$$U[PV_j] = U[PV_j](1 - \mu_j^* dt) + U[V]\frac{n_j^* \mu_j^* \nu_j}{q} dt. \tag{3.14}$$

Again assuming that dt is small, after substituting $U[V] = 1-\sum_j (U[IV_j] + U[PV_j])$ into (3.14) and using (3.12) to eliminate $U[IV_j]$, we find that

$$U[PV_j] \approx n_j^* \nu_j \frac{\left(1 - \sum_k U[PV_k]\right)}{q}.$$

Summing both sides of this gives an equation for $\sum_k U[PV_k]$, which can be used to give

$$U[PV_j] = \frac{n_j^* \nu_j}{q + \sum_k n_k^* \nu_k} \tag{3.15}$$

and

$$\Phi = \sum_j \frac{U[IV_j]}{dt} = \frac{\sum_j \mu_j^* n_j^* \nu_j}{q + \sum_k n_k^* \nu_k} \tag{3.16}$$

which is equivalent to (2.12).

Finally, we consider the two-compartment/one-cell type model using a retrospective approach (Figure 1(C)). We need to expand the collection of states to include virions in each compartment (V_i). Using the subscripts to denote compartment (1 or 2), the transition probabilities are

$$P[IV_i \to V_i] = 1 \tag{3.17}$$

$$P[V_i \to V_i] = 1 - A_i - B_i, \quad P[V_i \to V_j] = A_i, \quad P[V_i \to PV_i] = B_i \tag{3.18}$$

$$P[PV_i \to PV_i] = 1 - C_i - D_i, \quad P[PV_i \to IV_i] = C_i, \quad P[PV_i \to PV_j] = D_i \tag{3.19}$$

where

$$A_i = \frac{m_{ji} q_j}{q_i} dt, \quad B_i = \frac{n_i^* \mu_i^* \nu_i}{q_i} dt, \quad C_i = \frac{q_i \beta_i n_i}{n_i^*} dt, \quad D_i = \frac{n_j^* M_{ji}^*}{n_i^*} dt \tag{3.20}$$

and $j = 2$ if $i = 1$ and vice versa. The stationary probabilities satisfy the following equations:

$$U[V_i] = U[V_i](1 - A_i - B_i) + U[V_j]A_j + U[IV_i] \tag{3.21}$$

$$U[PV_i] = U[PV_i](1 - C_i - D_i) + U[PV_j]D_j + U[V_i]B_i \tag{3.22}$$

$$U[IV_i] = C_i U[PV_i]. \tag{3.23}$$

Solving,

$$\Phi = \frac{\sum_{i=1}^{2} \mu_i^* [(A_i + B_i)B_j D_j + A_j B_i (C_j + D_j)]}{\sum_{i=1}^{2} [B_i D_i (B_j + C_j) + A_i (B_j + C_j)(C_i + D_i) + A_i D_j (B_j + C_i)]}, \tag{3.24}$$

while (3.24) looks different from (2.17), they produce equivalent values when the system is in dynamical equilibrium. They appear different because there are different ways to use constraints of equations (2.6)–(2.7).

4. Discussion

An explicit linkage between dynamical and population genetic models of infection is useful for a number of reasons. For one, it illustrates the logical connection between ideas in two areas of study that are surprisingly disconnected, modeling of viral dynamics and evolution. Second, it may suggest ways to use virological data to address population genetic questions and vice versa [**7, 18, 22, 23, 24**]. The perspective can also guide the development of more realistic theoretical models. For example, there is an enormous literature concerning population structure in evolutionary genetics [**20**, Chapter 6], [**21, 55, 56**]. However, the great majority of models assume discrete, non-overlapping generations. While this assumption facilitates mathematical analysis, it clearly does not capture the demographic complexity of intra-host viral populations.

The model of equations (2.1)–(2.3) considers two distinct sorts of structure, cell types and compartments. From the standpoint of analysis, compartment structure is a bit more complicated. The model implicitly assumes that compartments are internally homogenous. Once a virion is released into a compartment, it can infect any cell type present, in proportion to their respective local abundances and infectivities. In contrast, we need three distinct parameter arrays, migration rates of virions and cells (both infected and uninfected), to describe movement among compartments. While these migration rates may be difficult to estimate, they are primary determinants of both dynamics and evolution within a structured population [**40**].

The effect of migration rate on the evolutionary generation time can be illustrated by considering the simplest case, two compartments with a single cell type. Even with these restrictions, the model has an eighteen dimensional parameter space, which is rather difficult to explore thoroughly. However, careful inspection of the equations for Φ ((2.17) and (3.24)) indicate that, like the simpler models (equations (1.11)–(1.12), (2.12)), this quantity can still be considered a weighted average of the death rates, μ_i^* and μ_i'. The dependence on μ_i' is less obvious, as its effect is subsumed by the relative values of n_i^* and q_i. The other dynamical parameters, particularly the migration rates, determine the weights given to each death rate.

Figure 2 illustrates predictions that emerge from many parameter sets in which one compartment is a source for viral production and the other is a sink. In this example, immune surveillance is much stronger in compartment 1 than compartment 2 resulting in larger values for μ^* and μ': $\mu_1^* = 1.0$, $\mu_2^* = 0.1$, $\mu_1' = 10$, $\mu_2' = 1$ (see Figure 2 legend for other parameter values). As a consequence, the viral population within compartment 1 persists only because of the continuous immigration from compartment 2. The total number of viral genomes $(q_1 + q_2 + n_1^* + n_2^*)$ is given as a function of migration rate in Figure 2(A). For the most part, migration among compartments reduces viral numbers. Greater migration moves a larger number of virions and/or infected cells into the sink compartment where they are more rapidly eliminated. However, when cells migrate among compartments, the highest viral densities obtain at intermediate migration rates. The asymmetry between virion and cell migration is due to the higher death rates for virions than for infected cells. If $\mu_i^* > \mu_i'$, then the highest viral densities occur at intermediate virion migration rates.

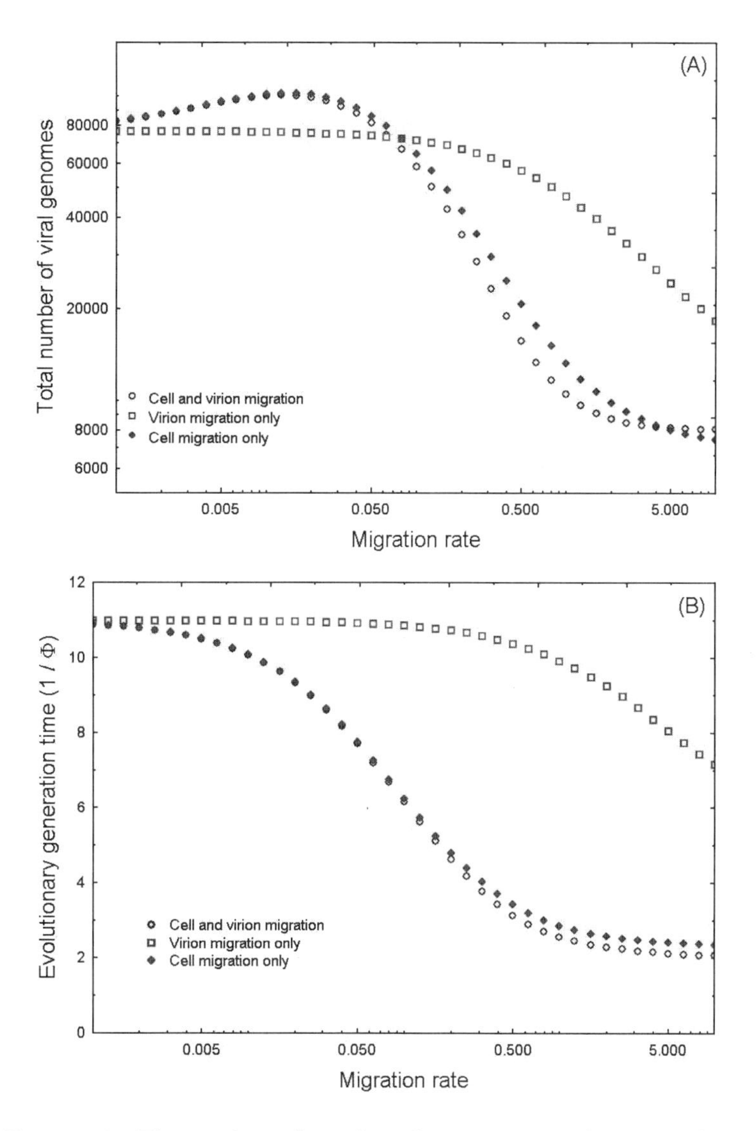

FIGURE 2. The total number of viral genomes as a function of the migration rate m. Here, migration is symmetric between the two compartments. With both cell and virion migration (open circles), $M_{12}^* = M_{21}^* = m_{12} = m_{21} = m$. With virion migration only, $m_{12} = m_{21} = 0$. The other parameter values are: $M_{12} = M_{21} = M_{12}^* = M_{21}^*$, $\mu_1^* = 1.0$, $\mu_2^* = 0.1$, $\mu_1' = 10$, $\mu_2' = 1$, $\nu_1 = \nu_2 = 50$, $\beta_1 = \beta_2 = 10^{-6}$, $\lambda_1 = \lambda_2 = 1500$, and $\mu_1 = \mu_2 = 0.01$.

Figure 2(B) considers the same parameter variations on $1/\Phi$, the evolutionary generation time. Migration among compartments dramatically reduces the mean generation time. All else equal, this will increase the rate of evolution. With low migration, the mean generation time for the entire population is very similar to what you would obtain for the source population considered as an isolated entity. The result from the simple model (equations (1.11)–(1.12)), where $\frac{1}{\Phi} = \frac{1}{\mu_2^*} + \frac{1}{\mu_2'} = 11.0$, predicts generation time at the left-hand side of Figure 2(B) very well. With little migration, the vast majority of viral genomes, at least those that leave descendents, spend their lifecycles within the source compartment. As migration increases, a larger fraction of viral genomes will necessarily reside in the sink compartment where both virions and infected cells have shorter lifespans. Viral lineages that accumulate replication cycles within the sink compartment must migrate back to the source compartment to persist. However, this is a frequent occurrence with high migration rates.

The paradox that emerges from the comparison of Figures 2(A) and 2(B) is that the existence of the sink compartment can greatly accelerate evolution even if it is a substantial drag on the number of viral genomes that reside in the host. Of course, this is only strictly true if the rate evolution maintains the same proportionality to the number of generations per unit time. This is expected with neutral evolution, where the predicted number of genetic substitutions within the population is equal to the mutation rate per replication cycle times the number of replication cycles (on average) that have accrued. With selection however, population size directly impacts the rate of evolution. The rate of fixation of deleterious mutations is inversely related to population size, whereas advantageous mutations should fix more frequently in larger populations [**16**, Chapter 3].

Internal structure is likely to have a number of evolutionary consequences, only one of which is altering the mean viral generation time. It will also impact the amount of gene sequence variation present in a population (polymorphism or nucleotide diversity) and the distribution of variation within and among compartments [**8, 12**]. If migration is sufficiently restricted, local adaptation to the conditions of a specific compartment may occur within the viral population [**48**]. Wade and Goodnight [**51**] review a range of evolutionary hypotheses associated with structured populations. I have focused on generation in this paper primarily for the purpose of illustration. It is possible to derive predictions for other genetic statistics, e.g. Wright's F statistics [**20**], for a population governed by equations (2.1)–(2.3). The algebra becomes increasingly cumbersome and closed-form results are difficult to obtain, but numerical analyses are straightforward.

There is currently a great deal of interest in using gene sequence data to draw dynamical and demographic conclusions about pathogen populations [**18**]. For example, Nickle et al. [**34, 35**] outline a series of predictions regarding the amount and pattern of gene sequence variation in HIV populations with both "reservoirs" and compartments. Virological reservoirs are cell types or tissues with low rates of viral replication. In the model of equations (2.1)–(2.3), this could correspond to either cell types with low death rates when infected (μ_{ji}^*), or to compartments with low (μ_{ji}^*) and μ_i' across cell types. Nickle et al. [**34, 35**] note that sequences sampled from a reservoir should be less divergent from the viral genotype(s) that initiated infection than sequences sampled from non-reservoirs, e.g. activated T cells in the blood. The analysis given here is fully consistent with this prediction. However, at least under the model of equations (2.1)–(2.3), the extent of the difference between compartments depends to a large extent on the migration parameters. A viral sequence sampled from a reservoir will not be that different (in terms of gene sequence divergence from the population founders) than a sequence from the blood if its ancestral lineage resided primarily in non-reservoir cell types or compartments.

The evolutionary generation time, $1/\Phi$, is certainly affected by differentiation among compartments, but it does not measure the extent of differentiation directly. The asymptotic results of equations (2.12) and (2.17)–(2.20) yield the same Φ for viral genomes in all subdivisions (compartments or cell types). This is because, over the long run, viral lineages will visit all subdivisions in according to the same probabilistic process. Viral genomes sampled from different compartments might be substantially different in their recent histories, with sequences drawn from reservoirs having fewer replication cycles in their ancestry. Such differences are reflected in differing values for the intercepts, the Δ terms of equations (2.10)–(2.10) and (2.13)–(2.16).

One final comment: the theory developed here was motivated by the study of persistent viral infection, mainly Lentiviruses. However, the basic issues emerge in other areas of epidemiology and population biology. The same sorts of "structure" are present in populations existing on larger scales. For example, there is an extensive literature considering the spread of epidemics within spatially structured host populations [**30, 31**, and references therein]. In translating predictions of the current theory, virions might correspond to the "free-living stage" a parasite. The cell types might correspond to alternative hosts or even alternative genotypes of the same host. Finally, compartments would naturally correspond to different subpopulations of the host or hosts. Of course, many details regarding infection and transmission would likely be different. Still, I predict that the same analytical approaches may prove useful.

Acknowledgements

I would like to thank B. Holt and M. Barfield for careful review of this paper.

References

[1] J. N. Blankson, D. Persaud, and R. F. Siliciano, The challenge of viral reservoirs in HIV-1 infection, *Annu. Rev. Med.* **53** (2002), 557–593.

[2] W. E. Boyce R. C. DiPrima, *Elementary differential equations and boundary value problems*, John Wiley and Sons, New York, 1986.

[3] D. S. Callaway and A. S. Perelson, HIV-1 infection and low steady state viral loads, *Bulletin of Mathematical Biology* **64** (2002), 29–64.

[4] T. W. Chun, L. Stuyver, S. B. Mizell, L. A. Ehler, J. A. Mican, M. Baseler, A. L. Lloyd, M. A. Nowak, and A. S. Fauci, Presence of an inducible HIV-1 latent reservoir during highly active antiretrovial therapy, *Proc. Nat. Acad. Sci. USA* **94** (1997), 13193–13197.

[5] R. Coombs, C. Speck, J. M. Hughes, W. Lee, R. Sampoleo, S. Ross, J. Dragavon, G. Peterson, T. Hooton, A. Collier, L. Cory, L. Koutsky, and J. Krieger, Association between culturable human immunodeficiency virus type 1 in semen and HIV RNA levels in semen and blood: evidence for compartmentalization of HIV-1 between semen and blood, *J. Infect. Dis.* **177** (1998), 320–330.

[6] K. Crandall, Multiple interspecies transmissions of human and simian T-cell leukemia/lymphoma virus type I sequences, *Molec. Biol. Evol.* **13** (1996), 115–131.

[7] K. Crandall, *The evolution of HIV*, Johns Hopkins University Press, Baltimore, 1999.

[8] S. Delassus, R. Cheynier, and S. Wain-Hobson, Nonhomogeneous distribution of human immunodeficiency virus type 1 proviruses in the spleen *J. Virol.* **66** (1992), 5642–5645.

[9] E. L. Delwart, J. I. Mullins, P. Gupta, J. J. Leary, M. Holodniy, D. Katzenstein, B. D. Walker, and M. Singh, Human immunodeficiency virus type 1 populations in blood and semen *J. Virol.* **72** (1998), 617–623.

[10] A. Drummond, R. Forsberg, and A. G. Rodrigo, The inference of stepwise changes in substitution rates using serial sequence samples *Molec. Biol. Evol.* **18** (2001), 1365–1371.

[11] W. Feller, *An introduction to probability theory and its applications*, John Wiley and Sons, New York, 1968.

[12] S. D. W. Frost, M.-J. Dumaurier, S. Wain-Hobson, and A. J. Leigh Brown, Genetic drift and within-host metapopulation dynamics of HIV-1 infection, *Pro. Natl. Acad. Sci. USA* **98** (2001), 6975–6980.

[13] S. D. W. Frost, H. F. Gunthard, J. K. Wong, D. Havlir, D. D. Richman, and A. J. Leigh Brown, Evidence for positive selection driving the evolution of HIV-1 env under potent antiviral therapy, *Virology* **284** (2001), 250–258.

[14] D. T. Gillespie, *Markov Processes, An Introduction for Physical Scientists*, Academic Press, San Diego, 1992.

[15] J. H. Gillespie, *The causes of molecular evolution*, Oxford University Press, New York, 1991.

[16] J. H. Gillespie, *Population genetics, a concise guide*, John Hopkins University Press, Baltimore, 2004.

[17] A. Granelli-Piperno, E. Delgado, V. Finkel, W. Paxton, and S. RM, Immature dendritic cells selectively replicate macrophagetropic (M-tropic) human immunodeficiency virus type 1, while mature cells efficiently transmit both M- and T-tropic virus to T cells, *J. Virol.* **72** (1998), 2733–2737.

[18] B. T. Grenfell, O. G. Pybus, J. R. Gog, J. L. N. Wood, J. M. Daly, J. A. Mumford, and E. C. Holmes, Unifying the epidemiological and evolutionary dynamics of pathogens, *Science* **303** (2004), 327–332.

[19] B. H. Hahn, G. M. Shaw, M. E. Taylor, R. R. Redfield, P. D. Markham, S. Z. Salahuddin, F. Wong-Stall, R. C. Gallo, E. S. Parks, and W. Parks, Genetic variation in HTLV-III/LAV over time in patients with AIDS or at risk for AIDS, *Science* **232** (1986), 1548–1553.

[20] D. L. Hartl and A. G. Clark, *Principles of population genetics*, Sinauer associates, Sunderland, Massachusetts, 1989.

[21] S. Karlin, A classification of selection-migration structures and conditions for protected polymorphism, in (M. K. Hecht, B. Wallace and G. T. Prance, eds), *Evolutionary Biology*, Plenum, New York, 1982, 61–204.

[22] J. K. Kelly, An Application of Population Genetic Theory to Synonymous Gene Sequence Evolution in the Human Immunodeficiency Virus (HIV), *Genetical Research* **64** (1994), 1–9.

[23] J. K. Kelly, Replication Rate and Evolution in the Human Immunodeficiency Virus, *J. Theor. Biol.* **180** (1996), 359–364.

[24] J. K. Kelly, S. Williamson, M. E. Orive, M. S. Smith, and R. D. Holt, Linking dynamical and population genetic models of persistent viral infection, *Amer. Nat.* **162** (2003), 14–28.

[25] T. B. Kepler and A. S. Perelson, Drug concentration heterogeneity facilitates the evolution of drug resistance, *Proc. Natl. Acad. Sci. USA* **95** (1998), 11514–11519.

[26] M. Kimura, *The Neutral Theory of Molecular Evolution*, Cambridge University Press, New York, 1983.

[27] J. F. C. Kingman, A simple model for the balance between selection and mutation, *J. Appl. Prob.* **15** (1978), 1–12.

[28] D. E. Kirschner, R. Mehr, and A. S. Perelson, Role of the thymus in pediatric HIV-1 infection *J. Acquir. Immune Defic. Syndr. Hum. Retrovirol.* **18** (1998), 95–109.

[29] L. M. Mansky, Forward mutation rate of human immunodeficiency virus type 1 in a T-lymphoid cell line *AIDS Res. Hum. Retrovir.* **12** (1996), 307–314.

[30] R. May and W. M. Anderson, Parasite-host coevolution, *Parasitology* **100** (1990), S89–S101.

[31] R. M. May, S. Gupta, and A. R. McLean, Infectious disease dynamics: what characterizes a successful invader? *Phil. Trans. R. Soc. Lond. B* **356** (1990), 901–910.

[32] A. Moya, S. F. Elena, A. Bracho, R. Miralles, and E. Barrio, The evolution of RNA viruses: A population genetics view *Pro. Natl. Acad. Sci. USA* **97** (2000), 6967–6973.

[33] M. V. Muse, Modelling the molecular evolution of HIV, in (K. Crandall, ed.), *The evolution of HIV*, John Hopkins, Baltimore, 1999, 122–152.

[34] D. C. Nickle, M. A. Jensen, D. Shriner, S. J. Brodie, L. M. Frenkel, J. E. Mittler, and J. I. Mullins, Evolutionary indicators of Human Immunodeficiency Virus Type 1 reservoirs and compartments, *J. Virol.* **77** (2003), 5540–5546.

[35] D. C. Nickle, D. Shriner, J. E. Mittler, L. M. Frenkel, and J. I. Mullins, Importance and detection of virus reservoirs and compartments of HIV infection, *Curr Opin Microbiology* **6** (2003), 410–416.

[36] M. Nowak, R. M. Anderson, A. R. McLean, T. F. Wolfs, J. Goudsmit, and R. M. May, Antigenic diversity thresholds and the development of AIDS *Science* **254** (1991), 963–969.

[37] M. Nowak and C. R. M. Bangham, Population dynamics of immune responses to persistent viruses, *Science* **272** (1996), 74–79.

[38] M. Nowak and R. M. May, *Virus Dynamics: The Mathematical Foundations of Virology and Immunology*, Oxford University Press, Oxford, 2000.

[39] T. Ohta, Role of very slightly deleterious mutations in molecular evolution and polymorphism, *Theor. Pop. Biol.* **10** (1976), 254–275.

[40] M. Orive, M. Stearns, J. Kelly, M. Barfield, M. Smith, and R. Holt, Viral infection in internally-structured hosts. I. Conditions for persistent infection, *J . Theor. Biol.* **232** (2005), 453–466.

[41] J. Overbaugh and C. R. M. Bangham, Selection forces and constraints on retroviral sequence variation, *Science* **292** (2001), 1106–1109.

[42] A. S. Perelson, Modeling the interaction of the immune system with HIV, in (C. Castillo-Chavez, ed.), *Mathematics and Statistical Approaches to AIDS Epidemiology*, Springer-Verlag, 1989, 350–370.

[43] M. Poss, A. G. Rodrigo, J. J. Gosink, G. H. Learn, D. de Vange Panteleeff, H. L. Martin, J. Bwayo, J. K. Kreiss, and J. Overbaugh, Evolution of envelope sequences from the genital tract and peripheral blood of women infected with clade A Human immuodeficiency Virus Type 1, *J. Virol.* **72** (1998), 8240–8251.

[44] B.D. Preston, B. J. Poiesz, and L. A. Loeb, Fidelity of HIV reverse transcriptase, *Science* **242** (1988), 1168–1171.

[45] C. R. Rinaldo, P. Gupta, X.-L. Huang, Z. Fan, J. I. Mullins, S. J. Gange, H. Farzadegan, R. Shankarappa, A. Munoz, and J. B. Margolick, Anti-HIV-1 memory cytotoxic T lymphocyte responses associated with changes in CD4+ T cell numbers in progression to HIV-1 infection, *AIDS Res. Hum. Retrovir.* **14** (1998), 1423–1433.

[46] A. G. Rodrigo and J. Felsenstein, Coalescent approaches to HIV population genetics, in (K. Crandall, ed.), *The evolution of HIV*, John Hopkins, Baltimore, 1999, 233–272.

[47] A. G. Rodrigo, E. G. Shpaer, E. L. Delwart, A. K. N. Iversen, M. V. Gallo, J. Brojatsch, M. S. Hirsch, B. D. Walker, and J. I. Mullins, Coalescent estimates of HIV-1 generation time in vivo, *Proc. Natl. Acad. Sci. USA* **96** (1999), 2187–2191.

[48] R. Sanjuan, F. Codoner, A. Moya, and S. F. Elena, Natural selection and the organ specific differentiation of HIV-1 V3 hypervariable region, *Evolution* **58** (2004), 1185–1194.

[49] R. Shankarappa, J. B. Margolick, S. J. Gange, A. G. Rodrigo, D. Upchurch, H. Farzadegan, P. Gupta, C. R. Rinaldo, G. H. Learn, X. He, X.-L. Huang, and J. I. Mullins, Consistent viral evolutionary changes associated with the progression of Human Immunodeficiency Virus Type 1 Infection, *J. Virol.* **73** (1999), 10489–10502.

[50] P. Simmonds, L. Q. Zhang, F. McOmish, P. Balfe, C. A. Ludlam, and A. J. L. Brown, Discontinuous sequence change of human immunodeficiency virus (HIV) type 1 env sequences in plasma viral and lymphocyte-associate proviral populations in vivo: implications for models of HIV pathogenesis, *J. Virol.* **65** (1991), 6266–6276.

[51] M. J. Wade and C. J. Goodnight, The theories of Fisher and Wright in the context of metapopulations: When nature does many small experiments, *Evolution* **52** (1998), 1537–1553.

[52] T. H. Wang, Y. K. Donaldson, R. P. Brettle, J. E. Bell, and P. Simmons, Identification of shared populations of human immunodeficiency virus type 1 infecting microglia and tissue macrophages outside the central nervous system, *J. Virol.* **75** (2001), 11686–11699.

[53] S. Williamson, Adaptation in the env gene of HIV-1 and evolutionary theories of disease progression, *Molec. Biol. Evol.* **20** (2003), 1318–1325.

[54] D. Wodarz and M. A. Nowak, Specific therapy regimes could lead to long-term immunological control of HIV, *Proc. Natl. Acad. Sci. USA* **96** (1999), 14464–14469.

[55] S. Wright, *Evolution and the genetics of populations. Vol. 3. Experimental results and evolutionary deductions*, University of Chicago Press, Chicago, 1977.

[56] S. Wright, *Evolution and the genetics of populations. Vol. 4. Variability within and among natural populations*, University of Chicago Press, Chicago, 1978.

[57] P. M. d. A. Zanotto, E. G. Kallas, R. F. de Souza, and E. C. Holmes, Genealogical evidence for positive selection in the nef gene of HIV-1, *Genetics* **153** (1999), 1077–1089.

Department of Ecology & Evolutionary Biology, University of Kansas, 1200 Sunnyside Ave., Lawrence, KS 66045-7534

E-mail address: jkk@mail.ku.edu

DIMACS Series in Discrete Mathematics
and Theoretical Computer Science
Volume **71**, 2006

Basic Methods for Modeling the Invasion and Spread of Contagious Diseases

Wayne M. Getz and James O. Lloyd-Smith

ABSTRACT. The evolution of disease requires a firm understanding of heterogeneity among pathogen strains and hosts with regard to the processes of transmission, movement, recovery, and pathobiology. In this and a companion chapter (Getz et al. this volume), we focus on the question of how to model the invasion and spread of diseases in heterogeneous environments, without making an explicit link to natural selection–the topic of other chapters in this volume. We begin in this chapter by providing an overview of current methods used to model epidemics in homogeneous populations, covering continuous and discrete time formulations in both deterministic and stochastic frameworks. In particular, we introduce Kermack and McKendrick's SIR (susceptible, infected, removed) formulation for the case where the removed (R) disease class is partitioned into immune (V class) and dead (D class) individuals. We also focus on transmission, contrasting mass-action and frequency-dependent formulations and results. This is followed by a presentation of various extensions including the consideration of the latent period of infection, the staging of disease classes, and the addition of vital and demographic processes. We then discuss the relative merits of continuous versus discrete time formulations to model real systems, particularly in the context of stochastic analyses. The overview is completed with a presentation of basic branching process theory as a stochastic generation-based model for the invasion of disease into populations of infinite size, with numerical extensions generalizing results to populations of finite size. In framework of branching process theory, we explore the question of minor versus major stochastic epidemics and illuminate the relationship between minor epidemics and a deterministic theory of disease invasion, as well as major epidemics and the deterministic theory of disease establishment. We conclude this chapter with a demonstration of how the basic ideas can be used to model containment policies associated with the outbreak of SARS in Asia in the early part of 2003.

1. Introduction

The pillars upon which modern epidemiological theory for directly-transmitted infectious microparasitic disease (primarily viral and bacterial) is built are the deterministic model of Kermack and McKendrick [**33**] (also see Hethcote [**28**]), the

Grants to the first author from the NSF/NIH Ecology of Infectious Disease Program (DEB-0090323), NIH-NIDA (R01-DA10135), and the James S. McDonnell Foundation 21st Century Science Initiative supported various components of the research reported in this chapter.

stochastic chain binomial models of Reed and Frost (unpublished lecture notes) and Soper [**51**], later enriched through the application of Galton-Watson branching process theory (see [**11**]). The field has only come of age over the past three decades, marked first by Bailey's seminal text [**5**], *The Mathematical Theory of Infectious Diseases and its Application*, later by Anderson and May's synthetic tome [**1**], *Infectious Diseases of Humans: Dynamics and Control*, and more recently with three research survey volumes by the broader community, arising from the Isaac Newton Institute's 1993 focus on infectious disease modeling [**24, 30, 45**]. Also, texts by Daley and Gani [**11**], Diekmann and Heesterbeek [**15**], and Thieme [**53**] provide well-crafted introductory and advanced mathematical presentations, while several edited volumes provide an overview of current interests and directions (e.g. [**14**] and [**29**], this volume).

An important current area of research is the development of theory and methods to model epidemics in heterogeneous systems. Heterogeneity due to spatial and other population structures, such as age, social groups, or genetic variation, can arise in many different ways, and modeling studies dealing with various kinds of heterogeneity are being published at an increasing rate. Heterogeneity is grist for the evolutionary mill. In this and the next chapter, however, we focus only on the development of more cohesive theoretical and methodological approaches to disease spread in heterogeneous systems, because further maturation of these fields are still needed to approach some of the most challenging problems in the coevolution of host-pathogen systems.

The primary goal of this chapter is to provide a didactic overview of the basic underpinnings of our own studies of HIV, TB and other diseases, presented in the next chapter [**23**].

2. SIR Models

Underlying all dynamical systems models of epidemiological processes is the S-I framework of Kermack and McKendrick [**33**] that was foreshadowed by the work of Enko [**18**]. Within this framework, a population infected by a microparasite (e.g. a bacterial or viral pathogen, or protist) is divided at its most basic level into susceptible and infected (assumed to also be infective) groups, with numbers or densities represented at time t by the continuous variables $S(t)$ and $I(t)$ respectively. At a trivial level, numbers are easily converted to densities once the area A confining the population is known; but, in general, aggregations in the density of populations across landscapes—or metapopulations when well-structured—has a critical influence on the epidemiology (see discussion in [**8, 12, 13, 43, 44**]). For deterministic models, a density representation is preferable because, as discussed in the next section on transmission, it focuses our attention on assumptions regarding the rate at which each susceptible individual acquires the disease as a function of the state of the system. In stochastic models, however, a numbers representation is often preferable. Descriptions of epidemics using continuous deterministic variables are only useful when the numbers in each disease class are relatively large. This, of course, is not the case in the initial stages of an epidemic when $I(t)$ is best described as a jump process on the integers.

At the heart of the original Kermack-McKendrick model is the *transmission function* βSI, with *transmission parameter* $\beta > 0$, which arises from the assumption that the rate at which susceptible individuals become infected is proportional to the densities or numbers of the susceptible and infected populations—i.e. transmission is a *mass action* process. (Note that some terminological confusion has arisen on this point through de Jong, Diekmann, and Heesterbeek [**13**], referring to βSI transmission as mass action when S and I are densities and pseudo mass action when S and I are numbers.) Many presentations of the Kermack-McKendrick model explicitly or implicitly assume that all individuals are infectious immediately on becoming infected (i.e. no latent period exists), and that infectious individuals are removed from the population at a rate α to enter a class of size $R(t)$ of *recovered* (assumed to also be immune) or *removed* (dead) individuals. Some ambiguity has arisen in the past in models where it has not been specified whether the R class individuals are immune or dead. To avoid this ambiguity in our presentation, we split the R class into a V (recovered and immune—i.e. naturally vaccinated) and a D (dead) class, with flows from the I class at rates α_V and α_D respectively. Immunity is life-long only for a subset of diseases, including many so-called childhood diseases such as measles or chickenpox. Thus, for generality, we assume that individuals in the V class lose their immunity at rate ρ (also known as the relapse rate), to return to the S class. Under these assumptions, the model has the form

$$\begin{aligned}
\frac{dS}{dt} &= -\beta SI + \rho V & S(0) &= S_0 \\
\frac{dI}{dt} &= \beta SI - (\alpha_V + \alpha_D)I & I(0) &= I_0 \\
\frac{dV}{dt} &= \alpha_V I - \rho V & V(0) &= V_0 \\
\frac{dD}{dt} &= \alpha_D I & D(0) &= 0.
\end{aligned} \tag{2.1}$$

In model (2.1), the total density of individuals alive at time t is $N(t) = S(t) + I(t) + V(t)$. The sum $N(t) + D(t)$ has the constant value $S_0 + I_0 + V_0$ throughout the epidemic because model (2.1) does not include a description of any demographic processes, which are assumed to be operating at much longer time scales than the epidemic itself. This model predicts that the invasion of a completely susceptible population by a single infected individual will give rise to an epidemic provided $S_0 > \frac{\alpha}{\beta} := S_T$ ($:=$ means "by definition"), where $\alpha = \alpha_V + \alpha_D$ and S_T is the conceptually appealing, although hard to demonstrate [**38**], threshold density of susceptibles that is required for a disease to be able to invade a population when the transmission term is assumed to have a mass-action form.

This threshold density is related to the more general threshold criterion $R_0 > 1$, where R_0 is the basic reproductive number defined as the expected number of individuals that a typical infectious case will infect in a wholly susceptible population [**1, 27**]. For the mass action model shown above, $R_0 = \frac{\beta S_0}{\alpha}$.

If an epidemic does occur then, in the absence of any new susceptibles coming into the population, the number of infectious cases will peak but ultimately die out leaving a proportion $s_\infty := \lim_{t\to\infty} \frac{S(t)}{N_0}$ of the population uninfected. This proportion is the solution to the equation

$$s_\infty = e^{R_0(1-s_\infty)}. \tag{2.2}$$

See [**41, 42**] and [**15**] for an illuminating discussion of conditions under which this expression holds true.

3. Transmission

At the core of any epidemic is the pathogen transmission process, modeled in the above SIR formulation by a mass-action process with resulting form βSI. A more refined analysis views transmission as the concatenation of two processes: a contact process focusing on the rate at which susceptible individuals encounter infected individuals, and a process relating the probability with which susceptible individuals become infected per susceptible-infective contact [**15, 26, 43**]. Let $C(N)$ be the rate at which each individual in a population contacts other individuals (assumed to depend on total population size N), and let p_T be the probability that a susceptible becomes infected during one such contact with an infective. In a randomly mixing population, the proportion of an individual's contacts that are with infectives is $\frac{I}{N}$. The expected transmission rate per-capita susceptible is then given by

$$\tau(I, N) = p_T C(N) \frac{I}{N}, \tag{3.1}$$

which is known in the epidemiology literature as the *force (or hazard) of infection.* Mass-action transmission (i.e. a transmission rate that is proportional to the product SI) arises when the rate at which susceptibles encounter infectives is *proportional* to the population density, that is $C(N) = cN$, with c the constant of proportionality, in which case the transmission parameter β is given by $\beta = p_T c$.

Although the mass-action formulation of transmission dominated epidemiological modeling into the early 1990's, by the end of the millennium most formulations were based on the assumption that contact rates are limited by behavioral or social factors and do not depend significantly on population density, with the simplest case being $C(N) = c'$. This assumption leads to so-called *frequency-dependent* transmission:

$$\tau(I, N) = p_T c' \frac{I}{N}, \tag{3.2}$$

with the interpretation that individuals contact each other at a fixed rate c', and if the density of all individuals is N, then a proportion $\frac{I}{N}$ of these contacts will be infective.

Use of the frequency-dependent rather than the mass-action formulation has profound implications for our understanding of epidemics [**22**]. For example, replacing mass action with frequency dependence in the SID model (2.1), we obtain

$$\begin{aligned}
\frac{dS}{dt} &= -\beta \frac{SI}{N} + \rho V & S(0) &= S_0 \\
\frac{dI}{dt} &= \beta \frac{SI}{N} - (\alpha_V + \alpha_D) I & I(0) &= I_0 \\
\frac{dV}{dt} &= \alpha_V I - \rho V & V(0) &= V_0 \\
\frac{dD}{dt} &= \alpha_D I & D(0) &= 0
\end{aligned} \tag{3.3}$$

where $N(t) = S(t) + I(t) + V(t)$. The concept of a threshold population density no longer applies in this model: on introduction of an infected individual into the population an epidemic can occur if and only if $\beta > \alpha = \alpha_V + \alpha_D$. Also, for the case where recovery from disease ($\alpha_V = 0$) does not occur (e.g. HIV infections), unlike the mass action SIR model all susceptibles become infected during the course of an epidemic in a well-mixed homogeneous population—that is, $s_\infty = 0$.

Mass-action and frequency-dependent transmission can be regarded as special cases of a more general transmission function [**43**]

$$\tau(I,N) = \left(\frac{p_T c'}{1+(K_T/N)}\right)\left(\frac{I}{N}\right), \tag{3.4}$$

where K_T is the population size or density at which transmission is half its maximum rate $p_T c'$ before accounting for the proportion of infective individuals in the population. Note that expression (3.4) is approximately mass action when $N \ll K_T$ (in this case $c \approx c'/K_T$ in the mass action expression given in equation (3.1) with $C(N) = cN$), is approximately frequency dependence when $N \gg K_T$ (cf. equation (3.2)), and is some interpolation of the two when N is the neighborhood of K_T. The influence of the interpolation can be controlled by a parameter $\gamma_T \geq 1$ using the expression

$$\tau(I,N) = \frac{p_T c'}{(1+(K_T/N)^{\gamma_T})^{1/\gamma_T}}\left(\frac{I}{N}\right), \tag{3.5}$$

where, for γ_T close to 1 (say, $1 \leq \gamma_T \leq 3$), the region of interpolation is quite extensive; and, for increasing values of γ_T, we get an increasingly abrupt switch from mass action when $N < K_T$ to frequency dependence when $N > K_T$. An alternative formulation of a saturating transmission function, based on mechanistic considerations, is given by Heesterbeek and Metz [**26**].

In a well-mixed population with no bounds on contact rates, mass action is likely to be superior to frequency dependence as a model of transmission of airborne or casual contact infectious diseases, such as tuberculosis or influenza. (Note that random mixing and unbounded contact rates are very strong assumptions that probably only apply within homogeneous subunits of a population.) In socially or spatially structured populations, or when the disease is only transmitted by close contact, the overall transmission may be better modeled using the more general force of infection function represented by expression (3.5). An open question remains regarding which values of K_T and γ_T best reflect the aggregated properties of different types of spatial structure. Many models implicitly assume that $N \gg K_T$ and apply frequency-dependent transmission.

For a sexually-transmitted disease (STD), K_T is likely to be quite low so the frequency of intimate contact is unlikely to depend on population density for many populations. In this case, transmission should be modeled using the pure frequency-dependent form expressed in (3.2). However, monogamy is a distinguishing feature of many (though not all) sexual relationships, and the finite duration of monogamous partnerships can have significant impacts on STD spread [**16**]. In a recent study, we examined the contact process for STDs at a finer resolution, formulating a model of pair formation and dissolution [**40**]. We showed that frequency-dependent transmission can be derived rigorously from the pair-formation mechanism, but only by applying a strong timescale assumption that pairing processes are much faster than disease processes. The derivation is exact only for instantaneous partnerships; for partnerships of finite duration epidemics progress more slowly and may saturate at lower levels. Using simulations we found that frequency-dependent transmission is a reasonable depiction of STD transmission via monogamous partnerships only for relatively promiscuous populations: for faster bacterial STDs, such as gonorrhea and chlamydia, partnerships had to last days on average, at most, for the approximation to be reasonable, while for slower viral STDs, such as HIV or herpes, partnerships needed to last a few months on average for the approximation to be reasonable.

Another important assumption made by most standard models of transmission is that the infection does not influence individuals' contact behavior. This is often not the case, because infected individuals may reduce their contacts due to physical effects of illness, social factors, or ethical concerns (e.g. [**34, 50**]), or the pathogen may manipulate host behavior to increase contact rates [**7**]. In the study described above, we extended the frequency-dependent transmission function to account for instances where individuals pairing behavior was a function of their disease class status (healthy (S) or sick (I)) [**40**]. In particular, if k_S and k_I are the relative rates at which healthy and sick individuals enter pairs and l_{SS}, l_{SI} and l_{II} are the relative rates at which SS, SI and II pairs break up (all rates on the same time scale, but the scale itself is arbitrary) then, assuming that infection is transmitted at the same average rate $p_T c'$ within all pairs, the frequency-dependent transmission expression (3.2) has the form

$$\tau(I, N) = p_T c' \frac{I}{N} \phi, \tag{3.6}$$

where the modifying factor ϕ depends only the proportions S/N and I/N, and the parameters k_S, k_I, l_{SS}, l_{SI} and l_{II}.

By way of example, if healthy and sick individuals pair up at different rates k_S and k_I, but all pairs break up at the same rate l irrespective of the disease status of the partners, then ϕ in expression (3.6) has the form

$$\phi = \frac{k_S k_I N}{k_S(k_I + l)S + k_I(k_S + l)I}.$$

The general case with different break up rates leads to more complex expressions [**40**].

4. Disease Class Extensions

The first obvious extension to the basic SIR model is to incorporate disease latency, defined as the period from when an individual is infected to when they become infectious. Note that while symptoms often aid transmission, as in the case of coughing (lung infections) or the formation of pustules (poxes), the latent period is distinct from the incubation period between infection and the appearance of symptoms [**19**]. Indeed, the proportion of transmission occurring before symptoms has a critical influence on options for disease control [**20, 48**]. In practice the onset of infectiousness is often difficult to measure, however, and the appearance of symptoms is sometimes used as a surrogate, particularly for novel diseases such as SARS when it emerged.

In a real population of individuals infected by a disease, the latent period will assume a distribution of values that is usually unimodal with a mode greater than zero [**19**]. The simplest approach conceptually is to model this with a fixed time delay. More generally, we can assume that the time from exposure to infectivity is characterized by a probability distribution $P_\Delta(t)$. In this more general case, just focusing on the transmission process terms, the equations for the susceptible and infective classes have the form

$$\begin{aligned}\frac{dS}{dt} &= -p_T C(N(t))S(t)\frac{I(t)}{N(t)} + \text{ non-transmission terms}\\ \frac{dI}{dt} &= p_T \int_0^t C(N(u))S(u)\frac{I(u)}{N(u)}P_\Delta(t-u)du + \text{ non-transmission terms.}\end{aligned}$$

This equation is an infinite-dimensional dynamical system with all the attendant numerical and analytical intricacies of such systems. A mathematically simpler approach is to add a disease class E of individuals that have been exposed to the disease and become infective at rate δ. In this case, ignoring equations and terms relating to removed or recovered individuals, the model takes the form

$$\tag{4.1} \begin{aligned}\frac{dS}{dt} &= -p_T C(N)S\frac{I}{N} & S(0) &= S_0\\ \frac{dE}{dt} &= p_T C(N)S\frac{I}{N} - \delta E & E(0) &= E_0\\ \frac{dI}{dt} &= \delta E - \alpha I & I(0) &= I_0.\end{aligned}$$

In this approach, just focusing on a fixed cohort of individuals E_0 that are exposed at time $t = 0$ (i.e. ignoring new infections from transmissions), then the rate at which these individuals advance to the infectious class is given by the following equation and its associated solution:

$$\frac{dE}{dt} = -\delta E, \quad E(0) = E_0 \quad \Rightarrow \quad E(t) = E_0 e^{-\delta t}.$$

In this case, the so-called *residence time* in the exposed class E is exponentially distributed with mean $1/\delta$ [**52**]. The exponential distribution is a poor match to biological distributions of latent periods, however, because its mode is at $t = 0$ rather than in the vicinity of its mean at $t = 1/\delta$.

This latter problem can be resolved using a *distributed delay* (i.e. staging or box car) approach [**47**] in which we have n classes of exposed individuals E_i, $i = 1, \ldots, n$, and assume that individuals transfer through each class at a rate $n\delta$. In this case, equations (4.1) are extended to:

$$\tag{4.2} \begin{aligned}\frac{dS}{dt} &= -p_T C(N)S\frac{I}{N} & S(0) &= S_0\\ \frac{dE_1}{dt} &= p_T C(N)S\frac{I}{N} - n\delta E_1 & E_1(0) &= E_{1\,0}\\ \frac{dE_i}{dt} &= n\delta(E_{i-1} - E_i) & E_i(0) &= E_{i\,0} \quad i = 2, \ldots, n\\ \frac{dI}{dt} &= n\delta E_n - \alpha I & I(0) &= I_0.\end{aligned}$$

The total residence time in the exposed class (i.e. the time from leaving S to entering I) is now gamma-distributed [**52**]. The mean residence time is still $1/\delta$, but the distribution is now modal near $1/\delta$ and has variance $1/(n\delta^2)$, which implies that as $n \to \infty$ the solution to model (4.2) approaches a fixed time lag (i.e. variance is 0) of duration $1/\delta$.

Of course, similar techniques can be employed to make the infectious period distribution more realistic as well. The dynamical consequences of using gamma-distributed latency and infectious periods were well-known in a queuing theory context in the mid-1900s but were first laid out in detail for an epidemiological audience by Anderson and Watson [**2**]. Subsequent to this, the implications of exponential versus fixed infectious periods have been explored in greater depth (e.g. [**31, 32, 36, 37**]).

5. Including Demography

Demographic considerations fall into two categories: 1.) flows in and out of the population due to births, deaths, and migration; and 2.) disease-independent internal population structure due to age, developmental stage, or group structure (all of these may also have a spatial component). The former is easy to incorporate in the absence of any internal structures and, thus, we deal with it first. Deterministic models (2.1) and (3.3) pertain purely to epidemic processes in a population that is initially of size $N(0) = S_0 + I_0 + V_0$ and declines due to the effects of the disease induced death rate α_D on individuals in class I. Demographic flows are easily included into such models, as follows. In the absence of disease, assume that the population of interest is subject to a total birth or recruitment rate $f_b(N)$ and a per-capita mortality rate μ. Then in the absence of migration and age structure the demographic equation for the population is simply

$$\frac{dN}{dt} = f_b(N) - \mu N.$$

In the presence of disease, we need to decide whether all individuals participate equally in the birth process or whether infected individuals have an altered reproductive capacity. The most appropriate assumption will depend on the disease in question. For simplicity of exposition, assume that susceptible and infected individuals contribute equally to births and that newborn individuals are susceptible (i.e. the disease is not vertically transmitted and there is no immunity due to maternal antibodies), and that individuals are removed from the infective population at a rate $(\mu + \alpha)$. In this case, using the general transmission function defined by expression (3.1), equations (2.1) and (3.3) can be written as (cf. [**1, 22**])

$$\begin{aligned}
\frac{dN}{dt} &= f_b(N) - \mu N - \alpha_D I & N(0) &= N_0 \\
\frac{dI}{dt} &= p_T C(N)(N - I - V)\frac{I}{N} - (\mu + \alpha_D + \alpha_V)I & I(0) &= I_0 \\
\frac{dV}{dt} &= \alpha_V I - \mu V & V(0) &= V_0.
\end{aligned} \tag{5.1}$$

In this model, written in a form that accentuates the dynamics of the population as a whole, the susceptible class is determined by $S(t) = N(t) - I(t) - V(t)$.

The two simplest forms for the "input" function $f_b(N)$ births, are the following. In the case where $f_b(N)$ is interpreted as recruitment (i.e. as is useful for short-term predictions when N is "distanced" from the birth process by a long time delay such as in HIV models where N represents individuals age 15 to 50 and predictions are made 10 to 20 years ahead) then $f_b(N)$ is assumed either to be constant or an externally determined time-varying input of the form $f_b(N) = \lambda(t)$. In the case where $f_b(N)$ arises from a constant per-capita birth rate, this rate must balance the death rate—that is $f_b(N) = \mu N$— otherwise the population will grow or decline exponentially (cf. [**28**]). This of course can be corrected by ensuring births are density dependent—as in the function

$$f_b(N) = \frac{bN}{1 + (N/K_b)^{\gamma_b}},$$

where we require $b > \mu$ (to ensure growth at low population densities) and $\gamma_b > 1$ (to ensure a sensible dependence on density—see [**21**]).

A treatment of age structure in the context of continuous time models is beyond the scope of this exposition because, as with the inclusion of time delays, continuous age-structure formulations are cast either in the context of McKendrick-von Foerster partial differential or integro-differential equations [**1, 10, 53**]. The mathematical properties and numerical solutions of such systems are much more difficult to obtain. Age structure, however, is easily incorporated in the context of discrete models. Further, as argued in subsequent sections, discrete time models better reflect the fact that empirical data are typically values obtained from averaging rates over predetermined discrete time intervals (e.g. daily, monthly, or annual birth, death and infection rates, and so on).

6. Discrete Time Formulations

We can find solutions to continuous time formulations of dynamic processes, such as the epidemiological models (2.1), (3.3) or (4.2), or we can, at least, analyze their behavior using the tools of calculus. Data used to estimate the parameters in these equations, however, are generally derived from events—such as births, deaths, new cases, cures, numbers vaccinated—recorded over appropriate discrete intervals of time (typically days for fast diseases such as SARS or influenza, weeks or months for slower diseases such as tuberculosis or HIV, and years for vital rates in seasonal breeders and long-lived species). Data reporting the proportion p_μ of individuals that die in a unit of time can be converted to a mortality rate parameter μ appearing in a differential equation model of the form $\frac{dN}{dt} = -\mu N$ by noting that the solution to this equation over any time interval $[k, k+1]$ is $N(k+1) = N(k)e^{-\mu}$. This implies that the proportion of individuals dying is

$$p_\mu = \frac{N(k) - N(k+1)}{N(k)} = \frac{N(k)(1 - e^{-\mu})}{N(k)} = 1 - e^{-\mu} \tag{6.1}$$

or, equivalently, $\mu = \ln\left(\frac{1}{1-p_\mu}\right)$.

It is often advantageous to formulate epidemiological and demographic models in discrete time. The primary advantage of differential equation models disappears once we resort to numerical simulation of systems rather than trying to obtain analytical results, which are difficult if not impossible to obtain for most detailed nonlinear models. Indeed, discrete time models can be implemented very naturally and easily in computer simulations, while numerical solutions of differential equations requires algorithms that use discretizing approximations. Also parameters in discrete time models can be more easily related to data that have been collated over discrete intervals (e.g. vital and transmission rates, etc.).

Discrete time equations, however, cannot properly account for the interactions of simultaneously nonlinear processes, such as individuals simultaneously subject to the processes of infection and death: in each time interval we either first account for infection and then natural mortality or vice versa. It does make a difference how we schedule things [**54**]. Alternatively we can treat the two processes simultaneously, but then cannot accurately depict both processes occurring in one time step if transition rates are state-dependent. A further challenge arises in discrete time disease models because the force of infection depends on the size of the infectious population at each moment, which cannot be updated over the course of a time step. Fortunately, a good approximation can be obtained by using a piecewise linear modeling approach, as follows.

First we write down the continuous time model of interest. Consider, for example, equations (4.1) with a constant total recruitment rate λ to the susceptible class and a constant per-capita natural mortality rate μ. Then, using the notation $\tau(I,N) := p_T C(N)I/N$ (see equation (3.1)), these equations become

$$\begin{aligned} \frac{dS}{dt} &= \lambda - \mu S - \tau(I,N)S & S(0) &= S_0 \\ \frac{dE}{dt} &= \tau(I,N)S - (\delta+\mu)E & E(0) &= E_0 \\ \frac{dI}{dt} &= \delta E - (\alpha+\mu)I & I(0) &= I_0. \end{aligned} \tag{6.2}$$

Now assume over a small interval $t \in [k, k+1]$ that the proportional change in τ over this interval due to a change in $I(t)$ is sufficiently small that $\tau(I(t), N(t))$ is well approximated by $\tau_k = \tau(I(k), N(k))$ (e.g. over the time interval the change in $\tau(I(t), N(t))$ may be a few percent, but our estimates of the parameters c' and p_T in $\tau(I(t), N(t))$, as defined by expression (3.2), may have uncertainties that are several times as large). Obviously this assumption will influence the choice of time step duration, with shorter time steps required for faster-growing epidemics. Then replacing $\tau(I(t), N(t))$ with the constant τ_k for $t \in [k, k+1]$, equation (6.2) is an inhomogeneous linear system of ordinary differential equations, with solution given by

$$\begin{pmatrix} S(k+1) \\ E(k+1) \\ I(k+1) \end{pmatrix} = \exp\{A_k\} \begin{pmatrix} S(k) \\ E(k) \\ I(k) \end{pmatrix} + \left(\int_0^1 \exp\{A_k t\} dt \right) \begin{pmatrix} \lambda \\ 0 \\ 0 \end{pmatrix}, \tag{6.3}$$

where

$$A_k = \begin{pmatrix} -(\mu+\tau_k) & 0 & 0 \\ \tau_k & -(\mu+\delta) & 0 \\ 0 & \delta & -(\mu+\alpha) \end{pmatrix}. \tag{6.4}$$

Calculation of the exponential matrix function $\exp\{A_k\}$ and its integral requires that we first find the eigenvalues and eigenvectors of the matrix A_k itself, which is cumbersome to calculate and will generally not have a closed form solution for systems with more than four disease classes (unless A_k is triangular, in which case the eigenvalues are the diagonal entries themselves). The matrix $\exp\{A_k\}$ and its integral can be calculated numerically, but this calculation will have to be performed at each time step k, because of the dependence of τ_k on the current values of the state vector $(S(k), E(k), I(k))'$ (here $'$ denotes vector transpose).

A discrete version of equation (6.2) can be argued directly from first principles under the assumptions that individuals are recruited at the beginning of time interval $[k, k+1]$ and that individuals die at the same constant rate μ throughout the time interval $[k, k+1]$. In this case we obtain the model

(6.5)
$$\begin{pmatrix} S(k+1) \\ E(k+1) \\ I(k+1) \end{pmatrix} = \begin{pmatrix} (1-p_\mu)(1-p_{\tau_k}) & 0 & 0 \\ (1-p_\mu)p_{\tau_k} & (1-p_\mu)(1-p_\delta) & 0 \\ 0 & (1-p_\mu)p_\delta & (1-p_\mu)(1-p_\alpha) \end{pmatrix} \times \begin{pmatrix} S(k) \\ E(k) \\ I(k) \end{pmatrix} + \begin{pmatrix} (1-p_\mu)\lambda \\ 0 \\ 0 \end{pmatrix},$$

which is iterated from the initial condition $(S_0, E_0, I_0)'$. The probabilities p_π are related to the corresponding rates π using the relationship expressed in (6.1), viz.

$$p_\pi = 1 - e^{-\pi} \quad \text{or} \quad 1 - p_\pi = e^{-\pi}, \quad \pi = \mu,\ \alpha,\ \delta,\ \tau_k. \tag{6.6}$$

While equations (6.5) are an approximation to exact solution (6.3), which itself is only exact if $\tau(I(t), N(t))$ is replaced by the constant $\tau_k = \tau(I(k), N(k))$, it is actually irrelevant how well equations (6.5) approximate equations (6.3) when solved precisely. The reason for this is that equations (6.3) are derived from a differential equation model that is not the "gold standard" for modeling epidemics; but, instead, equations (6.3) represent a highly simplified model that does not account for lags, latencies, or heterogeneity in the population being modeled—not to mention higher order processes taking place on faster time scales (one of which is the contact process discussed at the end of Section 3). No theoretical reason exists to prefer differential equation models over difference equation models: both have their strengths and weaknesses.

It is also worth noting at this point that all the deterministic models presented above can be interpreted as representing expected numbers of individuals in what are essentially stochastic epidemiological and demographic processes. Deterministic models provide reasonable realizations of stochastic models either when the size of the population is sufficiently large for the 'Law of Large Numbers' (proportions approach probabilities in the limit as population size approaches infinity) to prevail or they represent equations for the first order moments of the stochastic process in question when second and higher order moments are neglected (e.g. see [**11**, page 68]).

Discrete time models are more easily embedded in a Monte Carlo (i.e. stochastic) simulation framework than continuous time models. First the effects of demographic stochasticity can be simulated by treating the state variables as integers and then calculating the proportion that undergo each transition as one realization of a set of appropriate Bernoulli trials that will produce a binomial distribution of outcomes. The underlying probabilities p_π themselves can be subject to stochastic variation specified by some appropriate probability distribution, and Monte Carlo simulation methods can be used to generate the statistics of associated distributions of possible solutions to an equation (6.5) when the parameters are interpreted as random variables which themselves are drawn from statistical distributions.

Second, discrete time models are more flexible than ordinary differential equation models when it comes to fitting distributions reflecting the time spent in a particular disease stage. As we have seen, staging in continuous models can lead to gamma distributions on $[0, \infty)$ for the residence times of individuals in each disease class. Thus staging in continuous models allows us to construct a process that has a desired mean and variance, but also implies that the minimum time spent in a class is zero, even though residence times close to zero may be extremely unlikely for processes with variance much smaller than the mean.

Staging in discrete models, however, allows us to easily set a minimum and maximum time in a class, as well as a desired mean and variance, as illustrated by the following example of a staged exposed class (cf. equation (4.2)). In these equations, as previously defined, p_{τ_k} and p_α are respectively the probabilities over one time step of a susceptible becoming infectious and an infectious individual being removed. In addition, with regard to staging, we define p_{δ_i} as the probability that an individual in exposed class i, makes the transition to exposed class $i+1$ (except for the case $i = n$, from where individuals transition to the infectious class). For added flexibility we also introduce the probability p_{θ_i} of moving directly from exposed class i into the infectious class. Note that the probabilities are formulated so that we first account for the proportion that become infectious (via p_{θ_i}) and only then consider what proportion of the remainder make the transition to the next exposed class (via p_{δ_i}). Accordingly, for $i = n$ the proportion moving into the infectious class over one time step is $p_{\theta_i} + (1 - p_{\theta_i})p_{\delta_i}$. Under these assumptions, the staged model has the form:

(6.7)

$$
\begin{aligned}
S(k+1) &= S(k) - p_\tau(I(k), N(k))S(k) & S(0) &= S_0 \\
E_1(k+1) &= p_\tau(I(k), N(k))S(k) + (1 - p_{\theta_1})(1 - p_{\delta_1})E_1(k) & E_1(0) &= E_{1\,0} \\
E_{i+1}(k+1) &= (1 - p_{\theta_{i+1}})(1 - p_{\delta_{i+1}})E_{i+1}(k) + (1 - p_{\theta_i})p_{\delta_i}E_i(k), & E_i(0) &= E_{i\,0} \\
&\qquad i = 1, \dots, n-1 \\
I(k+1) &= \sum_{i=1}^{n} p_{\theta_i}E_i(k) + (1 - p_{\theta_n})p_{\delta_n}E_n(k) + (1 - p_\alpha)I(k) & I(0) &= I_0.
\end{aligned}
$$

Consider the fate of a cohort of $E_{1\,0}$ individuals entering the exposed class at time $t = 0$. The rate at which these individuals enter class I is the element $I(k)$ in the solution to equations (6.7) from initial conditions, $S_0 = 0$, $E_{1\,0} > 0$, $E_{i\,0} = 0$, $i = 2, \dots, n$, and $I_0 = 0$. In this case, it is easily shown that these $E_{1\,0}$ individuals will remain in the exposed class for a minimum of $k < n$ units of time whenever $p_{\theta_i} = 0$, $i = 1, \dots, k$, and will all have become infectious by time $n + 1$, provided $p_{\theta_n} = 1$ and $p_{\delta_i} = 1$ for $i = 1, \dots, n-1$. If $p_{\theta_n} < 1$, however, the final group of individuals will trickle into the infectious class at a geometrically decreasing total rate over time. The value for n and the parameters p_{θ_i} and p_{δ_i}, $i = 1, \dots, n$, can be selected to fit any empirically observed distribution.

7. Stochastic Branching Processes

During the invasion phase of an epidemic, the numbers of infected individuals are small and stochastic effects can play an important role. Branching processes are a family of stochastic models well-suited to modeling disease invasions in large populations. The theory of the branching process (also known as the Galton-Watson process: e.g. see [**9**]) was developed to explore the role of chance in demographic dynamics, originally to understand the extinction of notable family lines in England [**25**]. Consequently it is formulated in terms of generations, with its basic concept being the offspring distribution, which defines the probability that a given "parent" (here equivalent to an infectious index case) will give rise to a given number of "offspring" (here equivalent to new infections). The core assumption is that the numbers of offspring produced by different index cases are independent and identically distributed. In the context of epidemics, this amounts to assuming that the population is sufficiently large and well-mixed for depletion of the susceptible pool of individuals to be negligible, and that transmission conditions do not change with time. Thus a specific model may only be valid until control measures are introduced. The theory of branching processes is developed in detail elsewhere [**3, 4, 25**]; the treatment here is extracted essentially from Diekmann and Heesterbeek [**15**].

We define the offspring distribution $\{q_i\}_{i=0}^{\infty}$, where q_i is the probability that an infectious individual infects i other individuals. Thus we require $\sum_{i=0}^{\infty} q_i = 1$ and note that R_0, the mean number of cases contracting disease from each infective, is simply given by

$$R_0 = \sum_{i=0}^{\infty} i q_i. \tag{7.1}$$

A powerful tool for studying a branching process is the probability generating function $g(z)$ defined in terms of a dummy variable z by:

$$g(z) = \sum_{i=0}^{\infty} q_i z^i, \quad 0 \leq z \leq 1. \tag{7.2}$$

This power arises from the easily demonstrated properties

$$g(0) = q_0, \ g(1) = 1, \ \left.\frac{dg}{dz}\right|_{z=1} = R_0, \ \frac{d}{dz} g(z) > 0, \ \frac{d^2}{dz^2} g(z) > 0$$

and from the fact that the solutions z_k to the difference equation

$$z_k = g(z_{k-1}), \ z_0 = q_0, \ k = 1, 2, 3, \ldots,$$

are the probabilities that the disease traceable back to a single original infective dies out by generation k. Thus $z_\infty = \lim_{k\to\infty} z_k$ is the probability that an epidemic started by one individual will die out, and is equal to the smallest nonnegative root of the equation $z = g(z)$. It can also be shown that $z_\infty = 1$ if and only if $R_0 \leq 1$, in which case the epidemic will die out with certainty in a finite number of generations (essentially infecting a proportion of measure zero in an infinite population). When $0 < z_\infty < 1$ and there is initially one infective, then the disease causes a significant outbreak with probability $1 - z_\infty$.

Under the idealization of an infinite population, an epidemic that dies out in a finite number of generations is called a *minor epidemic*, while one that goes on to infect a positive proportion of individuals in our infinite population (i.e. measurable in mathematical terms) is called a *major epidemic*. Thus, in contrast to the deterministic case which guarantees invasion of a pathogen to epidemic proportions, with possible long term establishment, whenever $R_0 > 1$, in stochastic theory $R_0 > 1$ only implies that a *major* epidemic will occur with probability

$$P_{\text{epi}} = 1 - z_\infty, \tag{7.3}$$

whenever the initial number of infective individuals is one or, more generally, with probability $P_{\text{epi}}(a) = (1 - z_\infty)^a$ whenever the initial number of infective individuals is a.

In finite populations of size N, the distinction between minor and major epidemics can become less clear [**46**]. In very small populations such as households, the exact distribution of outbreak sizes can sometimes be calculated [**5**]. In larger

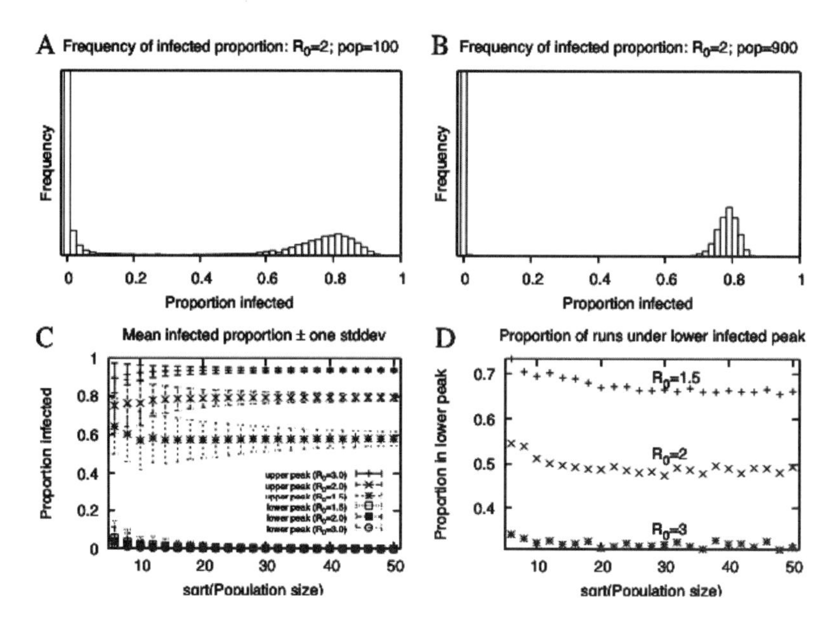

FIGURE 1. Results of individual-based stochastic simulations regarding the ultimate number of infections after the introduction of a single infected individual into a population of N-1 susceptible individuals (recall that $R_0 \approx \beta/\alpha$). **A.** $N = 100, \beta = 0.1, \alpha = 0.01$. **B.** $N = 900, \beta = 0.1, \alpha = 0.01$. **C.** The mean and standard deviation of the proportion of individuals in unsuccessful and successful invasions (*i.e.* relative areas in lower and upper peaks of histograms in **A** and **B**) are plotted as a function of $\sqrt{N}$ for the three cases $\beta = 0.075, 0.1$, and 0.15, with $\alpha = 0.05$ in all cases. **D.** Same three cases, but plotting the proportion of runs in the lower peak (10,000 simulations were used to generate each point). (We are indebted to Philip Johnson for producing this figure)

populations, a minor epidemic still goes extinct in a finite number of generations, while a major epidemic goes on to infect a proportion of the population approximated by the corresponding deterministic model, which for a basic SIR disease process without demographics is $i_\infty = 1 - s_\infty$, where s_∞ is the solution to equation (2.2). In this case, analytical results are difficult to generate, but the existence of a bimodal distribution of epidemic sizes can be demonstrated for sufficiently large N (see Figure 1A and B in [**55**]). Monte Carlo methods (discussed in the context of heterogeneous populations in the next chapter, [**23**]) can be used to generate simulated epidemics when N is finite. We can split the bimodal histograms obtained for sufficiently large N at the minimum frequency bin between the two modes and define p_{epi} and P_{epi} respectively to be the proportion of outbreaks in our population of size N falling to the left and to the right of this bin. Thus, as $N \to \infty$, $p_{\text{epi}} \to z_\infty$ and $P_{\text{epi}} \to 1 - z_\infty$ with modes of the proportion of individuals infected approaching 0 and i_∞ respectively (Figure 1).

Finally, we note that in a completely homogeneous infinite population where each individual has the same transmission rate and same fixed infectious period, all individuals are expected to transmit the same number R_0 of new infections in each generation. The actual number of infections will vary due to stochasticity in transmission events occurring over the fixed infectious period, and will follow the Poisson distribution arising from the purely random transmission process with intensity parameter R_0 [**15**]. The probability generating function for this Poisson distribution is:

$$\text{(7.4)} \qquad \underline{\text{Poisson}}: \quad g(z) = e^{R_0(z-1)}.$$

If we now apply our theory to a homogeneous population with no demography, subject to our standard SI process, we can show that the epidemic will die out if $R_0 \leq 1$, but will go on with probability $P_{\text{epi}} = 1 - z_\infty$ to infect a proportion $i_\infty = 1 - s_\infty$, where both z_∞ and s_∞ satisfy the same equation $x = e^{R_0(x-1)}$. (The root of this commonality can be explained via the theory of random graphs—see [**15**, Section 10.5.2].)

8. Group Structure and Containment of SARS

The assumption that a susceptible population is homogeneous is often too simplistic to address a number of important issues regarding the spread and containment of the spread of infectious disease. The theory we have outlined above is easily extended to a population in which the susceptible individuals are categorized into n (> 1) homogeneous subgroups. In this section, we develop this extension and then demonstrate its application to modeling the recent SARS outbreak in several cities in Asia, as well as Toronto, Canada.

We begin by considering a simple SIR process in a population composed of n subpopulations or groups. Let S_i and I_i respectively denote the density of susceptibles and infectives in population i, $i = 1, \ldots, n$. Let the constants c_{ij} denote the relative rates at which individuals in group i contact individuals in group j. Further, assume that the probability of transmission associated with each such contact is given by transmission parameter β_{ij}. Under the assumption of frequency-dependent transmission, the first two equations in system (3.3) generalize to

$$\text{(8.1)} \qquad \begin{aligned} \frac{dS_i}{dt} &= -\left(\frac{\sum_{j=1}^{n} c_{ij}\beta_{ij}I_j}{\sum_{j=1}^{n} c_{ij}N_j} \right) S_i + \text{ non-transmission terms} \\ S_i(0) &= S_{i\,0}, \\ \frac{dI_i}{dt} &= \left(\frac{\sum_{j=1}^{n} c_{ij}\beta_{ij}I_j}{\sum_{j=1}^{n} c_{ij}N_j} \right) S_i + \text{ non-transmission terms} \\ I_i(0) &= I_{i\,0}, \quad i = 1, \ldots, n. \end{aligned}$$

If we now discretize these equations, as discussed in Section 6 and use the vector notation $\mathbf{I} = (I_1, \ldots, I_n)'$ etc., equations (6.5) become

$$
\begin{aligned}
S_i(k+1) &= e^{-\tau_i(\mathbf{I}(k),\mathbf{N}(k))} S_i(k) + \text{ non-transmission terms} \\
S_i(0) &= S_{i\,0}, \\
I_i(k+1) &= (1 - e^{-\tau_i(\mathbf{I}(k),\mathbf{N}(k))}) S_i(k) + \text{ non-transmission terms} \\
I_i(0) &= I_{i\,0}, \quad i = 1, \ldots, n,
\end{aligned} \tag{8.2}
$$

where (cf. expression (3.2))

$$
\tau_i(\mathbf{I}, \mathbf{N}) = \frac{\sum_{j=1}^{n} c_{ij}\beta_{ij} I_j}{\sum_{j=1}^{n} c_{ij} N_j}. \tag{8.3}
$$

To avoid the notation becoming too complex, we have dropped the convention of using subscripted p's (cf. equation (6.6)) to emphasize that the transition parameters in the model are probabilities.

This approach was used in a recent model of the SARS epidemic that threatened to sweep through parts of Asia and Canada during the first half of 2003 [**39**]. Other models also examined this important example of a modern emerging infectious disease, as reviewed elsewhere [**6, 17**]; the primary focus of our study, though, was on structuring the population into health care workers (HCWs) in a hospital setting and the general community served by that hospital. Each individual in the model was designated as a HCW or community member, and their status did not change regardless of other transitions in the model (e.g. infection, hospitalization, etc.). The reason for this grouping was that it became evident soon after the start of the epidemic that HCWs were at much greater risk for SARS than other individuals in all cities where epidemics were threatening to flare out of control: as of mid-2003, HCWs comprised 63% of SARS cases in Hanoi, 51% in Toronto, 42% in Singapore, 22% in Hong Kong and 18% in mainland China (see [**39**] for specific references). It was therefore important to study the processes driving HCW infection explicitly, and to assess specific interventions to combat SARS spread within the hospital and back to the general community.

Moving to a specialized notation for this particular SARS model, the indices $i = h$, c, and m were respectively used to denote HCW, community, and case-managed (quarantined and isolated) individuals (Figure 2A). The model was iterated on a daily basis (i.e. the basic unit of time was one day) and included ten one-day exposed classes (Figure 2B), with probability p_j of progressing to the symptomatic (here assumed equivalent to infectious) class on the jth day since infection to simulate the empirically observed distribution of latent periods. The infectious state was broken into five subclasses with progression probabilities of $\{1, 1, r, r, r\}$ (Figure 2C) selected to fit empirical data on infectious periods. The parameters q_{ij} and h_{ij} denote the probabilities that exposed and infectious individuals are quarantined and hospitalized, respectively, where $i = c$ or h represents hosts from the community and HCW groups, and j represents the substage within the exposed or infectious class (because case management parameters may vary as a function of time since infection or appearance of symptoms).

(A) Overall Structure

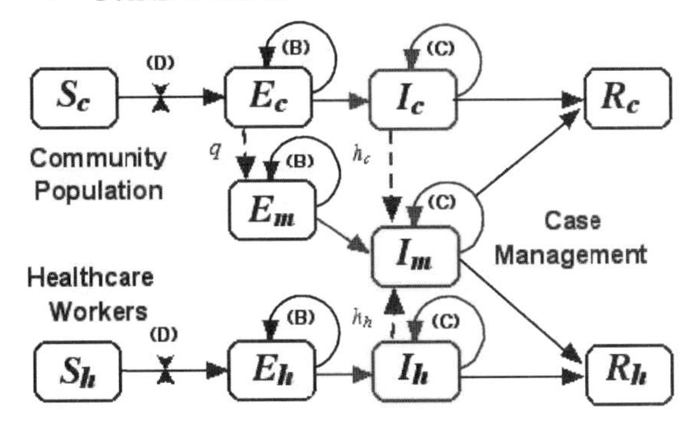

(B) Incubating Substructure

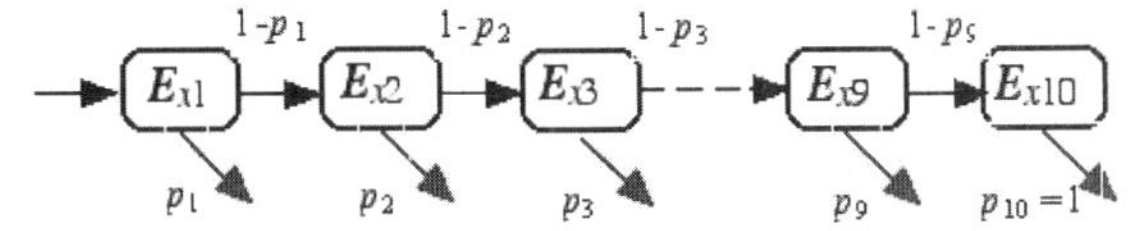

(C) Symptomatic Substructure

I_{x1} 1 I_{x2} 1 I_{x3} 1- r r I_{x4} 1- r r I_{x5} 1- r r

(D) Transmission substructure

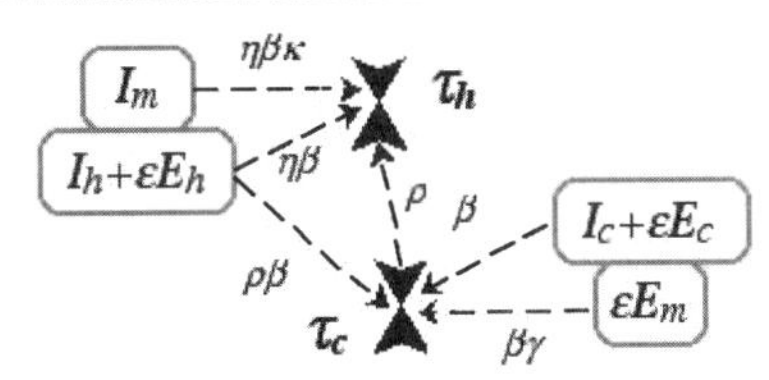

FIGURE 2. Flow diagram of the transmission dynamics of a SARS epidemic within a hospital coupled to that in a community (after Lloyd-Smith, Galvani and Getz, 2003). **(A.)** S: susceptible, E: incubating, I: symptomatic, R: removed. Subscripts h, c and m respectively represent individuals from healthcare worker, general community, and case-managed groups. E_m and I_m respectively are quarantined and isolated individuals. **(B.)** Incubating individuals (E_i, where $i = c, h, m$) were further structured into ten disease-age classes. Daily probabilities p_i of progressing to the symptomatic phase were linearly interpolated between $p_1 = 0$ and $p_{10} = 1$. **(C.)** For $0 < r < 1$, symptomatic individuals (I_i, where $i = c, h, m$) move through 5 stages to the final symptomatic class R_c or R_h, according to their group of origin. **(D.)** The transmission hazard rates for susceptible individuals S_i are denoted by $i(i = c, h)$, and depend on weighted contributions from community and HCW sources as described in the text.

The equations for both community (subscript c) and HCW (subscript h) pool are listed first, followed by the equations for the managed cases (subscript m). The managed case variables need a superscript $i = c, h$ in addition to the subscripts because recovered individuals are sent back to the pool of their origination. The variables and parameters used are depicted graphically in Figure 2. Transmission of the SARS coronavirus is represented by hazard rate functions τ_i for susceptible individuals in the i pool ($i = c, h$), which have the general form of expression (8.3). In particular, Lloyd-Smith, Galvani and Getz [**39**] formulated the parameters β_{ij} and c_{ij} in expression (8.3) in terms of the basic transmission rate β (not to be confused with the above transmission probabilities β_{ij}) and a collection of parameters modifying transmission for different settings: ε, η, γ, κ, and ρ (all on the interval $[0, 1]$). In particular, the reduced transmission rate of exposed (E) individuals (included because the extent of pre-symptomatic transmission of SARS was unknown when the model was created) is $\varepsilon\beta$. All transmission within the hospital setting occurs at a reduced rate $\eta\beta$ to reflect contact precautions adopted by all hospital personnel and patients, such as the use of masks, gloves and gowns. Additionally, quarantine of exposed individuals reduces their contact rates by a factor γ, yielding a total transmission rate of $\gamma\varepsilon\beta$, while specific isolation measures for identified SARS patients (I_m) in the hospital reduces their transmission by a further factor κ. Finally, we considered the impact of measures to reduce transmission rates between HCWs and community members by a factor ρ. Under these assumptions, the transmission hazards are:

$$\tau_c = \frac{\beta(I_c + \varepsilon E_c) + \rho\beta(I_h + \varepsilon E_h) + \gamma\beta\varepsilon E_m}{N_c}$$

and

$$\tau_h = \rho\tau_c + \frac{\eta\beta(I_h + \varepsilon E_h + \kappa I_m)}{N_h},$$

where E_i and I_i, $i = c, h$, represent sums over all sub-compartments in the incubating and symptomatic classes for pool j, and

$$N_h = S_h + E_h + I_h + V_h + I_m$$

and

$$N_c = S_c + E_c + I_c + V_c + \rho(S_h + E_h + I_h + V_h).$$

The detailed form of the SID equations that were formulated are:
Community and HCW equations:

$$\left.\begin{array}{l}
S_i(t+1) = \exp\left(-\tau_i(t)\right) S_i(t) \\
E_{i1}(t+1) = \left[1 - \exp\left(-\tau_i(t)\right)\right] S_i(t) \\
E_{ij}(t+1) = (1 - p_{j-1})(1 - q_{ij-1})E_{ij-1}(t) \quad j = 2, \ldots, 10 \\
I_{i1}(t+1) = \sum_{j=1}^{10} p_j(1 - q_{ij})E_{ij}(t) \\
I_{i2}(t+1) = (1 - h_{i1})I_{i1}(t) \\
I_{i3}(t+1) = (1 - h_{i2})I_{i2}(t) + (1 - r)(1 - h_{i3})I_{i3}(t) \\
I_{ij}(t+1) = r(1 - h_{i\,j-1})I_{i\,j-1}(t) + (1 - r)(1 - h_{ij})I_{ij}(t) \quad j = 4, 5 \\
V_i(t+1) = V_i(t) + rI_{i5}(t) + rI^i_{m5}(t)
\end{array}\right\} i = c, h,$$

$$\left.\begin{array}{l} E^i_{m,j}(t+1) = (1-p_{c\,j-1})\left(E^i_{m,j-1}(t) + q_{j-1}E_{c\,j-1}(t)\right) \quad j=2,\ldots,10 \\ I^i_{m1}(t+1) = \sum_{j=1}^{10} p_j\left(E^i_{mj}(t) + q_{ij}E_{ij}(t)\right) \\ I^i_{m2}(t+1) = h_{i1}I_{i1}(t) + I^i_{m1}(t) \\ I^i_{m3}(t+1) = h_{i2}I_{i2}(t) + I^i_{m2}(t) + (1-r)\left[h_{i3}I_{i3}(t) + I^i_{m1}(t)\right] \\ I^i_{mj}(t+1) = r\left[h_{i\,j-1}I_{i\,j-1}(t) + I^i_{m\,j-1}(t)\right] \\ \qquad\qquad +(1-r)\left[h_{ij}I_{ij}(t) + I^i_{mj}(t)\right] \; j=4,5 \end{array}\right\} i=c,h.$$

In the analysis presented here, the probabilities q_{ij} and h_{ij} vary between 0 and a fixed value less than 1 and account for delays in contact tracing or case identification. In addition, we did not analyze scenarios where health care workers are quarantined so that $q_{hj} = 0$ for all j. Deterministic solutions to this SARS model can be generated by directly iterating the above equations for specific sets of parameter values and initial conditions. However, because SARS outbreaks were invasion scenarios with initially small numbers of cases, stochastic simulations were required to incorporate the important influence of chance on outbreak dynamics.

For a range of R_0 values corresponding to conditions in different cities—but with emphasis on $R_0 \sim 3$ as reported for Hong Kong and Singapore [**35, 49**]—we performed sensitivity analyses (by calculating effective reproductive numbers under different control scenarios) and conducted stochastic simulations to explore the relative merits of different control measures for SARS. We assessed contributions of case management (i.e. isolation and quarantine) and contact precautions (such as masks, gowns, and hand-washing) to containment of a nascent SARS outbreak, and considered the extent to which one measure can compensate for another which is not available in a given setting.

A number of unintuitive and applicable conclusions arose from this analysis. For instance, hospital-wide contact precautions were identified as the single most potent containment measure—an encouraging finding since these are easily implemented and inexpensive. We investigated two types of delay in control efforts. At the individual level, delays of a few days in contact tracing and case identification severely degraded the utility of quarantine and isolation, particularly in high-transmission settings. Still more detrimental were delays at the population level between onset of an outbreak and implementation of control measures: for given control scenarios our model identified windows of opportunity beyond which efficacy of containment efforts was reduced greatly. In settings where hospital-based transmission was continuing, we showed that measures to reduce contact between healthcare workers and the community had dramatic benefits in preventing a widespread epidemic, emphasizing the importance of mixing restrictions (Figure 3) and, hence, considerations of heterogeneity in exposure to disease among individuals in the population.

9. Conclusion

Deterministic and stochastic theory of SIR processes in homogeneous populations has provided a solid foundation for the construction of modern epidemiological theory through the elaboration of the SIR framework. In this chapter we presented on overview of these SIR framework elaborated to include additional disease classes and demographic components in both discrete and continuous time equation formulations. We also demonstrated the extension of the discrete time setting to a population in which susceptible individuals are divided in n homogenous subgroups, with application to the recent SARS epidemics. In the next chapter [**23**], we discuss

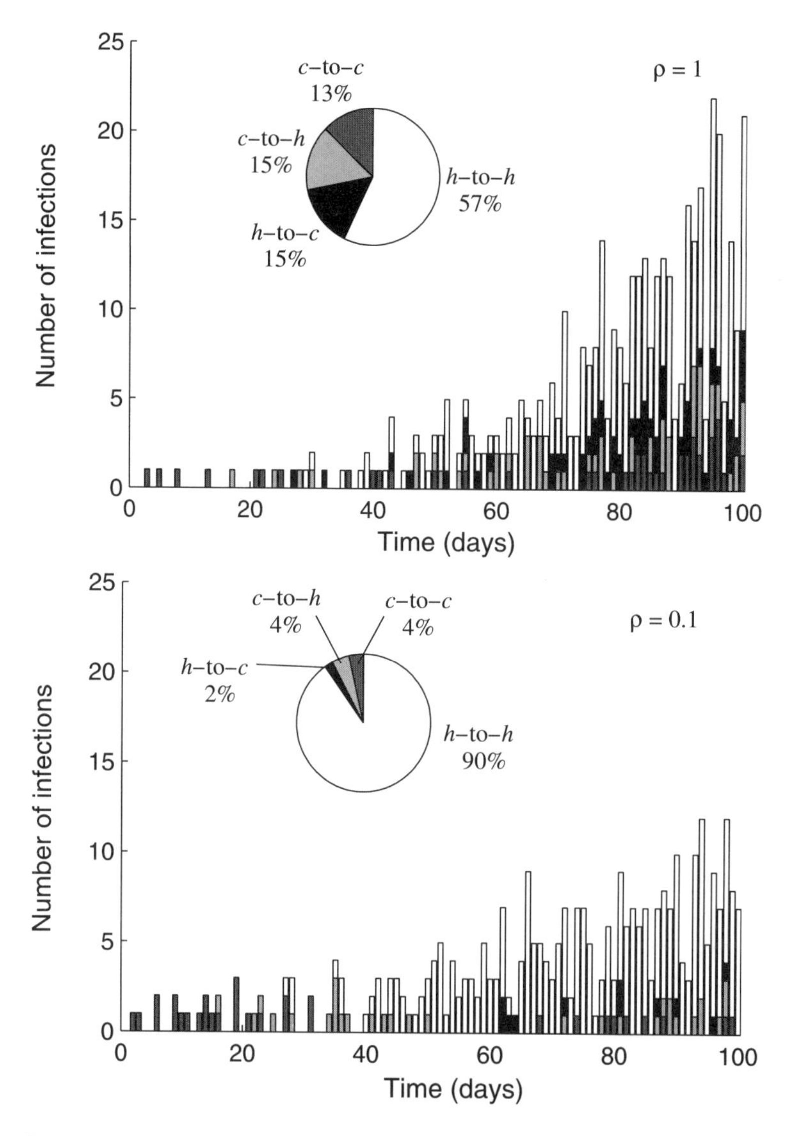

FIGURE 3. Daily incidence of infections are plotted for two stochastic epidemics with identical disease parameters ($R_0 = 3$) and control measures (no quarantine, but daily probability $h_c = 0.3$ of isolating an infectious community member, $h_h = 0.9$ of isolating an infectious HCW) implemented 14 days into the outbreak. Plots differ only in HCW(h) and community (c) mixing precautions (*i.e.*, $\rho = 1$ in **A** and $\rho = 0.1$ in **B**). Inset, pie-charts show average contributions of the different routes of infection for 500 stochastic simulations of each epidemic (after Lloyd-Smith, Galvani, and Getz, 2003).

how elaboration of the SIR framework can be used to structure populations into homogeneous subclasses of a more general nature than considered here.

Acknowledgements

We are indebted to Philip Johnson for writing the code needed to generate Figure 1 and producing the figure for us. We thank Shirli Bar-David, Paul Cross, Alison Galvani, Philip Johnson, Travis Porco, María S. Sánchez and an anonymous reviewer for comments that have lead to improvements of this chapter.

References

[1] R. Anderson and R. May, *Infectious Diseases of Humans: Dynamics and Control*, Oxford University Press, Oxford, 1991.

[2] D. Anderson and R. Watson, On the Spread of a Disease with Gamma-Distributed Latent and Infectious Periods, *Biometrika* **67** (1980), 191–198.

[3] K. B. Athreya and P. Jagers, *Classical and Modern Branching Processes*, Springer-Verlag, Berlin, 1997.

[4] N. T. J. Bailey, *The Elements of Stochastic Processes*, Wiley, New York, 1964.

[5] N. T. J. Bailey, *The Mathematical Theory of Infectious Diseases*, Griffin, London, 1975.

[6] C. T. Bauch, J. O. Lloyd-Smith, M. Coffee, and A. P. Galvani, Dynamically modeling SARS and respiratory EIDs: past, present, future, *Epidemiology* (in press) (2005).

[7] N. E. Beckage, *Parasites and Pathogens: Effects on Host Hormones and Behavior*, Chapman and Hall, New York, 1997.

[8] M. Begon, M. Bennett, R. G. Bowers, N. P. French, S. M. Hazel, and J. Turner, A clarification of transmission terms in host-microparasite models: numbers, densities and areas, *Epidemiol. Infect.* **129** (2002), 147–153.

[9] A. T. Bharucha-Reid, On the Stochastic Theory of Epidemics, in (J. Neyman, ed.), *Proceedings of the Third Berkeley Symposium on Mathematical Statistics and Probability* **4**, 1956, 111–120.

[10] S. N. Busenberg and K. P. Hadeler, Demography and epidemics, *Math. Biosci.* **101** (1990), 63–74.

[11] D. J. Daley and J. Gani, *Epidemic Modeling: An Introduction*, Cambridge University Press, Cambridge, 1999.

[12] M. C. M. De Jong, A. Bouma, O. Diekmann, and H. Heesterbeek, Modelling transmission: mass action and beyond, *Trends Ecol. Evol.* **17** (2002), 64–64.

[13] M. C. M. De Jong, O. Diekmann, and J. A. P. Heesterbeek, How Does Transmission of Infection Depend on Population Size? in (D. Mollison, ed.), *Epidemic Models : Their Structure and Relation to Data*, Cambridge University Press, Cambridge, 1995, 84–94.

[14] U. Dieckmann, J. A. J. Metz, M. W. Sabelis, and K. Sigmund, eds., *Adaptive Dynamics of Infectious Diseases: In Pursuit of Virulence Management*, Cambridge Series in Adaptive Dynamics, Cambridge University Press, Cambridge, 2002.

[15] O. Diekmann and J. A. P. Heesterbeek, *Mathematical Epidemiology of Infectious Diseases: Model Building, Analysis, and Interpretation*, Wiley, Chichester, 2000.

[16] K. Dietz and K. P. Hadeler, Epidemiological models for sexually transmitted disease, *J. Math. Biol.* **26** (1988), 1–25.

[17] C. A. Donnelly, A. C. Ghani, G. M. Leung, A. J. Hedley, C. Fraser, S. Riley, L. J. Abu-Raddad, L. M. Ho, T. Q. Thach, P. Chau, K. P. Chan, T. H. Lam, L. Y. Tse, T. Tsang, S. H. Liu, J. H. B. Kong, E. M. C. Lau, N. M. Ferguson, and R. M. Anderson, Epidemiological determinants of spread of causal agent of severe acute respiratory syndrome in Hong Kong, *Lancet* **361** (2003), 1761–1766.

[18] P. D. Enko, On the course of epidemics of some infectious diseases, *Vrac* **46** (1889), 1008–1010, **47**, 1039–43, **48**, 1061–1063 (For an abridged translation see: *International Journal of Epidemiology* **18**, 749–455).

[19] P. E. M. Fine, The interval between successive cases of an infectious disease, *Am. J. Epidemiol.* **158** (2003), 1039–1047.

[20] C. Fraser, S. Riley, R. M. Anderson, and N. M. Ferguson, Factors that make an infectious disease outbreak controllable, *Proc. Natl. Acad. Sci. USA* **101** (2004), 6146–6151.

[21] W. M. Getz, A hypothesis regarding the abruptness of density dependence and the growth rate of populations, *Ecology* **77** (1996), 2014–2026.

[22] W. M. Getz and J. Pickering, Epidemic models - thresholds and population regulation, *Am. Nat.* **121** (1983), 892–898.

[23] W. M. Getz, J. O. Lloyd-Smith, P. C. Cross, S. Bar-David, P. L. Johnson, T. C. Porco, M. S. Sánchez, Modeling the invasion and spread of contagious disease in heterogeneous populations, this volume.

[24] B. T. Grenfell and A. P. Dobson, eds., *Ecology of Infectious Diseases in Natural Populations*, Publications of the Newton Institute, Cambridge University Press, Cambridge, 1995.

[25] T. E. Harris, *The Theory of Branching Processes*, Springer, Berlin, 1963.

[26] J. A. P. Heesterbeek and J. A. J. Metz, The saturating contact rate in marriage– and epidemic models, *J. Math. Biol.* **31** (1993), 529–539.

[27] J. A. P. Heesterbeek, A brief history of R-0 and a recipe for its calculation, *Acta Biotheor.* **50** (2002), 189–204.

[28] H. W. Hethcote, The mathematics of infectious diseases, *SIAM Rev.* **42** (2000), 599–653.

[29] P. J. Hudson, A. Rizzoli, B. T. Grenfell, H. Heesterbeek and A. P. Dobson, eds., *The Ecology of Wildlife Diseases*, Oxford University Press, Oxford, 2002.

[30] V. Isham and G. Medley, *Models for Infectious Human Diseases*, Cambridge University Press, Cambridge, 1996.

[31] M. J. Keeling and B. T. Grenfell, Disease extinction and community size: Modeling the persistence of measles, *Science* **275** (1997), 65–67.

[32] M. J. Keeling and B. T. Grenfell, Effect of variability in infection period on the persistence and spatial spread of infectious diseases, *Math. Biosci.* **147** (1998), 207–226.

[33] W. O. Kermack and A. G. McKendrick, A contribution to the mathematical theory of epidemics, *Proc. Royal Soc. Lond. A* **115** (1927), 700–721.

[34] J. M. Kiesecker, D. K. Skelly, K. H. Beard, and E. Preisser, Behavioral reduction of infection risk, *Proc. Natl. Acad. Sci. USA* **96** (1999), 9165–9168.

[35] M. Lipsitch, T. Cohen, B. Cooper, J. M. Robins, S. Ma, L. James, G. Gopalakrishna, S. K. Chew, C. C. Tan, M. H. Samore, D. Fisman, and M. Murray, Transmission dynamics and control of severe acute respiratory syndrome, *Science* **300** (2003), 1966–1970.

[36] A. L. Lloyd, Destabilization of epidemic models with the inclusion of realistic distributions of infectious periods, *Proc. Royal Soc. Lond. B* **268** (2001), 985–993.

[37] A. L. Lloyd, Realistic distributions of infectious periods in epidemic models: Changing patterns of persistence and dynamics, *Theor. Popul. Biol.* **60** (2001), 59–71.

[38] J. O. Lloyd-Smith, P. C. Cross, C. J. Briggs, M. Daugherty, W. M. Getz, J. Latto, M. S. Sanchez, A. B. Smith, A. Swei, Should we expect population thresholds for wildlife disease? *Trends Ecol. and Evol.* (in press) (2005).

[39] J. O. Lloyd-Smith, A. P. Galvani, W. M. Getz, Curtailing SARS transmission within a community and its hospital, *Proc. Royal Soc. Lond. B* **270** (2003), 1979–1989.

[40] J. O. Lloyd-Smith, H. V. Westerhoff, and W. M. Getz, Frequency-dependent incidence in sexually-transmitted disease models: Portrayal of pair-based transmission and effects of illness on contact behaviour, *Proc. Royal Soc. Lond. B* **271** (2004), 625–634.

[41] D. Ludwig, Final size distributions for epidemics, *Math. Biosci.* **23** (1975), 33–46.

[42] D. Ludwig, Qualitative behavior of stochastic epidemics, *Math. Biosci.* **23** (1975), 47–73.

[43] H. McCallum, N. Barlow, and J. Hone, How should pathogen transmission be modelled? *Trends. Ecol. Evol.* **16** (2001), 295–300.

[44] H. McCallum, N. Barlow, and J. Hone, Modelling transmission: mass action and beyond – Response from McCallum, Barlow and Hone, *Trends Ecol. Evol.* **17** (2002), 64–65.

[45] D. Mollison, *Epidemic models: their structure and relation to data*, Cambridge University Press, Cambridge, 1995.

[46] I. Nåsell, Stochastic models of some endemic infections, *Math. Biosci.* **179** (2002), 1–19.

[47] R. E. Plant and L. T. Wilson, Models for age-structured populations with distributed maturation rates, *J. Math. Biol.* **23** (1986), 247–262.

[48] T. C. Porco, K. A. Holbrook, S. E. Fernyak, D. L. Portnoy, R. Reiter, and T. J. Aragón, Logistics of community smallpox control through contact tracing and ring vaccination: a stochastic network model, *BMC Public Health* **4** (2004), 34.

[49] S. Riley, C. Fraser, C. A. Donnelly, A. C. Ghani, L. J. Abu-Raddad, A. J. Hedley, G. M. Leung, L.-M. Ho, T.-H. Lam, T. Q. Thach, P. Chau, K.-P. Chan, S.-V. Lo, P.-Y. Leung, T. Tsang, W.

Ho, K.-H. Lee, E. M. C. Lau, N. M. Ferguson, and R. M. Anderson, Transmission dynamics of the etiological agent of SARS in Hong Kong: Impact of public health interventions, *Science* **300** (2003), 1961–1966.
[50] M. A. Schiltz and T. G. M. Sandfort, HIV-positive people, risk and sexual behaviour, *Social Sci. Med.* **50** (2000), 1571–1588.
[51] H. E. Soper, The interpretation of periodicity in disease prevalence, *J. Royal Statistical Society* **92** (1929), 34–73.
[52] H. M. Taylor and S. Karlin, *An Introduction to Stochastic Modeling*, third ed., Academic Press, San Diego, 1998.
[53] H. R. Thieme, *Mathematics in Population Biology*, Princeton University Press, Princeton, 2003.
[54] Y. H. Wang and A. P. Gutierrez, An assessment of the use of stability analysis in population ecology, *J. Animal Ecology* **49** (1980), 435–452.
[55] R. Watson, On the size distribution for some epidemic models, *J. Applied Probability* **17** (1980), 912–921.

Department of Environmental Science, Policy and Management, 140 Mulford Hall #3112, University of California, Berkeley, CA 94720-3112, United States and Mammal Research Institute, Department of Zoology and Entomology, University of Pretoria, Pretoria 0002, South Africa
E-mail address: getz@nature.berkeley.edu

Biophysics Graduate Group, University of California at Berkeley
E-mail address: jls@nature.berkeley.edu

Section II

Applications to Specific Diseases

DIMACS Series in Discrete Mathematics
and Theoretical Computer Science
Volume **71**, 2006

Modeling the Invasion and Spread of Contagious Diseases in Heterogeneous Populations

Wayne M. Getz, James O. Lloyd-Smith, Paul C. Cross, *Shirli Bar-David,
Philip L. Johnson, Travis C. Porco, and María S. Sánchez

Abstract. The evolution of disease requires a firm understanding of heterogeneity among pathogen strains and hosts with regard to the processes of transmission, movement, recovery, and pathobiology. In this chapter, we build on the basic methodologies outlined in the previous chapter to address the question of how to model the invasion and spread of diseases in heterogeneous environments, without making an explicit link to natural selection—the topic of other chapters in this volume. After a general introdution in Section 1, the material is organized into three sections (Sections 2–4). Section 2 covers heterogeneous populations structured into homogeneous subgroups, with application to modeling TB and HIV epidemics. Section 3 reviews a new approach to analyzing epidemics in well-mixed populations in which individual-level variation in infectiousness is represented by a distributed reproductive number [**51**]—in particular, the expected number of secondary cases due to each individual is drawn from a gamma distribution, yielding a negative binomial offspring distribution after stochasticity in transmission is taken into account. In Section 3, we discuss ideas relating to superspreading events, as well as the best way to characterize the heterogeneity associated with transmission in real epidemics, including SARS, measles, and various pox viruses. Section 4 deals with individual-based approaches to modeling the spread of disease in finite populations with group structure, focusing on several issues including interactions among movement, transmission, and demographic time-scales, the effects of network connectivity on the spread of disease, and the spread of disease in invading or colonizing hosts. The applications in Section 5 focus on bovine TB (BTB) in an African buffalo population and the potential for BTB to invade a colonizing Persian fallow deer population.

1. Introduction

In the previous chapter [**35**], a set of methods for modeling epidemics in homogeneous populations was presented. In this chapter, we use these methods to address theoretical and applied problems on the invasion and spread of contagious diseases in heterogeneous populations.

Grants to the first author from the NSF/NIH Ecology of Infectious Disease Program (DEB-0090323), NIH-NIDA (R01-DA10135), and the James S. McDonnell Foundation 21st Century Science Initiative supported various components of the research reported in this chapter.

*Authorship is alphabetical from here onwards.

Population heterogeneity can often be represented by dividing a host population into homogeneous subgroups based on spatial location, sex, behavior, genetics or other factors. Analysis of basic epidemic properties in such multi-group or multi-type populations is well-developed [**2, 5, 25, 26**], but more advanced questions and applications continue to motivate research in this important area.

Heterogeneous populations with homogeneous subgroups can broadly be divided into those with and those without inter-group transitions on time scales relevant to the analysis. Another important distinction is whether transmission occurs among individuals in different groups or only among individuals within the same group. These two criteria define a basic taxonomy of multi-group disease models, for which we provide several examples below. For instance, a population may be structured into groups according to some unchanging social categorization (hence no transitions among groups) or all individuals are potentially able to interact with another (hence transmission among groups is possible), but interaction rates are much greater within than among groups.

In this chapter, we present three approaches to dealing with heterogeneity, using several of our own recent studies as illustrative examples. The first is the relatively simple approach of dividing a heterogeneous population into a finite number of homogeneous subpopulations and then modeling the dynamics of the epidemic using a system of discrete difference equations with application to tuberculosis and HIV/AIDS in humans. The second approach is an application of stochastic branching process theory where the *individual reproductive number* associated with each infectious case (i.e. the expected number of new infections caused by each infected individual) is itself a random variable rather than a constant. We apply this approach to analyzing outbreak data from several important diseases including SARS, measles, smallpox, pneumonic plague, and other viral diseases.

The third approach is the use of individual-based discrete time stochastic simulation models. We consider their application to investigating how the timescales of host mixing and recovery from disease interact to determine the probability of a pandemic. We also discuss their use in modeling the spread of disease in a network of individuals characterized by an empirically derived association matrix for the purposes of obtaining insight into the spread of bovine tuberculosis in the African buffalo (*Syncerus caffer*) in the Kruger National Park, South Africa.

In this chapter, for convenience of presentation, we refer to the 25 equations presented in the previous chapter [**35**] using the numbers they have been designated in that chapter, such as [**35**, Equation number].

2. Interconnected homogeneous subgroups: TB and HIV

Classification of populations into groups is almost always problematic because group boundaries are ad-hoc. In modeling the spread of HIV, for example, it would be useful to be able to organize individuals into groups based on sexual preference, practice, and level of promiscuity. In many ways, an individual's behavior is better described by continuous rather than categorical variables. Nevertheless, categorical approaches are most often taken [**64**] because of their relative simplicity compared to approaches using a continuous descriptive variable. For example, the Actuarial Society of South Africa's (ASSA) official model for projecting the HIV/AIDS epidemic in South Africa divides the population into four risk groups based on categorization of sexual behavior, without allowing movement among groups. In reality, some individuals are bound to change their behavior as they learn more about the epidemic, as they age through time, or as their disease symptoms progress (e.g. [**37, 42, 43**]).

A critical source of heterogeneity in a population challenged with HIV is the presence of other diseases that can act as cofactors, particularly venereal diseases that enhance opportunities for the spread of HIV. Conversely, because it impairs immune function, HIV infection can have dramatic impacts on host response to other diseases, particularly the so-called opportunistic infections [**44**], one of the most important of which is tuberculosis (TB) [**41**]. For many TB-infected individuals the infection remains latent and has little effect on their health, but infections become acute and deadly once individuals are immunocompromised. Thus, the course of the disease is going to be vastly different in individuals infected with HIV than in otherwise healthy individuals [**1, 20, 52, 69**]. Accordingly, when considering the interaction of these two epidemics, we can formulate a full TB-HIV model in a homogeneous population, where we include the transmission and progression dynamics of both diseases. The caveat for this framework is its high level of complexity, which is compounded by the fact that there is a great level of uncertainty in the parameter values characterizing the epidemiological features of both TB and HIV.

Alternatively, we can begin by modeling TB in a population in which a certain proportion of individuals have HIV, without explicitly modeling the HIV epidemic itself [**66**]. This approach is justified by the fact that the impact of TB on the epidemiology of HIV appears to be less dramatic than that of HIV on TB [**4, 21, 24, 53**]. Under these assumptions, the background HIV structure is not static because we allow individuals in different HIV groups to progress according to the World Health Organization disease staging system (susceptible, clinical stages I to IV, and dead). The essential simplification is that we model the transition of susceptibles to HIV stage I using a recruitment process, based on historical patterns obtained from empirical data, rather than modeling HIV transmission in detail. Essentially, we have a TB only epidemic model, embedded within a population that is heterogeneous with respect to HIV stages [**22, 29, 57, 62**]. This kind of framework may prove to be just as informative as a full TB-HIV model if we are trying to understand the course of the TB epidemic within a relatively short period of time (e.g. 10–20 years), because input of the best estimates of HIV incidences from current data are likely to be as reliable as predictions from an HIV transmission model over the period of concern.

Here we sketch out how we use an incidence input approach to develop a model of a TB epidemic occurring in a population that is heterogeneous with respect to HIV status. The model is constructed as outlined in Section 6 of the previous chapter [**35**], but it has many more disease classes because the aetiology and clinical presentations of TB are very complex (Figure 1A). One of our motivations for developing a TB-HIV model is to determine the potential impact of reducing treatment duration in TB infected patients in areas of high HIV prevalence. Thus the model outlined below has a strong focus on TB treatment classes progressing over a monthly time scale. This level of time resolution entails 42 different categories of disease/treatment classes for the TB epidemic alone.

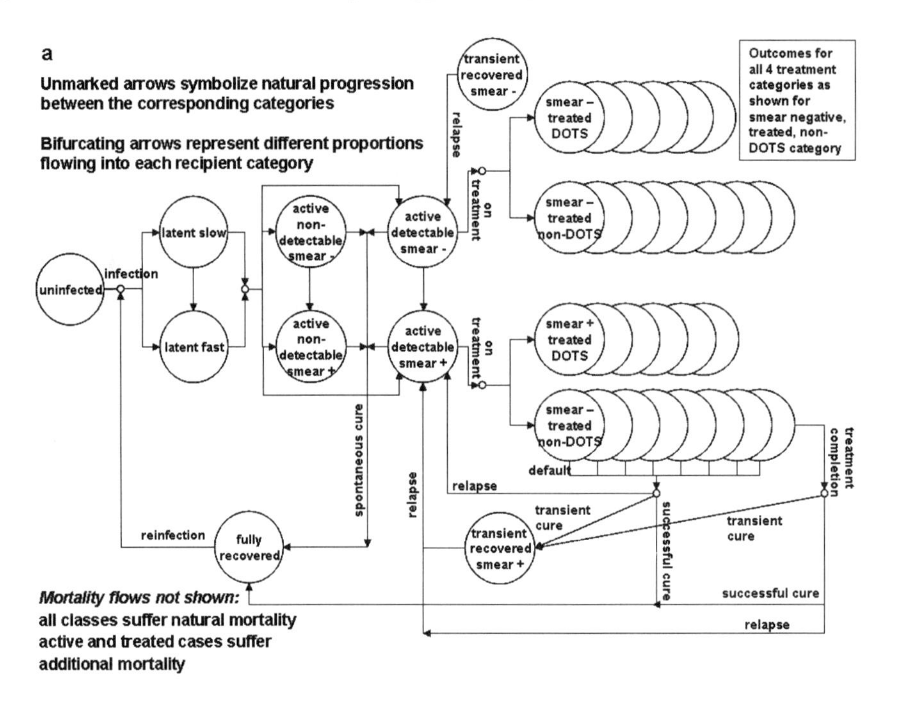

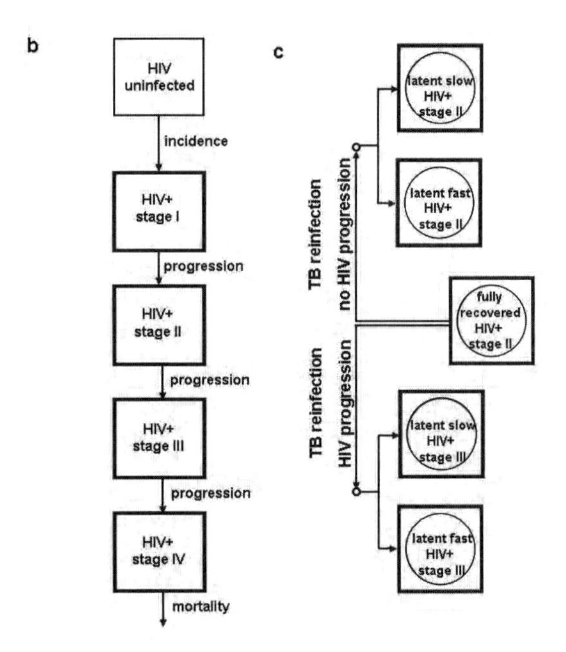

FIGURE 1. **a.**) TB transmission, treatment, and mortality in an HIV-free population is represented by a progression through 42 categories. **b.**) In an HIV infected population, HIV incidence, progression, and mortality is accounted for by 5 stages. **c.**) Details of the TB reinfection flow from the "fully recovered HIV+ stage III. This is a pattern that applies for other HIV stages as well. Each HIV stage includes the 42 TB categories for a total of 210 combined TB/HIV categories.

Specifically, individuals newly infected with TB can progress to active disease at a slow or fast rate, and those with an active infection can be classified as sputum-smear positive or negative. Among those individuals with an active infection, a certain fraction can be detected and placed on treatment, others are undetectable (due to subclinical symptoms or lack of access to health care) and will never become treated, while some can recover completely without ever being treated. At each time step, individuals within the detectable category are placed on treatment at a given rate, and can then enter either a DOTS or a non-DOTS regimen (DOTS stands for Directly Observed Treatment Short-course, but represents many more aspects of a well-run governmental TB control program—see [**23**]). We model the impact of treatment regimens that take different lengths of time to cure TB. Treated individuals can recover transiently or completely, or relapse to active disease. Flows among these categories were chosen to reflect the most critical processes determining TB incidence, prevalence, and mortality under different treatment regimens, paying particular attention to case detection, default, relapse, and reinfection.

In the combined TB-HIV model, we assume HIV-negative individuals become infected with HIV at a fixed input level given by reported incidence levels, and flow into the corresponding TB category of individuals that are in stage I of their HIV infection. HIV-positive individuals then progress through the four HIV stages, for a total of 42×5 different TB/HIV categories (Figure 1B). Individuals in HIV stage IV die of AIDS according to a fixed HIV mortality rate. We allow for HIV to affect TB infection, progression, and mortality rates—and vice-versa. All TB processes, together with the background mortality and HIV mortality, are modeled as competing rates within a continuous and deterministic framework that is updated on a discrete monthly time step. This is an approximation to the exact solution of a deterministic model [**35**, cf. Equations (6.3) and (6.4)] which allows all processes to occur simultaneously but restricts each individual to undergo one state transition per time step (unless the model is specifically adjusted to allow multiple transitions). HIV progression from stages I through IV occurs simultaneously with progression through various TB stages. At each time step, any given individual can progress in both diseases, in only one, or in neither.

Preliminary results indicate that a 2- compared with a 6-month treatment regimen may offer important benefits that appear to deminish when HIV prevalence is high. The model is being used to investigate this reduction in benefits in more detail under new drug scenarios that include increased treatment compliance by patients [**9, 73**], reduced relapse after treatment completion [**30**], and enhanced case detection [**10, 31**].

3. Migrants and the spread of disease

In Section 8 of the previous chapter and Section 2 of this chapter, we presented models in which individuals respectively 1.) do not make transitions among non-disease categories of individuals and 2.) make transitions along a unidirectional chain of categories. Here we generalize these approaches by, in theory, permitting any individual in any category to move to any other category, where movement in the absence of disease is defined by a *migration matrix* $M(k)$ with elements $M_{ij}(k)$ representing the probability that an individual in category i moves to category j during time interval $[k, k+1]$, $i, j = 1, \ldots, n$ [**65**]. This approach generalizes the preceeding sections, and it is applicable to modeling epidemics over a region consisting of several urban areas or in animal populations that have an identifiable group (e.g. troop, pod, colony, herd) or metapopulation structure [**34, 36, 38, 40, 81**].

We begin by considering a population structured into n_g identifiable groups. For generality, we also divide the population by gender and include age structure, because movement rules among animal groups are very often influenced by gender and age of individuals. This level of generality, however, can turn into a notational nightmare. Thus one of our goals in this section is to present a notation that does not obscure the structure of the equations themselves. There are many ways to do this—our goal is to preserve the macrostructure in the equations while choosing representations that are mnemonic where possible.

The notation is built around the symbol Z, where $Z = X$ denotes females and $Z = Y$ males (the obvious mnemonic being the genetic context of X and Y chromosomes). A superscript $U = S, E, I, R$ (D, V) is used to denote disease class [**35**] (where E, I, V etc. may be staged: e.g., I_l, $l = 1, \ldots, n_l$). Also, a subscript $a = 1, \ldots, n_a$ is used to denote age and a subscript $i = 1, \ldots, n_g$ is used to denote group (e.g. herd etc.). Thus the variables Z represent the number or density of individuals of gender Z, disease class U, age a, and group i. As in the previous section, we avoid the additional level of subscripting, by using $\pi = \alpha, \gamma, \mu$ etc., rather than p_π to represent transition probabilities (or proportions in large populations represented by deterministic models). We also use the same convention to refer to the migration parameters, but with an additional superscript needed to denote gender and another subscript needed to denote the proportion of individuals leaving group i that move to group j—that is, the migration parameters have the form $M^{ZU}_{aij}(k)$, which can be viewed as an $n_g \times n_g$ migration matrix for each gender, age and disease class of host st time k. With this notation, using EPI to denote terms controlling disease class transitions and DEMOG to denote terms controlling demographic transitions (aging, births, deaths) all equations have the generic form

$$Z^U_{a+1\ i}(k+1) = \mathrm{EPI}^{ZU}_{ai}(k) + \mathrm{DEMOG}^{ZU}_{ai}(k) + \sum_{i=1,\ i\neq j}^{n_g} \left(M^{ZU}_{aji}(k) Z^U_{aj}(k) - M^{ZU}_{aij}(k) Z^U_{ai}(k)\right). \tag{3.1}$$

We will not elaborate further on the structure of the terms in EPI and DEMOG in this section. We already have a sense from our models in Section 2 what form the EPI terms might take, while DEMOG terms follow the type of structure found in Leslie matrix formulations of age-structured models (e.g. see [**17**] for an example in the context of vaccinating African buffalo to control bovine TB). In terms of the movement process, we generally expect the coefficients $M^{ZU}_{aij}(k)$ to depend on the state of each group in our population (e.g. [**81**]), and to reflect a spatial topology typically represented by a set of parameters ξ_{ij} characterizing the "distance" (Euclidean or otherwise) between groups i and j over the time interval $[k, k+1)$. Dependence on group size might be quite complex, as discussed more fully in [**50**].

The primary issue we want to focus on in this section is the fact that in group structured systems of the type modeled by equation (3.1), we can define a matrix $\mathbf{R}_0$ of elements R^{ij}_0 defined to be the expected number of individuals in group j that will be infected directly (i.e. in the next generation) by an individual infected in group i. We will return to the question of how to calculate $\mathbf{R}_0$, but once calculated it can be used to derive the expected number infected in the offspring generation for different situations. For example, if an infective is introduced to group j at time $k = 0$, then, in an otherwise susceptible population, the expected number infected in the offspring generation is

$$R_0^{\bullet j} := \begin{pmatrix} R_0^{11} & \cdots & R_0^{1j} & \cdots & R_0^{1n} \\ \vdots & \vdots & \vdots & \vdots & \vdots \\ R_0^{n1} & \cdots & R_0^{nj} & \cdots & R_0^{nn} \end{pmatrix} \begin{pmatrix} 0 \\ \vdots \\ 1 \\ \vdots \\ 0 \end{pmatrix} = \sum_{i=1}^{n} R_0^{ij}.$$

Similarly, if an infective is introduced into group j with probability w_j at time $k = 0$ (this probability could be proportional to the initial group size or determined by its location on a landscape) then, in an otherwise susceptible population, the expected number infected in the offspring generation is

$$R_0^w := \sum_{i=1}^{n} \sum_{j=1}^{n} w_j R_0^{ij}.$$

Diekmann and Heesterbeek [**26**] have developed methods for generating the matrix $\mathbf{R}_0$, which they have dubbed the *next-generation matrix*. This matrix can also be generated numerically through simulation; but no example of this for a relatively detailed system has been published to date. If the next-generation matrix is irreducible and acyclic, its dominant eigenvalue is the basic reproduction number intrinsic to the system (as opposed to that defined above for the vector $\mathbf{w}$ of introduction probabilities) [**26**]. A more detailed discussion of the properties of the next-generation matrix can be found in [**60**].

An SIR model of the form (3.1), in a system without age or gender structure, has been analyzed by Hagenaars et al. [**38**] to consider how disease persistence is influenced by the tradeoff between the number of groups n_g and the initial group size $N_i = \hat{N}$, $i = 1, \ldots, n_g$, when the total initial population size $N_{\text{Tot}} = n_g \hat{N}$ is fixed. This model assumes that individuals in one group contact individuals in any other group at a relative rate ε, which implies that $f(\varepsilon) = \frac{1}{1+\varepsilon(n_g-1)}$ is the fraction of contacts with individuals from one's own group, with the rest equally distributed among the other groups. They also assume a relatively simple demographic component in which 1.) the birth and immigration rates are balanced by death and emigration rates, 2.) the migration is to a population that is external to the structured group of interest, and 3.) the external population is at constant disease prevalence. Their stochastic analysis of this model reveals that: "... if the overall transmission potential is kept fixed, increasing the level of spatial heterogeneity typically results in a decrease in disease persistence. For weak spatial coupling between subpopulations, the persistence changes as a function of coupling can be understood in terms of rescue effects. For intermediate and strong spatial couplings, coherence effects become important." Here, rescue effects imply that fade out of disease in one group is followed by reinfection from another group [**14**]. Also coherence in the context of relatively slow diseases implies similar levels of incidence in all groups, which for fast diseases can take the form of synchronized oscillations.

4. Individual heterogeneity in well-mixed populations

4.1. The individual reproductive number. For the past 25 years, analysis of epidemic dynamics has centred on the basic reproductive number, R_0, which is the expected number of new infections due to one infectious individual in a wholly susceptible population [**39**]. Models of homogeneous populations usually use a point estimate of R_0, implicitly assuming that every individual has the same degree of infectiousness. In reality, however, the infectiousness of each individual (as manifested by the number of secondary cases they cause) varies due to a complex blend of host, pathogen and environmental factors. The joint action of these factors leads to continuous variation in infectiousness and distinct risk groups often cannot be recognized *a priori*, thereby hampering our ability as modelers to represent this heterogeneity using group structure as described in the previous section.

To account for this variation, we introduce the individual reproductive number, ν, defined as the expected number of new infections that a given individual in the current generation will cause in the next generation [**51**]. At the population level, ν has some probability distribution that can be fitted to datasets describing the observed distribution of secondary cases caused by each individual (i.e. the empirical realization of the offspring distribution introduced in Section 7 in the previous chapter). The actual number of cases caused by each individual will vary stochastically around ν, so the realized offspring distribution will be a compound distribution with the form Poisson(ν) (i.e., a Poisson with parameter ν that is itself distributed). In a completely homogeneous population, where all individuals have identical infectiousness $\nu = R_0$, the offspring distribution will be Poisson, as discussed in Section 7 of the previous chapter [**35**]. If all individuals transmit at the same rate (i.e. they have equal transmission coefficients β) and exponentially-distributed infectious periods (i.e. they have constant per capita rates of recovery or death), then ν is distributed exponentially with mean R_0 and the offspring distribution is geometric [**71**]. These two scenarios represent the standard assumptions used in homogeneous population models.

To allow for a more flexible degree of individual variation in infectiousness, we propose a model with a gamma-distributed ν. Gamma distributions are a useful two- parameter family of distributions for developing epidemic theory in heterogeneous populations, because they are unimodal and non-zero only on $(0, \infty)$. More important for the development of branching process theory, however, the offspring distribution arising from a population with gamma-distributed individual reproductive number ν is a negative binomial distribution. Further, this model is a generalization of the two conventional models (Poisson and geometric offspring distributions), as described below. The negative binomial distribution is typically expressed in terms of a scale parameter p and a dispersion parameter k. To emphasize the link with epidemiological models, we deviate from this practice and use the notation $\phi^{\mathrm{NB}}_{R_0,k}$ to denote a negative binomial offspring distribution with mean R_0 related to p and k by the equation

$$R_0 = k\left(\frac{1}{p} - 1\right). \tag{4.1}$$

The probability generating function for negative binomial distribution has the form

$$\underline{\text{Negative Binomial}}\ \ \phi^{\mathrm{NB}}_{R_0,k} :\ g(z) = \left(1 + \frac{R_0}{k}(1-z)\right)^{-k}.$$

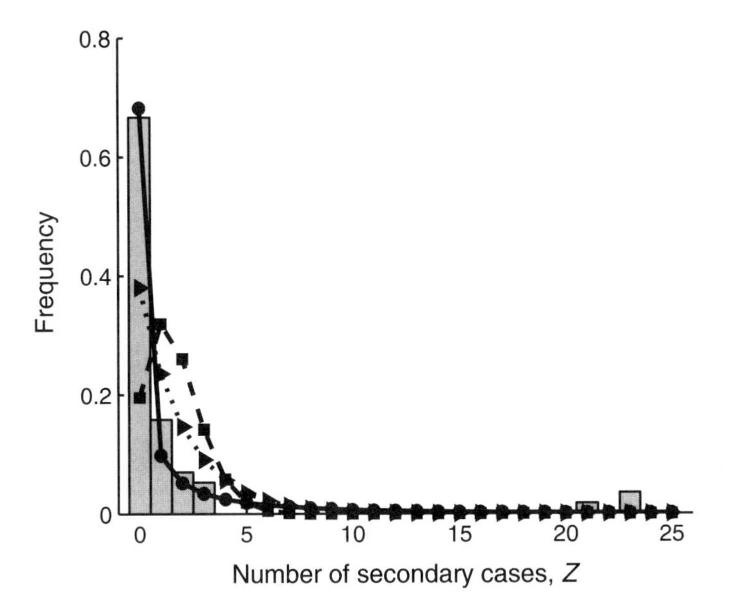

FIGURE 2. Empirical offspring distribution from the 2003 SARS outbreak in Singapore [**48**]. Bars show observed frequency of Z, the number of individuals infected by each case; lines show maximum likelihood fits for $Z \sim$ Poisson (squares), $Z \sim$ geometric (triangles), and $Z \sim$ negative binomial (circles).

As mentioned in Section 7 of the previous chapter, the probability q_0 that an infectious individual in the parent generation will not transmit to anyone in the offspring generation is $g(0)$, from which it follows for the negative binomial that

$$q_0 = \left(1 + \frac{R_0}{k}\right)^{-k}. \tag{4.2}$$

The Poisson and geometric distributions are special cases of the negative binomial distribution with $k \to \infty$ and $k = 1$, respectively.

During a number of recent epidemics, contact tracing of cases provided sufficient data for the construction of empirical offspring distributions to which Poisson, geometric and binomial distributions could be fitted. An example of this data for the SARS outbreak in Singapore, 2003 [**48**], is illustrated in Figure 2. Maximum likelihood methods can be used to find parameters of the negative binomial, Poisson, and geometric distributions that best fit these data. Of course, because of its extra parameter, the negative binomial distribution will always provide a better fit but the Akaike information criterion can be used to assess whether the improvement in fit over the one-parameter Poisson or geometric models is sufficient to merit inclusion of the extra parameter (e.g. see [**13**]). Further, bootstrap methods [**32**] can be used to estimate confidence intervals for the parameters characterizing these best-fitting offspring distributions. The results of such an exercise for 14 different disease datasets are summarized in Table 1, where we use $\hat{}$ to denote estimated values of the parameters R_0 and k. The Akaike weights reported in Table 1 represent the probability that each model is the best choice to represent the data of the three candidate models considered. The datasets include well-traced single outbreaks, combinations of data from multiple outbreaks, and surveillance data tracking the first generation of many introductions of a disease. Note that many of the datasets, particularly measles and smallpox, pertain to populations with high levels of vaccination, so observed heterogeneity may reflect differences among vaccinated and unvaccinated individuals rather than intrinsic host or pathogen characteristics. More details pertaining to data, methods and results are given in Lloyd-Smith et al. [**51**].

A large degree of individual variation in infectiousness is evident in almost all of the 14 datasets analyzed in Table 1. For five of the disease datasets, the 90% confidence intervals for the negative binomial dispersion parameter k are bounded below 1, indicating that only the negative binomial offspring distribution can represent the observed patterns. The best estimate of k for 11 of the datasets suggests that heterogeneity is either greater or much greater than that arising from an exponential distribution of infectiousness (i.e. best-fit $k < 1$). Only one dataset exhibited sufficient homogeneity in infectiousness that the Poisson model was favored by the Akaike weight model-selection technique. This was Ebola hemorrhagic fever with the most likely value for k estimated to be $\hat{k} = 5.1$; this dataset had only 13 index cases and contact tracing was probably incomplete, so our confidence in this result is limited.

In half of the epidemics the best estimate for k was $\hat{k} < 1/3$. The estimated values of R_0 are also quite low, and likely biased because the detailed contact tracing data required for this analysis is much more difficult to obtain when disease spread is very rapid. The low values for k, together with relatively low values for R_0, suggest that the great majority of individuals are unlikely to infect any other individual. For example, if $k = 1/3$, then from equation (4.2) it follows that for each individual to have a greater than 50% chance of infecting another individual, R_0 would have to exceed

$$R_0 > \frac{(2)^3 - 1}{3} = 2.33,$$

which is outside the 90% confidence interval for all the datasets analyzed except for smallpox in Europe over the period 1958 to 1973.

Similarly, our methods allow us to estimate that 73% of SARS cases in Singapore were below the critical infectious level of $\nu = 1$, while only 6% had infectiousness of $\nu > 8$ [**51**]. This result is consistent with field reports from SARS-afflicted regions [**48, 67**] indicating that infectiousness is highly overdispersed, but contrasts sharply with many published SARS models that do not take heterogeneity into account (reviewed in [**8**] and [**28**]).

4.2. Characterizing heterogeneity in infectiousness. The dispersion parameter k has little intuitive value as a measure characterizing the heterogeneity of infectiousness among individuals, particularly as the mean infectiousness R_0, expressed in equation (4.1), is not itself independent of k. One way to characterize heterogeneity is to ask what percentage of infectious cases are responsible for a given percentage of all on-going transmission. In a recent publication, Woolhouse et al. [**80**] proposed a general "20/80" rule: 20% of infectious individuals cause 80% of all infections for vector-borne and sexually transmitted diseases. In a homogeneous population the rule would be 20/20 or 80/80.

In general for a heterogeneous population we expect a "20/P" rule to arise, where increasing levels of heterogeneity in ν lead to increasing values for P. From

TABLE 1. Parameter estimation and for the Poisson (P), geometric (G) and negative binomial (NB) models of the offspring distribution and statistical support.

Datasets	Model	Akaike weight	$\hat{R}_0$ (90% CI for NB)	$\hat{k}$ (90% CI for NB)	$P\%$ ("20/P" rule) (90% CI for NB)
SARS	P	0	1.63	0.16	88%
Singapore 2003	G	0	(0.54,2.65)	(0.11,0.64)	(60,94)
$N = 57$	NB	1			
SARS	P	0	0.94	0.17	87%
Beijing 2003	G	0	(0.27,1.51)	(0.10,0.64)	(60,95)
$N = 33$	NB	1			
Measles	P	0	0.63	0.23	81%
US 1997-9	G	0.01	(0.47,0.80)	(0.13,0.40)	(70,92)
$N = 165$	NB	0.99			
Measles	P	0	0.82	0.21	83%
Canada 1998-2001	G	0.15	(0.72,0.98)	(0.09,0.52)	(44,86)
$N = 49$	NB	0.85			
Smallpox	P	0	3.19	0.37	71%
Europe 1958-73	G	0.02	(1.66,4.62)	(0.26,0.69)	(59,79)
$N = 32^*$	NB	0.98			
Smallpox	P	0	0.80	0.32	74%
Benin 1967	G	0.45	(0.32,1.20)	(0.16,1.76)	(44,88)
$N = 25$	NB	0.55			
Smallpox	P	0	1.49	0.72	58%
W. Pakistan	G	0.71	(*)	(0.32,2.23)	(41,74)
$N = 47$	NB	0.29			
Variola minor	P	0	1.60	0.65	60%
England 1966	G	0.71	(0.88,2.16)	(0.34,2.32)	(41,73)
$N = 25$	NB	0.29			
Monkeypox	P	0	0.32	0.58	62%
Zaire 1980-84	G	0.62	(0.22,0.40)	(0.32,3.57)	(36,74)
$N = 147$	NB	0.37			
Pneumonic plague	P	0	1.32	1.37	0.47%
6 outbreaks	G	0.67	(1.01,1.61)	(0.88,3.53)	(37,54)
$N = 74$	NB	0.33			
Avian influenza	P	0.17	0.06	0.026	100%
S.E. Asia 2004	G	0.32	(0,0.18)	(0.026,∞)	(20,100)
$N = 33$	NB	0.51			
Rubella	P	0	1.00	0.032	100%
Hawaii 1970	G	0	(0.0,1.95)	(0.013,∞)	(20,100)
$N = 19$	NB	1			
Hantavirus	P	0.31	0.70	1.66	45%
Argentina 1996	G	0.52	(0.20,1.05)	(0.24,∞)	(20,80)
$N = 20$	NB	0.17			
Ebola HF	P	0.56	1.50	5.10	34%
Uganda 2000	G	0.28	(0.85,2.08)	(1.46,∞)	(20,46)
$N = 13$	NB	0.17			

*neither the raw data nor a fitted confidence interval were available.

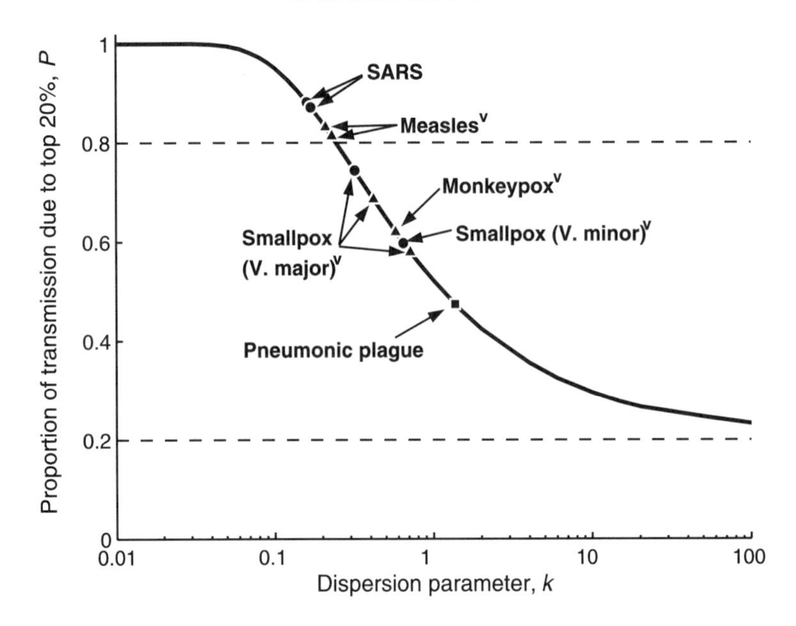

FIGURE 3. Proportion of transmission expected from the most infectious 20% of cases, for data drawn from single outbreaks (circles), multiple outbreaks (squares), and long-term surveillance (triangles). Dashed lines show proportions expected under the 20/80 rule (top) and in a homogeneous population (bottom). The superscript v indicates a highly-vaccinated host population. See details in Lloyd-Smith et al. [**51**].

the last column in Table 1, we see that "20/80" rule could well be a slight underestimate of the variation observed for SARS, fairly accurate for measles in highly vaccinated populations, and a slight overestimate for smallpox, monkeypox and pneumonic plague. For other diseases broad confidence intervals prevent firm conclusions, but best-fit parameters indicate that the "20/80" rule may seriously underestimate the heterogeneity of infectiousness for H5N1 avian influenza and rubella, and overestimate the heterogeneity for hantavirus and Ebola hemorrhagic fever.

The theory allows us to construct a curve denoting the expected proportion of transmission due to the most infectious 20% of transmitting individuals, as a function of the dispersion parameter of the negative binomial offspring distribution, k. (Of course, 20% is arbitrary and we can construct curves for any percentage we choose.) In Figure 3 such a curve is depicted along with the locations on this curve predicted by best-fit parameters for some of the disease datasets listed in Table 1.

From a technical point of view, particularly when complete contact tracing and construction of a reliable histogram (such as in Figure 2) is difficult to achieve, a crude estimate of the best fitting negative binomial distribution can be obtained from estimates of the mean number of offspring R_0 and proportion q_0 of non-transmitting infected individuals. In this case the value for k is calculated by solving equation (4.2) implicitly using an appropriate numerical method.

In conventional theory for homogeneous populations, R_0 is the only statistic needed to calculate the probability P_{epi} of a major epidemic, as discussed in Section

7 of the previous chapter [**35**]. The same branching process theory applies to heterogeneous populations, in that the value z_∞ to be used in [**35**, Equation (7.3)] is still the solution to the probability generating function equation $z = g(z)$, but now the probability generating function is for the negative binomial rather than Poisson distribution. Thus z_∞ in this case is the implicit solution of the equation

$$z = \left(1 + \frac{R_0}{k}(1-z)\right)^{-k}.$$

Using this equation it becomes clear that the probability of a major epidemic is now critically dependent on both k and R_0, and that greater degrees of individual heterogeneity in infectiousness lead to higher probabilities of stochastic extinction in the early phase of disease establishment (Figure 4). Thus, along with R_0, it should be very useful to specify P_{epi} as well. The latter informs us of the probability with which disease will invade the population (i.e. occurrence of an epidemic), while the former informs us on relatively how fast it will spread, if the disease should invade.

4.3. Superspreading events (SSEs) and loads (SSLs). In all epidemics, whether in homogeneous or heterogeneous populations, some infectious cases will not infect any individuals while others will infect many more than the expected number R_0. In heterogeneous populations this greater degree of transmission may arise from biological or social properties of the host individual, while in homogeneous populations it arises from random circumstance; that is, the host individual happens to spend time in close confinement with other individuals (e.g. in a crowded hospital ward or aboard a commuter train) while highly infectious. Occurrences of this kind have been referred to as superspreading events (SSEs) [**48, 49, 63, 67**], even though the individual involved may not have been more infectious than the average infective individual.

Despite numerous published accounts of SSEs in the literature, 37 of which are summarized by Lloyd-Smith et al. [**51**], no coherent approach to their quantitative analysis has emerged until recently. In the case of SARS, for example, using the random variable Z to represent the actual number of individuals infected by a known infectious case, SSEs have been defined as $Z \geq 8$, $Z \geq 10$, $Z > 10$, and "many more than the average number" [**48, 63, 67, 74**]. We propose a general protocol for defining an SSE that scales with the infectiousness of different diseases and naturally incorporates the influence of stochasticity [**51**]. The definition centers on using the Poisson distribution with mean R (where R is the effective reproductive number for the disease and population in question, taking immunization levels into account) as the expected range of Z due to stochasticity in the absence of individual heterogeneity. An SSE is then any case causing more secondary cases than would occur in at least 99% of infectious histories in a homogeneous population; that is, $Z \geq Z_R^{99}$, where Z_R^{99} is the first integer value for which the cumulative proportion of cases for the Poisson distribution is at least 0.99 (i.e., because the Poisson distribution is discrete Z_R^{99} will, in general, not correspond exactly to the 99th percentile: it is first integer at or beyond this percentile point).

From this definition it follows that in a homogeneous population, at most 1% of infectious cases causes SSEs. From a tabulation of the cumulative Poisson distribution, one finds for example that $Z > 6$ is an SSE when $R = 2$, while an SSE for $R = 5$ requires $Z > 11$. In a heterogeneous population, for an epidemic with

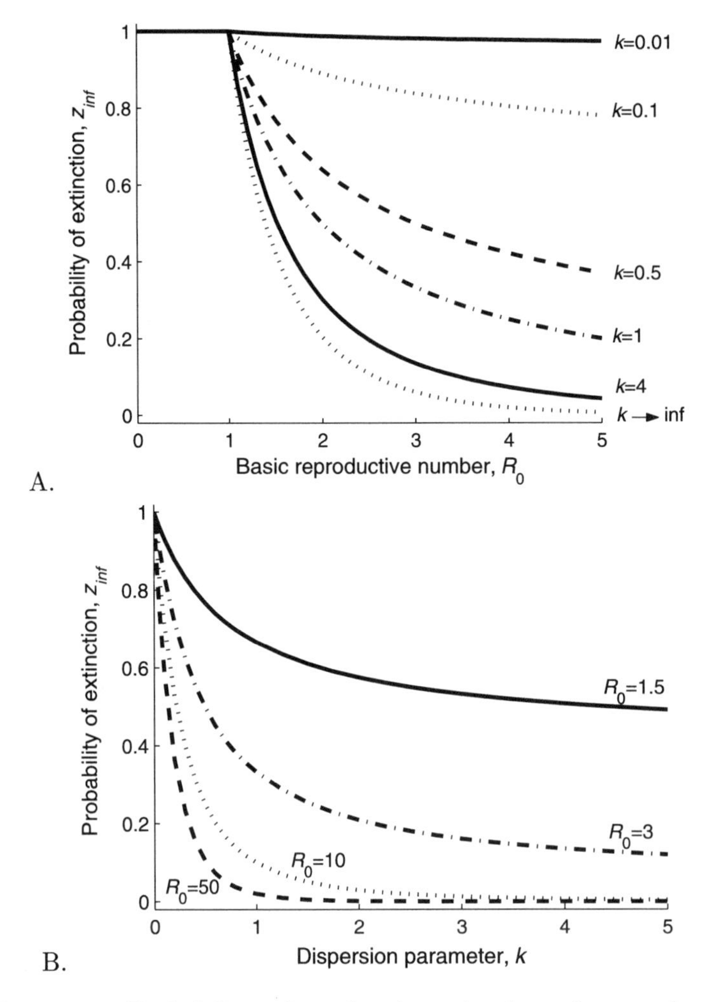

FIGURE 4. Probability of stochastic extinction of an outbreak beginning with a single infectious case for a branching process epidemic with negative binomial offspring distribution (i.e., z_∞, cf. [**35**, Equation (7.3)]). (A.) Extinction probability versus population-average reproductive number R_0 for different dispersion parameters k. (B.) Extinction probability versus dispersion parameter k for different values of R_0.

effective reproductive number R, the tail defined by $Z \geq Z_R^{99}$ will contain an increasing proportion of the population with increasing heterogeneity. We define the magnitude of this increase as the superspreading load (SSL) associated with a heterogeneous population. Specifically, if the tail defined by $Z \geq Z_R^{99}$ contains $r\%$ of events, then we define SSL$= r$. In the context of the negative binomial distribution we can plot the value of SSL as a function of k, as illustrated in Figure 5. In this figure we have selected values of R for which Z_R^{99} defines a tail with precisely 1% of events. As noted above, this is not the case for arbitrary values of R due to the discreteness of the Poisson distribution, which will cause discontinuous jumps in

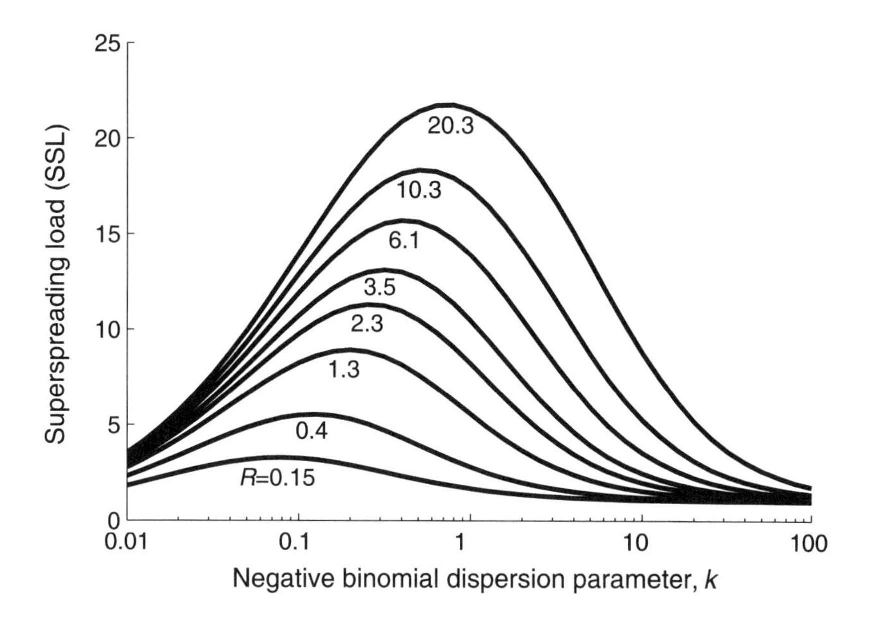

FIGURE 5. Superspreading load (SSL) versus dispersion parameter k for outbreaks with negative binomial offspring distributions. The SSL reflects the percentage of index cases expected to cause 99th-percentile SSEs. Values of R were selected such that $\Pr\{Z \geq Z_R^{99} \mid Z : \text{Poisson}(R)\}$ is precisely 0.01, as described in the text.

SSL as R changes. One way to deal with this problem is to re-define SSL as

$$r = 100 \left(\theta \left(\phi_{R,k}^{\text{NB}} \left(Z_R^{99} - 1 \right) \right) + \sum_{Z=Z_R^{99}}^{\infty} \phi_{R,k}^{\text{NB}}(Z) \right), \tag{4.3}$$

where $\phi_{R,k}^{\text{NB}}(Z)$ is the cdf of the negative binomial distribution, and $0 \leq \theta < 1$ is selected so that in the case of the Poisson ($k \to \infty$) with mean R the value for the SSL is precisely 1%. That is, some proportion of the density in the next-lower bin of the Poisson cdf is added to bring SSL to exactly 1%, and the identical proportion of that bin is added when calculating the SSL for negative binomial offspring distributions.

Note from Figure 5 that for R in the range $[1, 20]$, the SSL is maximized when k lies in the range $[0.1, 1]$. In particular, for $R = 6.1$ ($Z_{R=6.1}^{99} = 13$), the SSL is maximized at $k \approx 0.5$. The loads are not quite as severe for lower values of R, but even for the case $R = 1.3$ ($Z_{R=1.3}^{99} = 5$), the SSL rises to a maximum of approximately 9 at $k \approx 0.2$. Thus levels of heterogeneity observed in real epidemics (e.g. SARS or smallpox, Table 1) will result in an order of magnitude more SSEs than would be expected if the epidemic were in a population that is homogeneous with respect to infectiousness.

5. Individual-based models in heterogeneous populations

Individual-based models, also referred to as agent-based models or microsimulations, depend on the computational power of computers to represent large numbers of individuals each with its own independent record containing information relevant to the questions being addressed (e.g. age, sex, location, disease status, and so forth). This information is then updated either at stochastically determined times (event driven models) or periodically (discrete time or 'clock-driven' models) (e.g. see [**12**]). The amount of detail that can be put into such models is limited only by the size of the population to be simulated and the power of the computer used to carry out the simulations. Currently, several groups are developing general agent-based modeling software suitable for simulating epidemics in spatially and demographically structure populations (e.g. [**33, 46, 72**]). The three studies outlined in this section used computer programs written in Matlab® and C++ specifically for the problems at hand. The material presented in the first subsection extends our intuition relating to the spread of epidemics in highly idealized populations consisting of n homogeneous groups, initially of size m, linked to each other by migration patterns representing particular spatial configurations. The material presented in the second subsection uses association data obtained from empirical studies to explicitly characterize the contact process when modeling the spread of disease in a population where the movements of each individual are known. Finally, in the third subsection we introduce the question of what happens when a disease invades a host population that is itself invading and colonizing a region, with an illustration on a realistic heterogeneous landscape.

5.1. Epidemics in a group structured population with movement among groups. The formulation in this section is based on the assumption that a population consists of m identifiable subpopulations—a classic metapopulation structure—which is fixed throughout time, but the number of individuals N_i in each subpopulation varies with time and can become 0. In this case, the total population size is $N = \sum_{i=1}^{m} N_i$. The subpopulation in each group is further structured into S and I disease classes. The model reflects two processes: transmission within groups and movement between groups [**65**]. Each group is regarded as being homogeneous with respect to the hazard of transmission, where for the ith group with transmission parameter β_i and frequency-dependent transmission, it follows that the probability that a susceptible in group i will become infected over the time interval $[k, k+1]$ is (cf. [**35**, Expression (6.6) and (8.1)])

$$p_{\mathcal{T}_{ik}} = 1 - \exp\left(-\beta_i \frac{I_i(k)}{N_i(k)}\right).$$

For simplicity, assume that movement between groups is not influenced by the size of any group other than the group from which an individual is departing. Our assumption that the subgroups are homogenous thus implies that any individual in group i has the same probability, $\mu_{ik}(N_i(k))$, of leaving its current group during time step $[k, k+1]$. The group that a departing individual then joins can reflect both spatial (e.g. distance among groups) and recipient group factors (e.g. size of groups). Ways of characterizing these movements have received considerable attention in the ecological literature [**11, 15, 34, 50, 81**], leading to movement matrices of the form considered in Section 3.3, with elements M_{ij} that may depend in relatively complex ways on the group size vector $\mathbf{N} = (n_1 \ldots, n_m)'$. Here we only consider relatively simple rules that permit us to focus on epidemiological rather than demographic questions.

Cross et al. [**19**] used an individual based model of a metapopulation, structured at the start of the simulation into equally sized groups spatially organized on a square lattice, to investigate the question of how SIV epidemics (recall V is recovered with immunity) depend on relative rates influencing the spread of disease within groups versus the movement of infected individuals among groups. The relative rate of the spread of disease within groups was controlled by the value of a transmission parameter β, which was assumed to be the same in all groups. The relative rate of the spread of disease among groups was controlled by a departure probability parameter μ that was assumed to be the same value for all groups and independent of both time k and departing group size $N_i(k)$.

Three different sets of rules were used to assess the effects of spatial connections on the epidemic in question. First, departing individuals were assumed to move to only one of their four nearest-neighboring groups, each with probability 1/4, and, to avoid boundary effects, opposite edges of the array were connected so that effectively, the population was moving on a torus. Second, individuals were assumed to move with equal probability to one of only two nearest-neighbors, an analysis that is topologically equivalent to populations confined to a one-dimensional loop (e.g. villages located around the circumference of a lake). Third, individuals were permitted to join with equal probability any other group, which is equivalent to an "island" model previously used by Hess [**40**] and Fulford et al. [**34**] to study the spread of disease in metapopulations. Finally, Cross et al. [**19**] assume that all infected individuals recover to an immune class at a constant probability per time step, $p_\alpha = \alpha$. (Note, p_α is only approximately equal to the recovery rate α when α is small: more precisely $p_\alpha = 1 - e^{-\alpha}$). Although all groups begin in this particular study with the same number of individuals, group sizes do change over the course of simulations. The symmetries within the model with respect to group structure (e.g. lack of boundaries on a torus or loop), however, ensured that group sizes remained relatively equitable during the course of each simulation.

Cross et al. [**19**] compared the dynamics of two diseases with the same R_0 ($\approx \beta/\alpha$ when time steps are relatively small), but one disease was slow with a relatively long infectious period (e.g. $\beta = 0.05$, $\alpha = 0.01$) while the other was an order of magnitude faster (e.g. $\beta = 0.5$, $\alpha = 0.1$). The probability p_μ for an individual to move in one time step was the same for both cases: thus the analysis essentially compares diseases with equal values of R_0 that are fast and slow relative to the time scale of movement. The spread of diseases was compared for four different host population structures varying in levels of heterogeneity: specifically, a single homogeneous group with 1000 individuals, 10 homogeneous groups each of 100 individuals, 25 homogeneous groups each of 40 individuals, and 100 groups each of 10 individuals.

As expected [**40, 78**], the probability of invasion decreases as the population is divided into more smaller groups because the number of intergroup transfers needed for the disease to penetrate the entire population must increase (Figure 6). Also slower diseases are more likely to invade structured populations than faster diseases with the same R_0 because, in the former case, the mean period of infectiousness ($1/\alpha$) is longer, thereby providing more time for inter-group movement of infectious individuals. The simulated epidemics in the single group and 100 group populations were far more similar for the slow than the fast disease. In other words, slow diseases are better approximated by a single homogeneously mixed group than are fast

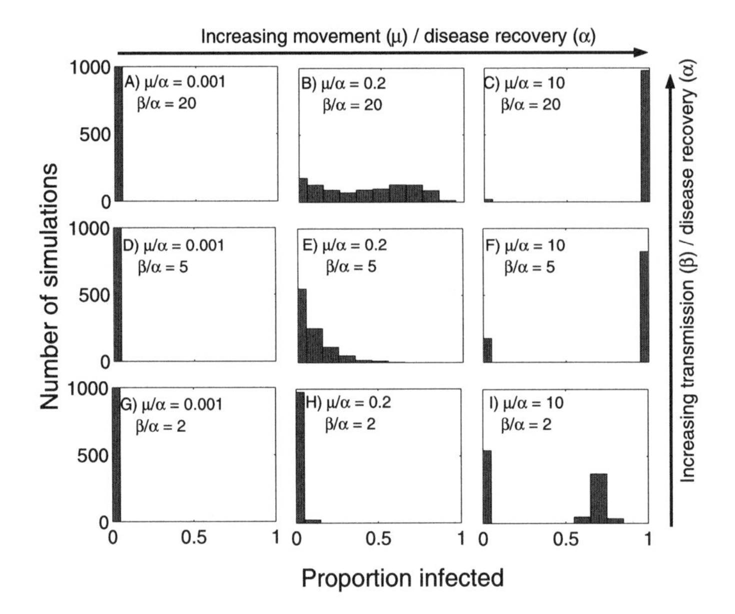

FIGURE 6. Histograms of the proportion of individuals infected during an epidemic for different transmission (β) and movement (μ) values scaled by the probability of disease recovery (α). Each parameter set was simulated 1000 times on an 11×11 toroidal array of groups with 10 individuals each and a recovery probability of 0.1.

diseases (Figure 7). The mean infectious period ($1/\alpha$) defines the natural disease timescale. When movement occurs on roughly the same timescale or longer, mixing among groups should be modeled mechanistically with explicit host movement (as presented in this and the next subsection) rather than implicitly as a between-group transmission rate that operates simultaneously with within-group transmission.

In socially or spatially structured populations, invasion of disease may depend more on the rate of movement between groups p_μ divided by the mean period of infectiousness than on R_0 itself (where the latter is approximate by the transmission rate β divided by the recovery rate α). In Figure 6, R_0 ($\approx \beta/\alpha$) is 20 for all cases. When p_μ/α is small, however, the disease does not invade the metapopulation. As a rule of thumb, a disease will invade the metapopulation if p_μ/α is greater than the reciprocal of the expected number of individuals that will be infected within a single group [**19**]. This makes intuitive sense because in this model system p_μ/α is the expected number of between-group movements made by each infectious individual. Thus p_μ/α multiplied by the expected number of infected individuals is the expected number of infected dispersers per group, which must exceed 1 for a pandemic to occur. When R_0 is high almost all individuals in a group will be infected. Thus, for a pandemic to occur, p_μ/α should be greater than the reciprocal of the average group size. This rule of thumb, however, also depends upon the average number of neighboring groups. The completely connected topology has a lower movement to recovery threshold than a more restrictive loop topology (Figure 8).

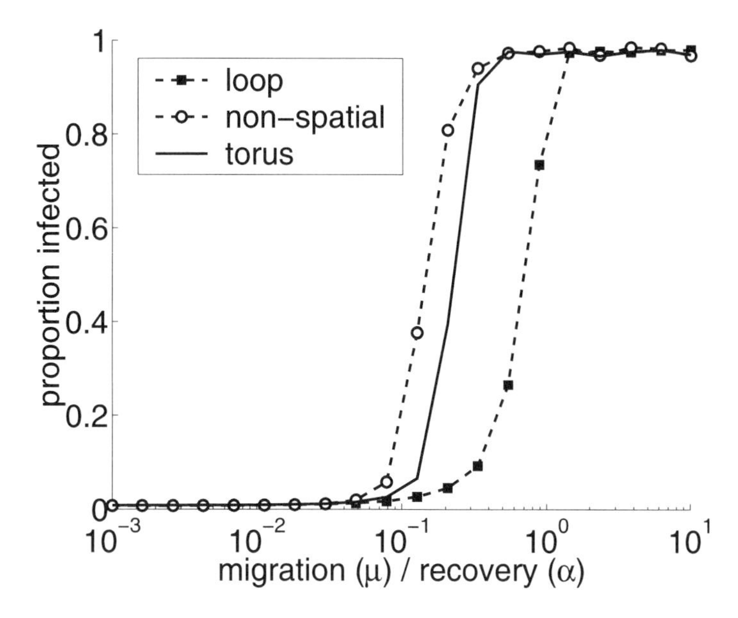

FIGURE 7. The average proportion of the population that becomes infected depends on the spatial configuration: torus, loop, or non-spatial. In the non-spatial array individuals could move to any other group during a single time step. All simulations have 121 groups. Each parameter set was simulated 1000 times, with a group size of 10 and a recovery probability of 0.1.

5.2. Epidemics on dynamic social networks: Populations structured by an association matrix. In Section 5.1 we selected three canonical topologies to describe the movement process in a group structured population. Each of these topologies is an idealization, none of which may fit a particular system very well. Further, in real populations movement is influenced by seasons and space, and movement rates may be time dependent. As a result, individual contacts in real populations are most accurately represented by a dynamic network of social interactions (described in more detail below). Network models provide a much-needed alternative to the assumption of random mixing. Consequently, they have been the focus of much recent research in epidemic theory [**33, 54, 58, 76**]. Network researchers, however, are challenged by depiction of changing social contacts (often resorting to using static, unchanging networks) and by the unavailability of data from which to estimate network parameters.

One approach to these challenges is to apply established methods from studies of animal behavior by constructing an *association matrix* depicting the changing social interactions of all studied individuals [**77**]. A demonstration of this method, discussed below, is a study reported by Cross et al. [**16**]) on the spread of disease in an African buffalo population in the Kruger National Park, South Africa. We tracked buffalo with radio collars for a two-year period, noting how often different individuals were seen together in the same herd. These data were then used to construct a time-dependent set of association matrices $\mathbf{A}(k)$ with entries $a_{ij}(k)$ that represent the proportion of time individuals i and j (dyad i-j) spent together

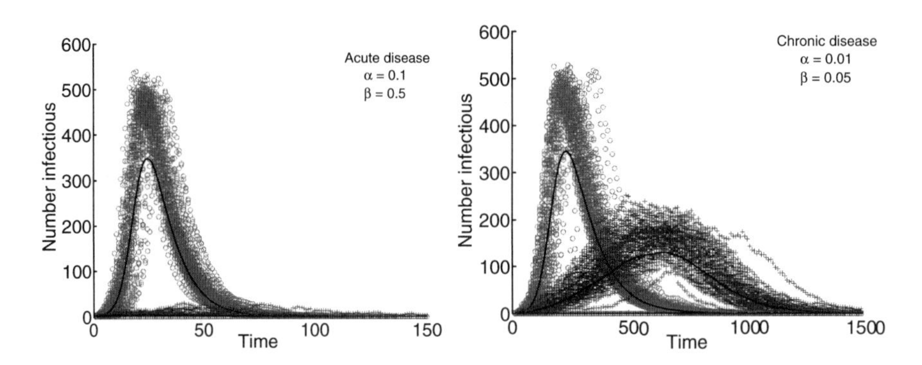

FIGURE 8. Disease invasion depends upon population structure (green circles: 1 group of 1000 individuals; red points: 25 groups of 40 individuals; blue crosses: 100 groups of 10 individuals) and the duration of the infectious period. A mean-field model of one group is a worse approximation for an acute disease with $\alpha = 0.1$ (left panel) than a chronic disease with $\alpha = 0.01$ (right panel). For both diseases $\beta/\alpha = 5$, but the slow disease is more likely to invade the structured population. Lines represent the mean of 100 simulations with frequency-dependent transmission and 1000 individuals. Simulations with 25 or 100 groups were run on a toroidal spatial structure with a movement probability μ of 0.01.

(i.e. were located in the same herd) over the kth time interval $[k-1, k]$, $k = 1, \dots, T$ [**16, 18, 70, 77, 79**].

Assuming that association indices $a_{ij}(k)$ for all dyads in the population have been measured, the matrices $\mathbf{A}(k)$, $k = 1, \dots, T$, can be used in an individual-based model to investigate how a disease introduced at time $t = 0$ into this population might spread through it. Specifically, let the vector $X_i(k) = \binom{a}{b}$, $a, b = 0$ or 1, represent the disease state of individual i at time k, where $x_i(k)' = (0,0)$, $(1,0)$ and $(0,1)$ ($'$ is the transpose of the vector) respectively represent a susceptible, infected, and recovered individual at time t. Then over a time interval $[0, T]$, every individual that becomes infected has a profile of the form

$$
\begin{array}{lll}
 & X_i(l) = (0,0)' \ (S \text{ class}) & l = 0, \dots, k_1 - 1 \leq T \\
(5.1) & X_i(l) = (1,0)' \ (I \text{ class}) & l = k_1, \dots, k_2 - 1 \leq T \text{ provided } k_1 \leq T \\
 & X_i(l) = (0,1)' \ (R \text{ class}) & l = k_2, \dots, T \text{ provided } k_2 \leq T,
\end{array}
$$

where k_1 and k_2 are the points in time where the individual made the transition from $S \to I$ and $I \to R$ respectively. Thus summing over the first and second entries for each individual at time k immediately tells us how many individuals are infected and how many removed at time k. Now assume that the hazard rate for the transmission of disease between an infected individual j and a susceptible individual i, $i, j = 1, \dots, N$, is determined by a common underlying transmission coefficient β multiplied by the association coefficient $a_{ij}(k)$ (recall that $a_{ij}(k)$ is an estimate of the proportion of time individuals i and j spent together over the period $[k-1, k]$.) These pairwise transmission hazards are then summed over all j to obtain the total hazard rate for susceptible i. Additionally, assume that infected individuals enter a V class at a rate α independent of time. Under these assumptions, recalling [**35,**

Expressions (6.6) and (8.3)], the probabilities of infection and recovery take the form (note the use of vector dot product in the first expression):

$$(5.2)\qquad \begin{aligned} p_{i\tau_k} &= 1 - e^{-\beta \sum_{j=1}^{n} a_{ij}(k)((1,0)\cdot X_j(k))}, \quad k = 0, 1, 2, \ldots, T-1, \\ p_\alpha &= 1 - e^{-\alpha}. \end{aligned}$$

Note that the ith individual's probability of infection differs from other individuals because the ith individual has its own set of association coefficients $\{a_{ij}(k), j = 1, \ldots, N\}$. This approach essentially elaborates the average contact rate in a mass action expression of transmission with individual level details (the association matrix $\mathbf{A}(k)$ with elements $(\mathbf{A}(k))_{ij} = a_{ij}(k)$, $i, j = 1, \ldots, N$, determines the actual contact rates for each individual).

In most cases, the number of individuals monitored to obtain the association matrix $\mathbf{A}$ is only a small fraction of the population, perhaps as low as 5%. Thus the question in the context of individual-based models is how to statistically create the other 95% of individuals in the population by scaling up the matrix $\mathbf{A}$ from N^2 to $(20N)^2$ entries, while retaining the emergent group structures (i.e. herd structure in the case of African buffalo) evident in the sampled subset of the population. A quick and dirty solution in the case of a 5% sampling is to assume that the every individual would have 20 times more contacts than indicated by the matrix $\mathbf{A}$ arising from the sample; in which case, we might simply assume that

$$p_{i\tau_k} = 1 - e^{-20\beta \sum_{j=1}^{n} a_{ij}(k)((1,0)\cdot X_j(k))}.$$

This simplification is adequate when the number of infected individuals is relatively large, but not for invasion analysis. In this latter case, introducing one infective into the model is not equivalent to introducing 20 infectives into the population because the effects of demographic stochasticity cannot simply be scaled up. Further, this kind of scaling ignores the network correlations between infected individuals that inevitably build up in the early stages of disease invasion [**45**]. If we are only interested in a comparative analysis, however, of the course of an epidemic for different values of β and α that constitute the same ratio $R_0 = \beta/\alpha$, then an individual based simulation using an association matrix constructed from a population sample provided some insights into the effects of the relative differences in the disease and movement time scales on epidemics.

Associated with every $N \times N$ matrix $\mathbf{A}(k)$ is a network of N nodes, where the ith and j^{th} nodes are connected at time k if and only if $a_{ij} > 0$. The importance of this network topology on the spread of disease can be investigated by randomly rewiring a proportion δ of all connections to obtain versions of the matrices $\mathbf{A}(k)$ that represent the same number of connections. Any non-random group structure inherent in the topology at time k would be progressively destroyed as δ increased from 0 to 1. In particular, randomizing all connections ($\delta = 1$) destroys all group structures while preserving the average degree of connectedness of the network. Two types of rewiring simulations can be conducted. In *static rewiring* simulations, a proportion δ of the connections are rewired at the beginning of the simulation with the same reorganized matrix used throughout the rest of the simulation. In *dynamic rewiring* simulations, in each time step a proportion δ of the connections of $\mathbf{A}$ used in the previous time step are rewired so that even for small δ the matrix $\mathbf{A}$ becomes progressively more randomized over the course of the simulation. (Examples of these two procedures are shown below.)

The importance of variation in connection frequency or strength across the values $a_{ij}(k)$, $k = 1, \ldots, T$, obtained for each dyad i-j, $i, j = 1, \ldots, N$, can also be investigated by increasing the variance over time while keeping the time-averaged association value $\bar{a}_{ij} = \sum_{k=1}^{T} a_{ij}(k)$ among dyads constant. High variance among a_{ij} values corresponds to situations where individuals have two sets of associates, those that they spend most of their time with and those that they rarely encounter. Low variance implies a well-mixed population. One might hypothesize that weak connections are less significant for disease transmission, and hence that systems with a high variance in connection strength may be less permeable to disease spread. On the other hand, the disease may spread more rapidly amongst those individuals that are tightly associated.

The variance can be manipulated while preserving the mean of the associations of each dyad i-j by defining a new set of elements a_{ij}^{γ} of an association matrix $\mathbf{A}^{\gamma}$ modified using the following algorithm: a value z of a uniformly distributed random variable Z on $[0, 1]$ is drawn and then

$$a_{ij}^{\gamma} = \begin{cases} \bar{a}_{ij} + \gamma(1 - \bar{a}_{ij}) & \text{if } \bar{a}_{ij} > z \\ \bar{a}_{ij} - \gamma \bar{a}_{ij} & \text{otherwise} \end{cases}$$

is calculated with the process repeated for each $i, j = 1, \ldots, N$. Clearly, for the extreme cases $\gamma = 0$ and 1, this procedure produces $a_{ij}^{0} = \bar{a}_{ij}$ and $a_{ij}^{1} = 0$ or 1. Note that the algorithm preserves the topology of the connections except when $\alpha = 1$, when some connections are entirely removed and the total proportion of connections is reduced to $\bar{a}_{ij}$.

Cross et al. [**16**] applied the above methods to association data obtained for African buffalo from multi-week observations over a two-year period. From these data, they constructed 24 monthly association matrices for 64 individuals. The networks associated with May 2002 and the entire study period are illustrated in Figure 9. These 24 matrices were then used to investigate questions regarding the spread of disease in this population. Repeating the sequence of 24 association matrices twice (i.e. $\mathbf{A}(24 + k) = \mathbf{A}(k)$, $k = 1, \ldots, 24$) to construct a 48 month period, Cross et al. simulated the spread of a slow moving ($\beta = 0.4$, $\alpha = 0.3$) and a fast moving ($\beta = 0.04$, $\alpha = 0.03$) SIV disease (Figure 10). Note that both epidemics have the same basic reproductive number $R_0 = 4/3$, so that any differences that arise between sets of simulations of epidemics are due to time scale differences in the disease dynamics (the first is ten times faster than the second) relative to the rate of mixing as determined by the association matrices. These simulations indicate that a faster-moving disease is more likely to fade out than a slower-moving disease because the latter integrates over a longer time period; and, the longer the time period of integration, the more likely it becomes that any two individuals make contact (i.e. the underlying contact network becomes more fully connected). These simulations based on empirical social networks thus yield conclusions regarding timescales that are consistent with findings presented in Section 5.1 for an idealized metapopulation.

Cross et al. [**16**] also showed, using the same association matrix at each timestep (specifically, buffalo data from November 2001) that partial random rewiring had a non-linear effect upon disease dynamics. In Figure 11 we see that small increases in the proportion δ of nodes rewired randomly at each step (i.e. the dynamic rewiring

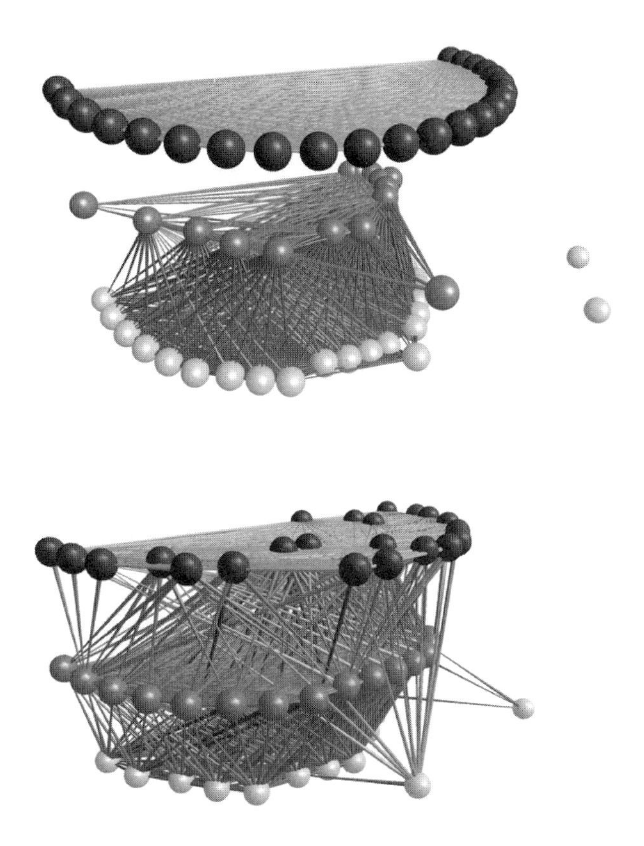

FIGURE 9. Network graphs of the buffalo association data for (A.) May 2002 and (B.) November 2001 through October 2003. Balls represent individual buffalo and the lines represent all non-zero association values. Individuals are distributed vertically according to herd membership, which was determined by UPGMA cluster analysis.

simulation described above) lead to solutions that rapidly approach those of a completely random network ($\delta = 1.0$) by the time $\delta = 0.2$.

Finally, Cross et al. found that variance in the connection strength among dyads had a substantial effect only under certain circumstances. Using the same association matrix at each timestep, they found that increasing the variance (i.e. increasing the value of the parameter γ) had less effect upon slower than faster diseases. Further, increasing variance had little effect when the particular association matrix used in the simulation represented either a very well or very weakly connected group of individuals. This effect was investigated further in association matrices made more sparse by dropping a proportion of connections at random from an empirical association matrix at the beginning of the simulation. Reducing the number of connections between individuals had a relatively large impact compared with the effect of increasing the variance in connection strength (Figure 12).

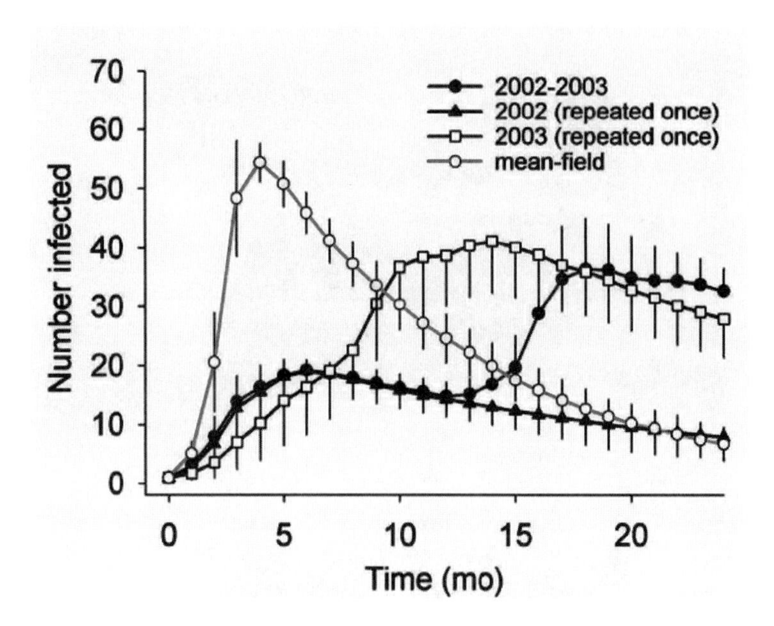

FIGURE 10. Mean and standard deviations (over 50 stochastic simulations) of the number of infected individuals using monthly association data from the entire study period (closed circles), 2002 (closed triangles), 2003 (open squares), or a mean-field model (open circles). All simulations used a transmission coefficient β of 0.3 and recovery probability γ of 0.1. For simulations using one year of data, the same association matrices were used again the second year.

5.3. Epidemics in colonizing populations. Our final example considers the spread of disease in a population that is itself expanding over a given landscape. The particular setting for this problem pertains to the reintroduction of Persian fallow deer (*Dama mesopotamica*) in northern Israel [**6**] and the assessment of how this reintroduction might have been affected if one of the founding individuals had been infected with a transmissible disease [**7**]. A disease of potential concern for Persian fallow deer is bovine tuberculosis (BTB), which occurs in the European fallow deer, *Dama dama* [**3, 56**]. BTB is endemic in wildlife species such as brushtail possums, badgers, African buffalo, and white-tailed deer in numerous countries worldwide, with serious economic, ecological and public health consequences [**59**]. Therefore it is essential to understand the basic processes underlying its possible spread in this reintroduced species, and in evaluating the potential efficacy of different management strategies should BTB be detected in the northern Israel Persian fallow deer population. We have undertaken such an evaluation using a range of parameter values that have been measured with regard to the transmission of BTB in other deer populations (unfortunately all captive, though—see Wahlström et al. [**75**]).

In a homogeneous landscape, simple models predict that a colonizing host population will spread out in a radially symmetric fashion, with a wavefront that has a characteristic velocity $V_p(t)$ [**68**]. Depending on details of the population's dispersal process, this velocity may be constant or increase with time [**47, 55**] until

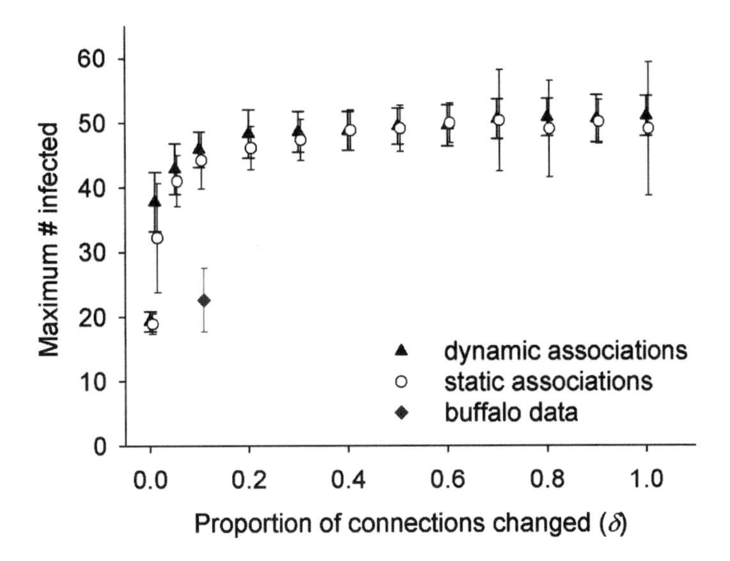

FIGURE 11. The maximum number of individuals infected at any point in time after 50 time steps depends upon the amount δ of random rewiring of the association network at the beginning of each simulation (static) or cumulatively every time-step (dynamic). Dynamic and static simulations started with association data from November 2001; the point pertaining to the buffalo data was generated using all of the association data (i.e. unmanipulated). Disease parameters were $\beta = 0.3$, $\alpha = 0.2$. Error bars represent the standard deviations from 50 stochastic simulations. (For clarity, δ values of the static simulations were increased slightly before plotting.)

the population runs out of space to expand. The rate at which disease spreads in a colonizing population obviously depends on the value of R_0 for that disease (as would be measured in population that has a relatively large number of susceptibles). From the material that has been presented thus far, we expect that the disease will fade out unless R_0 is comfortably above 1. If the disease invades successfully, it can also be regarded as a colonizing process that in a homogeneous population has its own radially expanding wavefront with velocity $V_d(t)$ [**55**].

In a homogeneous population we expect the following three scenarios [**7**]: 1.) The disease fades out if R_0 is less than or equal to 1, or fades out with high probability if R_0 is slightly greater than 1; 2.) If $V_d(t) < V_p(t)$ for all t during the population colonization phase, then the disease wavefront lags behind the population wavefront and the disease only pervades the entire population some time after the population runs out of space; 3.) If $V_d(t) > V_p(t)$ during the initial phase of the disease, then the disease wavefront follows close behind the population wavefront until colonization is complete.

Here we illustrate these three possible outcomes for the situation of Persian fallow deer colonizing a heterogeneous landscape. The model is run on a GIS

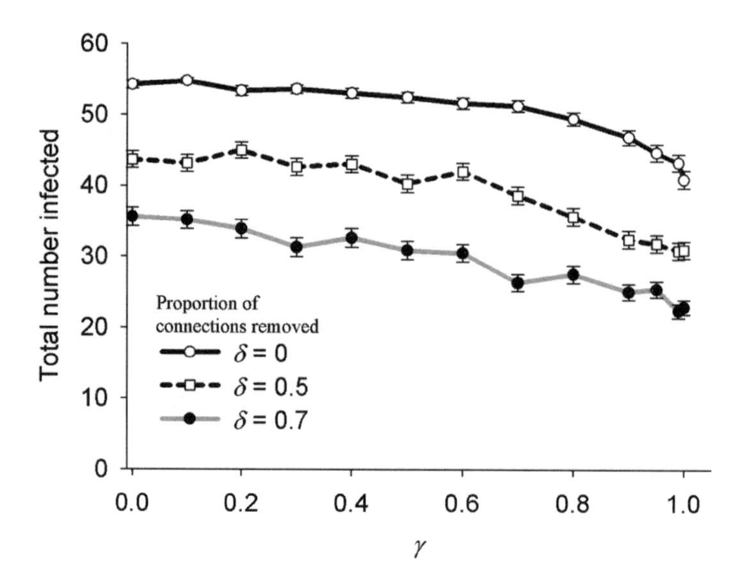

FIGURE 12. The total number of individuals infected after 50 time-steps decreases with increasing variability in the time-averaged connection strength between pairs and decreasing temporal variability of connections within pairs (γ) and the proportion of connections that are removed from the network. See the text for a description of how γ increases the variance in association indices. Error bars indicate the standard errors of 200 simulations using September 2003 as the association matrix, $\beta = 0.3$, and $\alpha = 0.1$.

landscape template that maps a 630 km^2 region of northern Israel into 300×213 1-hectare pixel elements, each of which has been rated with respect to its quality as fallow deer habitat. Details of how animals move around and establish territories are given elsewhere [**6**]. Transmission is modeled by assuming that individuals make contact with other individuals in proportion to the degree to which their respective home ranges overlap [**61**]. Individuals are assumed to have a fixed time budget for interaction with other deer (cf. the density-independent contact rate of frequency-dependent transmission), so the hazard rate of infection scales with the proportion of all territory overlaps that are occupied by infectious deer (rather than the total overlap, which would correspond to mass-action transmission with density-dependent contact rates).

To implement this assumption, the area of overlap a_{ij} for any two individuals i and j needs to be calculated, after which in the notation of expression (5.1) and the following frequency dependent analog of equation (5.2) can be applied to calculate the probability that any individual in the model contracts the disease over a given period of time:

$$p_{i\tau_k} = 1 - e^{-\beta \left(\sum_{j=1}^{n} a_{ij}(k)((1,0)\cdot X_j(k)) \right) \Big/ \sum_{j=1}^{n} a_{ij}(k)}$$

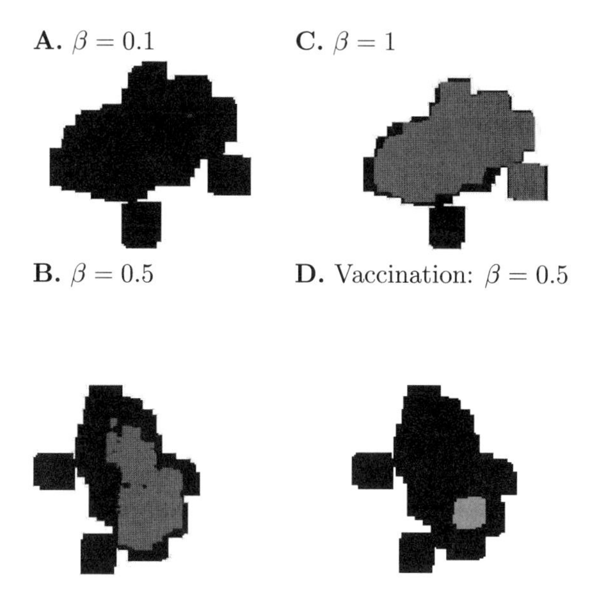

FIGURE 13. Host population density (black) and disease prevalence (purple) are plotted at the end of 10 year projections from onset of the reintroduction of deer into northern Israel (one infected individual at start or reintroduction), for the cases (A.) $\beta = 0.1$, (B.) $\beta = 0.5$, (C.) $\beta = 1$ and (D.) $\beta = 0.5$ with vaccination (100% effective with life-long protection in all released individuals from the third year onwards and all wild-born young). Results are averaged over 250 simulations runs, and pixels were colored if their average density of host individuals (black) or infected individuals (purple) over all runs was at least 0.5.

In stochastic simulations of the model, the heterogeneous landscape affects the pattern of the frontal wave of the population expansions, and induces the establishment of population activity centers in preferred habitats [**6, 7**] and disease centers within them. The disease range expansion within the colonizing population follows patterns similar to our predictions for a homogeneous landscape: the transmission coefficient values (β) have a major effect on the velocity of disease expansion and the distance at which it follows the range expansion of the population (Figure 13A–C). When β is low, unsurprisingly, the disease has a high probability of fading out of the population, even without any management interventions (Figure 13A).

For cases in which intervention efforts are needed, simulations of such a model may help managers to set targets and predict management outcomes (Figure 13D). Field tests in which host individuals are examined for their disease status along transects radiating outward from the introduction site are important to ascertain the relative positions of the disease and population wavefronts. This information will help managers to distinguish between the three idealized scenarios described above, and to evaluate the management options aimed at containing the spread of disease.

6. Conclusion

In this chapter, we have provided a synthetic overview of connections among the various models we have used to address a range of theoretical and applied problems, by showing how they are outgrowths of canonical deterministic and branching process SIR methods presented in the previous chapter [**35**]. Along the way, we have endeavoured to emphasize new properties that emerge as one moves beyond the structurally homogeneous theory into the more realistic world of heterogeneity with regard to demographic, epidemiologic, behavioural and spatial structure.

From the material we present, it is clear that assumptions of homogeneity oversimplify the analysis of most epidemics to the point where estimates of R_0 alone do not adequately describe disease dynamics. Heterogeneity has a profound effect on epidemics and ignoring it leads to substantial bias in estimating the probability with which a disease will invade (Figure 4) or the proportion of individuals that will ultimately become infected (Figure 6). This is why it is important to characterize a disease in heterogeneous populations using appropriate measures, such as R_0 and P_{epi} (cf. [**35**, Equation (7.3)]) when heterogeneity exists with regard to the rates at which diseases are transmitted from one individual to another or, in the case of metapopulations, group size and relative time-scales of movement and transmission/demographic processes.

Much work remains to be done to provide a more coherent theory on the spread of disease in heterogeneous populations. Of course, it will be difficult to generalize the effects that the idiosyncratic spatial structures found in real systems will have on epidemics. Other more generally defined processes, however, such as time scale relationships in canonical metapopulation formulations, or descriptions of heterogeneity using negative binomial, gamma, or other well-known distributions, provide opportunities for gaining new insights into the spread and control of disease in heterogeneous populations.

Acknowledgements

We thank Alison Galvani and an anonymous reviewer for comments that have lead to improvements of this chapter.

References

[1] L. Aaron, D. Saadoun, I. Calatroni, O. Launay, N. Memain, V. Vincent, G. Marchal, B. Dupont, O. Bouchaud, D. Valeyre, and O. Lortholary, Tuberculosis in HIV-infected patients: a comprehensive review, *Clinical Microbiology and Infection* **10** (2004), 388–398.

[2] R. M. Anderson and R. M. May, *Infectious Diseases of Humans: Dynamics and Control*, Oxford University Press, Oxford, 1991.

[3] A. Aranaz, L. de Juan, N. Montero, C. Sanchez, M. Galka, C. Delso, J. Alvarez, B. Romero, J. Bezos, A. I. Vela, V. Briones, A. Mateos, and L. Dominguez, Bovine tuberculosis (Mycobacterium bovis) in wildlife in Spain, *Journal of Clinical Microbiology* **42** (2004), 2602–2608.

[4] M. Badri, R. Ehrlich, R. Wood, T. Pulerwitz, and G. Maartens, Association between tuberculosis and HIV disease progression in a high tuberculosis prevalence area, *Int. J. Tuberc. Lung Dis.* **5** (2001), 225–232.

[5] F. Ball and O. D. Lyne, Stochastic multi-type SIR epidemics among a population partitioned into households, *Adv. Appl. Prob.* **33** (2001), 99–123.

[6] S. Bar-David, D. Saltz, and T. Dayan, Predicting the spatial dynamics of reintroduced populations—the Persian fallow deer, *Ecological Applications*, in press, 2005.

[7] S. Bar-David, . O. Lloyd-Smith, and W. M. Getz, Infectious disease in colonizing populations: simultaneous expansion processes and implications for conservation, in review, 2005.

[8] C. T. Bauch, J. O. Lloyd-Smith, M. Coffee, and A. P. Galvani, Dynamically modeling SARS and respiratory EIDs: past, present, future, *Epidemiology* (2005).

[9] S. M. Blower, P. M. Small, and P. C. Hopewell, Control strategies for tuberculosis epidemics: new models for old problems, *Science* **273** (1996), 497–500.

[10] S. M. Blower and T. Chou, Modeling the emergence of 'hot zones': tuberculosis and the amplification dynamics of drug resistance, *Nat. Med.* **10** (2004), 1111–1116.

[11] J. M. Bossenbroek, C. E. Kraft, and J. C. Nekola, Prediction of long-distance dispersal using gravity-models: Zebra mussel invasion of inland lakes, *Ecol. Appl.* **11** (2001), 1778–1788.

[12] P. Bratley, B. L. Fox, and L. E. Schrage, *A Guide to Simulation*, Springer-Verlag, New York, 1987.

[13] K. P. Burnham and D. R. Anderson, *Model Selection and Inference: a Practical Information-Theoretic Approach*, Springer-Verlag, New York, 1998.

[14] A. D. Cliff and P. Haggett, Disease diffusion: The spread of epidemics as a spatial process, in (M. Pacione, ed.), *Medical Geography: Progress and Prospect*, Croom Helm., London, 1986.

[15] J. Clobert, E. Danchin, A. A. Dhondt, and J. D. Nichols, eds., *Dispersal*, Oxford University Press, Oxford, 2001.

[16] P. C. Cross, J. O Lloyd-Smith, J. Bowers, C. T. Hay, M. Hofmeyr, and W. M. Getz, Integrating association and disease dynamics: an illustration using African buffalo data, *Annals Zoologici Fennici* **41** (2004), 879–892.

[17] P. C. Cross and W. M. Getz, Assessing vaccination as a control strategy in an ongoing epidemic: Bovine Tuberculosis in African buffalo, *Ecological Modelling*, in review, 2005.

[18] P. C. Cross, J. O Lloyd-Smith, and W. M. Getz, Disentangling association patters in fission-fusion societies using African buffalo as an example, *Animal Behavior* **69** (2005), 499–506.

[19] P. C. Cross, J. O. Lloyd-Smith, P. L. Johnson, and W. M. Getz, Dueling time scales of host mixing and disease recovery determine invasion of disease in structured populations, *Ecology Letters* **8** (2005), 587–595.

[20] E. L. Corbett, C. J. Watt, N. Walker, D. Maher, B. G. Williams, M. C. Raviglione, and C. Dye, The growing burden of tuberculosis: global trends and interactions with the HIV epidemic, *Arch. Intern. Med.* **163** (2003), 1009–1021.

[21] M. Cruciani, M. Malena, O. Bosco, G. Gatti, and G. Serpelloni, The impact of human immunodeficiency virus type 1 on infectiousness of tuberculosis: a meta-analysis, *Clin. Infect. Dis.* **33** (2001), 1922–1930.

[22] C. S. Currie, B. G. Williams, R. C. Cheng, and C. Dye, Tuberculosis epidemics driven by HIV: is prevention better than cure? *AIDS* **17** (2003), 2501–2508.

[23] P. D. Davies, The role of DOTS in tuberculosis treatment and control, *Am. J. Respir. Med.* **2** (2003), 203–209.

[24] J. H. Day, A. D. Grant, K. L. Fielding, L. Morris, V. Moloi, S. Charalambous, A. J. Puren, R. E. Chaisson, K. M. De Cock, R. J. Hayes, and G. J. Churchyard, Does tuberculosis increase HIV load? *J. Infect. Dis.* **190** (2004), 1677–1684.

[25] O. Diekmann, J. A. P. Heesterbeek, and J. A. J. Metz, On the definition and computation of the basic reproduction ratio R_0 in models for infectious diseases in heterogeneous populations, *J. Math. Biol.* **28** (1990), 365–382.

[26] O. Diekmann and J. A. P. Heesterbeek, *Mathematical Epidemiology of Infectious Diseases: Model Building, Analysis, and Interpretation*, Chichester, Wiley, New York, 2000.

[27] K. Dietz and K. P. Hadeler, Epidemiological models for sexually transmitted disease, *J. Math. Biol.* **26** (1988), 1–25.

[28] C. A. Donnelly, A. C. Ghani, G. M. Leung, A. J. Hedley, C. Fraser, S. Riley, L. J. Abu-Raddad, L. M. Ho, T. Q. Thach, P. Chau, K. P. Chan, T. H. Lam, L. Y. Tse, T. Tsang, S. H. Liu, J. H. B. Kong, E. M. C. Lau, N. M. Ferguson, and R. M. Anderson, Epidemiological determinants of spread of causal agent of severe acute respiratory syndrome in Hong Kong, *Lancet* **361** (2003), 1761–1766.

[29] C. Dye, G. P. Garnett, K. Sleeman, and B. G. Williams, Prospects for worldwide tuberculosis control under the WHO DOTS strategy. Directly observed short-course therapy, *Lancet* **352** (1998), 1886–1891.

[30] C. Dye, C. J. Watt, and D. Bleed, Low access to a highly effective therapy: a challenge for international tuberculosis control, *Bulletin of the World Health Organization* **80** (2002), 437–444.

[31] C. Dye, C. J. Watt, D. M. Bleed, and B. G. Williams, What is the limit to case detection under the DOTS strategy for tuberculosis control? *Tuberculosis* **83** (2003), 35–43.

[32] B. Efron and R. J. Tibshirani, *An introduction to the bootstrap*, Chapman and Hall, London, 1993.

[33] S. Eubank, H. Guclu, V. S. A. Kumar, M. V. Marathe, A. Srinivasan, Z. Toroczkai, and N. Wang, Modelling disease outbreaks in realistic urban social networks, *Nature* **429** (2004), 180–184.

[34] G. R. Fulford, M. G. Roberts and J. A. P. Heesterbeek, The metapopulation dynamics of an infectious disease: Tuberculosis in possums, *Theor. Popul. Biol.* **61** (2002), 15–29.

[35] W. M. Getz and J. O. Lloyd-Smith, Basic methods for modeling the invasion and spread of contagious diseases, this volume.

[36] B. Grenfell and J. Harwood, (Meta)population dynamics of infectious diseases, *Trends Ecol. Evol.* **12** (1997), 395–404.

[37] K. P. Hadeler and C. Castillo-Chavez, A core group model for disease transmission, *Math. Biosci.* **128** (1995), 41–55.

[38] T. J. Hagenaars, C. A. Donnelly, and N. M. Ferguson, Spatial heterogeneity and the persistence of infectious disease, *J. theor. Biol.* **229** (2004), 349–359.

[39] J. A. P. Heesterbeek, A brief history of R-0 and a recipe for its calculation, *Acta Biotheor.* **50** (2002), 189–204.

[40] G. Hess, Disease in metapopulation models: Implications for conservation, *Ecology* **77** (1996), 1617–1632.

[41] C. B. Holmes, E. Losina, R. P. Walensky, Y. Yazdanpanah, and K. A. Freedberg, Review of Human Immunodeficiency Virus Type 1-Related Opportunistic infections in Sub-Saharan Africa, *Clin. Inf. Dis.* **36** (2003), 652–662.

[42] Y. H. Hsieh and S. P. Sheu, The effect of density-dependent treatment and behavior change on the dynamics of HIV transmission, *J. Math. Biol.* **43** (2001), 69–80.

[43] J. M. Hyman and J. Li, Behavior changes in SIS STD models with selective mixing, *SIAM J. Appl. Math.* **57** (1997), 1082–1094.

[44] J. E. Kaplan, D. Hanson, M. S. Dworkin, T. Frederick, J. Bertolli, M. L. Lindegren, S. Holmberg, and J. L. Jones, Epidemiology of human immunodeficiency virus-associated opportunistic infections in the United States in the era of highly active antiretroviral therapy, *Clin. Infect. Dis.* **30** (2000), S5–14.

[45] M. J. Keeling, The effects of local spatial structure on epidemiological invasions, *Proceedings of the Royal Society of London Series B-Biological Sciences* **266** (1999), 859–867.

[46] E. L. Korenromp, R. Bakker, S. J. de Vlas, R. H. Gray, M. J. Wawer, D. Serwadda, N. K. Sewankambo, and J. D. F. Habbema, HIV dynamics and behaviour change as determinants of the impact of sexually transmitted disease treatment on HIV transmission in the context of the Rakai trial, *AIDS* **16** (2002), 2209–2218.

[47] M. Kot, M. A. Lewis, and P. van den Driessche, Dispersal data and the spread of invading organisms, *Ecology* **77** (1996), 2027–2042.

[48] Y. S. Leo, M. Chen, B. H. Heng, C. C. Lee, N. Paton, B. Ang, P. Choo, S. W. Lim, A. E. Ling, M. L. Ling, B. K. Tay, P. A. Tambyah, Y. T. Lim, G. Gopalakrishna, S. Ma, L. James, P. L. Ooi, S. Lim, K. T. Goh, S. K. Chew, and C. C. Tan, Severe acute respiratory syndrome – Singapore, 2003, *Morbidity and Mortality Weekly Report* **52** (2003), 405–411.

[49] M. Lipsitch, T. Cohen, B. Cooper, J. M. Robins, S. Ma, L. James, G. Gopalakrishna, S. K. Chew, C. C. Tan, M. H. Samore, D. Fisman, and M. Murray, Transmission dynamics and control of severe acute respiratory syndrome, *Science* **300** (2003), 1966–1970.

[50] J. O. Lloyd-Smith, Scaling of resource-driven density dependence in heterogeneous landscapes: dispersal as a case study, *Ecology* (2005).

[51] J. O. Lloyd-Smith, S. J. Schreiber, P. E. Kopp, and W. M. Getz, Superpreading and the impact of individual variation on disease emergence, *Nature*, in review, 2005.

[52] L. G. Louie, W. E. Hartogensis, R. P. Jackman, K. A. Schultz, L. S. Zijenah, C. H. Y. Yiu, V. D. Nguyen, M. Y. Sohsman, D. K. Katzenstein, and P. R. Mason, Mycobacterium tuberculosis/HIV-1 coinfection and disease: Role of human leukocyte antigen variation, *J. Infect. Dis.* **189** (2004), 1084–1090.

[53] E. Mañas, F. Pulido, J. M. Pena, R. Rubio, J. González-García, R. Costa, E. Pérez-Rodríguez, and A. Del Palacio, Impact of tuberculosis on the course of HIV-infected patients with a high initial CD4 lymphocyte count, *Int. J. Tuberc. Lung Dis.* **8** (2004), 451–7.

[54] L. A. Meyers, M. E. J. Newman, M. Martin, and S. Schrag, Applying network theory to epidemics: Control measures for Mycoplasma pneumoniae outbreaks, *Emerg. Infect. Dis.* **9** (2003), 204–210.

[55] D. Mollison, Dependence of epidemic and population velocities on basic parameters, *Math. Biosci.* **107** (1991), 255–287.

[56] R. S. Morris, D. U. Pfeiffer, and R. Jackson, The epidemiology of Mycobacterium-Bovis infections, *Veterinary Microbiology* **40** (1994), 153–177.

[57] C. J. L. Murray and J. A. Salomon, Modeling the impact of global tuberculosis control strategies, *Proceedings of the National Academy of Sciences of the United States of America* **95** (1998), 13881–13886.

[58] M. E. J. Newman, The structure and function of complex networks, *SIAM Rev.* **45** (2003), 167–256.

[59] L. M. O'Reilly and C. J. Daborn, The epidemiology of Mycobacterium bovis infection in animals and man: A review, *Tuber. Lung. Dis.* **76** (1995), 1–46.

[60] O. T. Ovaskainen and B. T. Grenfell, Mathematical tools for planning effective intervention scenarios for sexually transmitted diseases, *Sex Transm Dis* **30** (2003), 388–394.

[61] A. Perelberg, D. Saltz, S. Bar-David, A. Dolev, and Y. Yom-Tov, Seasonal and circadian changes in the home ranges of reintroduced Persian fallow deer, *Journal of Wildlife Management* **67** (2003), 485–492.

[62] T. C. Porco, P. M. Small, and S. M. Blower, Amplification dynamics: Predicting the effect of HIV on tuberculosis outbreaks, *J. Acquir. Immune Defic. Syndr.* **28** (2001), 437–444.

[63] S. Riley, C. Fraser, C. A. Donnelly, A. C. Ghani, L. J. Abu-Raddad, A. J. Hedley, G. M. Leung, L.-M. Ho, T.-H. Lam, T. Q. Thach, P. Chau, K.-P. Chan, S.-V. Lo, P.-Y. Leung, T. Tsang, W. Ho, K.-H. Lee, E. M. C. Lau, N. M. Ferguson, and R. M. Anderson, Transmission dynamics of the etiological agent of SARS in Hong Kong: Impact of public health interventions, *Science* **300** (2003), 1961–1966.

[64] M. S. Sánchez, R. M. Grant, T. C. Porco, K. L. Gross, and W. M. Getz, Could a decrease in drug resistance levels of HIV be bad news? *Bull. Math. Biol.* (2005).

[65] L. Sattenspiel and C. P. Simon, The spread and persistence of infectious diseases in structured populations, *Math. Biosci.* **90** (1988), 341–366.

[66] M. T. Schechter, N. Le, K. J. P. Craib, T. N. Le, M. V. O'Shaughnessy, and J. S. G. Montaner, Use of the Markov model to estimate the waiting times in a modified WHO staging system for HIV infection, *Journal of Acquired Immune Deficiency Syndromes and Human Retrovirology* **8** (1995), 474–479.

[67] Z. Shen, F. Ning, W. G. Zhou, X. He, C. Y. Lin, D. P. Chin, Z. H. Zhu, and A. Schuchat, Superspreading SARS events, Beijing, 2003, *Emerg. Infect. Dis.* **10** (2004), 256–260.

[68] N. Shigesada and K. Kawasaki, *Biological Invasions: Theory and Practice*, Oxford University Press, New York, 1997.

[69] P. Sonnenberg, J. R. Glynn, K. Fielding, J. Murray, P. Godfrey-Faussett, and S. Shearer, How soon after infection with HIV does the risk of tuberculosis start to increase? A retrospective cohort study in South African gold miners, *J. Infect. Dis.* **191** (2005), 150–158.

[70] M. Szykman, A. L. Engh, R. C. Van Horn, S. M. Funk, K. T. Scribner, and K. E. Holekamp, Association patterns among male and female spotted hyenas (Crocuta crocuta) reflect male mate choice, *Behavioral Ecology And Sociobiology* **50** (2001), 231–238.

[71] H. M. Taylor and S. Karlin, *An Introduction to Stochastic Modeling*, third ed., Academic Press, San Diego, 1998.

[72] C. P. B. Van der Ploeg, C. Van Vliet, S. J. De Vlas, J. O. Ndinya-Achola, L. Fransen, G. J. Van Oortmarssen, and J. D. F. Habbema, STDSIM: A microsimulation model for decision support in STD control, *Interfaces* **28** (1998), 84–100.

[73] J. Volmink, P. Matchaba, and P. Garner, Directly observed therapy and treatment adherence, *Lancet* **355** (2000), 1345–1350.

[74] J. Wallinga and P. Teunis, Different epidemic curves for severe acute respiratory syndrome reveal similar impacts of control measures, *Am J Epidemiol* **160** (2004), 509–516.

[75] H. Wahlström, L. Englund, T. Carpenter, U. Emanuelson, A. Engvall, and I. Vagsholm, A Reed-Frost model of the spread of tuberculosis within seven Swedish extensive farmed fallow deer herds, *Preventive Veterinary Medicine* **35** (1998), 181–193.

[76] D. J. Watts and S. H. Strogatz, Collective dynamics of 'small-world' networks, *Nature* **393** (1998), 440–442.

[77] H. Whitehead, Testing association patterns of social animals, *Anim Behav* **57** (1999), 26–29.
[78] E. B. Wilson and J. Worcester, The spread of an epidemic, *Proc. Natl. Acad. Sci. USA* **31** (1945), 327–333.
[79] G. Wittemyer, I. Douglas-Hamilton, and W. M. Getz, The socio-ecology of elephants: analysis of the processes creating multi-tiered social structures, *Animal Behavior*, in press, 2005.
[80] M. E. J. Woolhouse, C. Dye, J. F. Etard, T. Smith, J. D. Charlwood, G. P. Garnett, P. Hagan, J. L. K. Hii, P. D. Ndhlovu, R. J. Quinnell, C. H. Watts, S. K. Chandiwana, and R. M. Anderson, Heterogeneities in the transmission of infectious agents: Implications for the design of control programs, *Proc. Natl. Acad. Sci. USA* **94** (1997), 338–342.
[81] Y. Xia, O. N. Bjornstad, B. T. Grenfell, Measles metapopulation dynamics: a gravity model for epidemiological coupling and dynamics, *Am. Nat.* **164** (2004), 267–281.

Department of Environmental Science, Policy and Management, 140 Mulford Hall #3112, University of California, Berkeley, CA 94720-3112, U.S.A., and Mammal Research Institute, Department of Zoology and Entomology, University of Pretoria, Pretoria 0002, South Africa
E-mail address: getz@nature.berkeley.edu

Biophysics Graduate Group, University of California at Berkeley
E-mail address: jls@nature.berkeley.edu

Department of Environmental Science, Policy and Management, 140 Mulford Hall #3112, University of California, Berkeley, CA 94720-3112, U.S.A., and USGS, Northern Rocky Mountain Science Center, Bozeman, MT 59715, U.S.A.
E-mail address: pcross@nature.berkeley.edu

Department of Environmental Science, Policy and Management, 140 Mulford Hall #3112, University of California, Berkeley, CA 94720-3112, U.S.A.
E-mail address: bardavid@nature.berkeley.edu

Biophysics Graduate Group, University of California at Berkeley
E-mail address: plfjohnson@berkeley.edu

California Department of Public Health
E-mail address: TPorco@dhs.ca.gov

Department of Environmental Science, Policy and Management, 140 Mulford Hall #3112, University of California, Berkeley, CA 94720-3112, U.S.A.
E-mail address: msanchez@nature.berkeley.edu

DIMACS Series in Discrete Mathematics
and Theoretical Computer Science
Volume **71**, 2006

A cophylogenetic perspective on host-pathogen evolution

Michael A. Charleston and Alison P. Galvani

ABSTRACT. Cophylogeny reconstruction is an increasingly important component of the study of evolution, as it pertains to the relationships among ecologically linked organisms, particularly those of hosts and pathogens. Cophylogenetic models have enhanced our understanding of the evolution of many pathogens, including malaria, adenoviruses and lyssaviruses. These models have also been used to infer the conditions surrounding the emergence of HIV. We compare the primary techniques used in cophylogenetic analysis, and survey some of the insights that cophylogenetic analysis has yielded about host-pathogen interactions and the emergence of new diseases.

1. Introduction

Cophylogeny, the study of linked phylogenies, can be used to reveal macroevolutionary patterns and dynamics of host-pathogen coevolution, and to identify ancient associations between host and pathogen species. Cophylogenetic analysis can be used as a predictive tool: for example, missing lineages can reveal the occurrence of previously unidentified pathogen species [**29**], particularly as some host species will be more highly sampled, such as humans. These unidentified pathogens may have potential for zoonosis into human or other host species. Cophylogenetic methods can also be employed to identify lineages from which zoonosis is disproportionately frequent. Cophylogenetic analysis thus has public health significance in terms of focusing on areas for vigilance and protection against zoonosis. In addition to inferring the relationship between hosts and their pathogens, cophylogenetic models provide a quantitative measure of the degree of divergence between the hosts and their pathogens.

Strong congruence between host and parasite phylogenies has historically been interpreted as evidence for cospeciation. However, analytical methods have proliferated recently to resolve the interplay of cospeciation, host switching, extinction and duplication, which together determine degree of congruence. Cophylogenetic methods have been applied to a diverse array of host-pathogen associations, including copepods and teleost fish [**41**], birds and tapeworms [**26**], clams and sulphur oxidizing endosymbionts [**44**], ascouracarid mites and megapodes [**45**], bacterial endosymbionts and aphids [**17**], crocodiles and helminths [**4**] and lice and their hosts [**23**, page 1991].

Key words and phrases. Host-parasite coevolution, Cophylogeny, Modeling.

Most emerging diseases arise through zoonosis, equivalent to host-switching in cophylogenetic models. Recent zoonotic diseases include Ebola, West Nile virus, monkeypox, hantavirus, HIV and new subtypes of influenza A. The emergence of a disease can be separated into two steps: introduction into a new host population and perpetuated transmission within the new host species [**2**]. Cophylogenetic analysis can provide remarkably detailed insights into the events that have generated specific emergence events. For example, Caribbean lizards form distinct species isolated from each other on the islands of Eastern Caribbean. These lizard species are infected by lizard malaria *Plasmodium azurophilum*, the vector of which has not been identified. There is evidence for codivergence between these hosts and their malarial parasites: mapping the malarial trees into the *Anolis* host species shows significant agreement, consistent with a history of codivergence (Figure 3). Intriguingly, the Pareto-optimal maps include host-switching events from the southern to the northern islands, which is in the direction of prevailing winds. This suggests a rafting vector, possibly another *anolis* species [**12**]. Thus, knowledge of the relationship between the lizard species and *P. azurophilum* cophylogenetic histories throws light on the unknown vector for the malarial species *P. azurophilum.*

Cophylogenetic methods have also furthered our understanding of the coevolution between snakes and retroviruses. Retroviruses are obligate parasites, and as such are likely codiverge with their hosts, yet the degree of coherence between retrovirus and host trees is highly variable [**33, 34**]. However as Jackson and Charleston [**31**] found, there are many factors that must coincide if codivergence is to occur to an extent that is recoverable. Martin's studies revealed cases in which there was highly significant codivergence between retroviruses and their snake hosts, whereas Kossida et al.'s study showed a significant *lack* of codivergence. These studies demonstrate that satisfaction of established requirements for codivergence to take place and form coevolutionary patterns, is not necessarily sufficient. Obligate parasites, highly host-specific, vertically transmitted, may yet not show significant codivergence with their hosts.

2. Basic terminology

A phylogeny is typically represented by a *rooted binary tree*, the tips of which are labeled with extant taxa. A tree is a type of *graph*, which is a set of *vertices*, connected by *edges*. Edges are also known by biologists as *branches*. The vertices of T are denoted $V(T)$, and the edges are denoted $E(T)$. An edge e of T is often written as a pair of vertices (u, v). Trees can be either directed or undirected: in directed trees the edges are called *arcs* and can only be traversed along a particular direction. Thus, arcs (u, v) and (v, u) are not the same. If a tree is *rooted*, then its *root* is a special vertex, and the canonical orientation from the root to the arcs gives directionality to the tree: in phylogeny the direction of the arcs is away from the root, forward in time toward the tips of the tree. Here phylogenies are represented by rooted binary trees, the tips of which are labeled with the names of extant taxa, and the internal branches of which may be weighted with estimated times. Throughout this discussion H will mean the host tree, and P will mean the pathogen or parasite tree. H can be thought of as the independent variate, and P as the dependent variate. For example, (H, P) pairs are (host, pathogen), (species, gene), and (area, organism). In addition, h/p will be used to indicate a particular association between host and parasite/pathogen species.

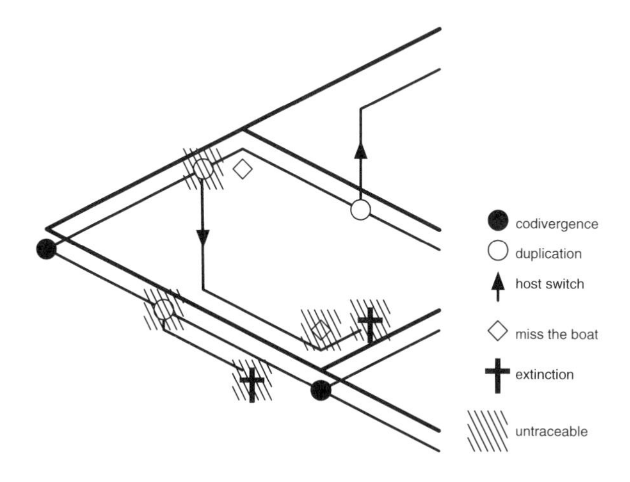

FIGURE 1. Cophylogenetic events. This figure shows the different types of cophylogenetic events that can take place. The host tree is in heavy lines; the pathogen/parasite tree in lighter lines, and time is from left to right. Codivergence and duplication events are shown by solid and open circles respectively, host switches by arrows to contemporaneous host lineages, lineage sorting by diamonds and extinction by termination of the pathogen line. Certain events are untraceable from the surviving phylogeny; these are hashed.

Cophylogenetic events

Linked phylogenies be subject to a number of types of cophylogenetic events (Figure 1). In host-pathogen/parasite systems, coevolutionary patterns are determined by underlying epidemiological dynamics. In this context, *codivergence* is defined as the joint speciation of both host and pathogen/parasite [**39**]. Alternatively, if a pathogen or parasite is distributed over only part of the host's distribution, divergence of the host lineage may lead to descendants that are not infected by the species in question. This process has been termed '*missing the boat*' [**43**] or, sometimes, as *lineage sorting*. A *duplication* event refers to the speciation of the pathogen that occurs within a single host lineage [**39**]. *Host switching* is the transmission and establishment of a pathogen lineage from one host to another host that is not itself the immediate descendent of the original host [**39**]. For instance, a combination of cophylogenetic analysis and historical biogeographical data on Alcidae auks and *Alcataenia* cestodes indicated that sequential host switching events provided a more likely explanation than scenarios that involved codivergence [**26**]. A *loss* is simply the absence of a pathogen where we would expect it to be, e.g., in the descendants of a host that previously had the pathogen. Losses can be caused by the extinction of the pathogen and "missing the boat", or just failing to observe pathogens. Losses in reconciled trees may often represent undiscovered pathogen species and not necessarily extinction, particularly as some host species will be more highly sampled, such as humans. More exhaustive sampling helps to resolve whether the losses are genuine extinctions.

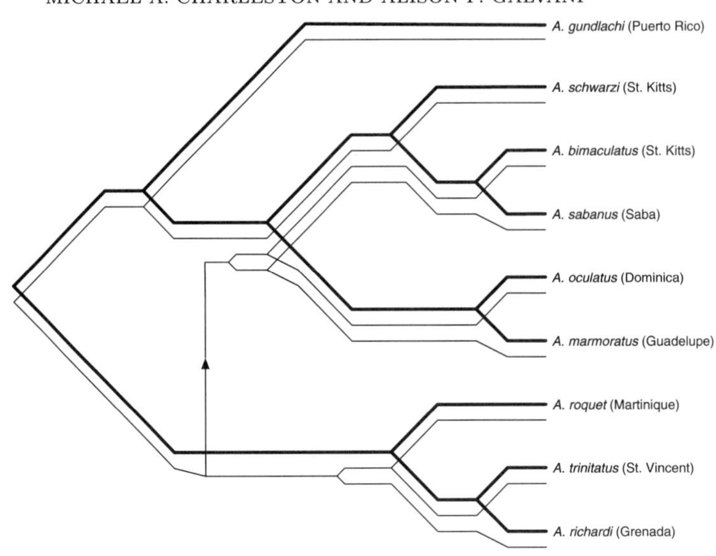

FIGURE 2. Mapping SIV into the primate tree. In Charleston and Robertson [**13**] several analyses were carried out with different plausible resolutions of the SIV phylogeny, one of which is shown above. All showed the characteristic significant congruence in terms of numbers of codivergence events, which could mostly be attributed to the similarity of the trees near the tips.

Linked trees are *reconciled* by inducing cophylogenetic events to explain the differences between them, and *maps* from P into H are most intuitive method of representing this reconciliation. A *reconciled tree* has come to be a commonly used term in this sense, and while initially it referred to a specific method of Page which prohibited host switching [**36**], we use it more generally, as above.

3. Methodology

The inclusion of host switching events dramatically increases the difficulty of cophylogeny problems, and initial methods (and some persisting ones) do not deal with them correctly. One method for dealing with host switching uses a mathematical structure called a *jungle* [**8**] and implemented it in the "TreeMap" program [**11**]. The jungle is a graph containing a set of feasible maps from P into H, permitting the set of all possible recoverable events as described above. The implicit assumption of most, if not all, methods in cophylogeny of independence of pathogen lineages means that without stretching our assumptions we may treat the arcs of a (mapped) pathogen tree independently: because each such arc may occur in more than one map, the jungle can be constructed to contain just one copy of each such arc. Even so, the number of feasible reconstructions grows rapidly both with the size of host and pathogen trees [**10**], and so the jungle becomes unwieldy for moderately large problems, even with the bounding methods implemented in TreeMap [**11**].

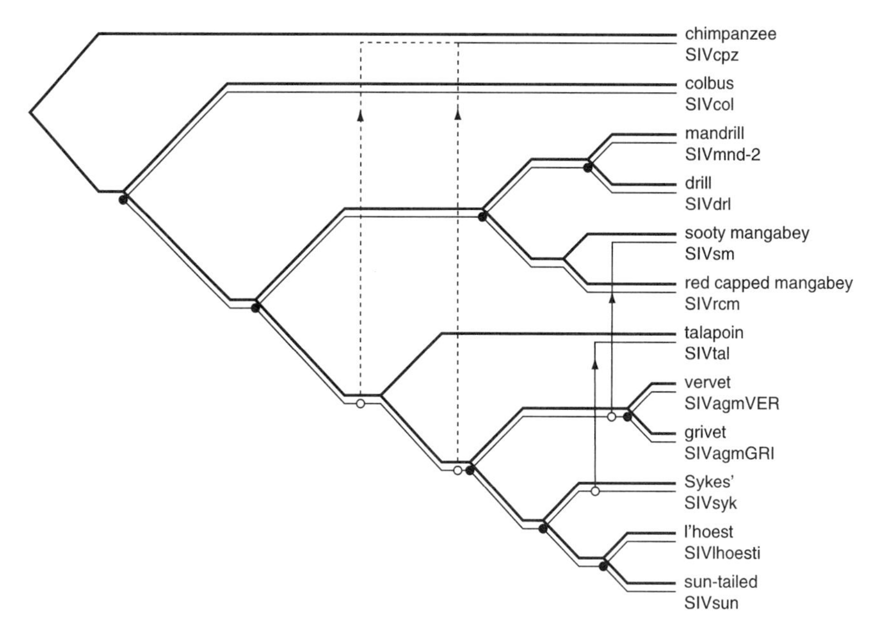

FIGURE 3. Lizard malaria in the Eastern Caribbean. This cophylogeny map of malaria (*Plasmodium azurophilum* (Red) types) into the *Anolis* species in the Eastern Caribbean archipelago is adapted from Charleston and Perkins [**12**]. The map is typical of Pareto-optimal maps that were found in that analysis and shows the host switch from the southern islands (Martinique, St. Vincent and Grenada) to the northern islands in that chain.

The feasible solutions can be compared either by the approach of assigning costs to each type of distinguishable event (codivergence, duplication, loss and host switching) and summing the cost of events in each map, or by taking a much more conservative approach using the concept of *Pareto optimality.* A solution is Pareto optimal if it is not *dominated* by any other solution. Map X dominates map Y in this case if no matter what possible event costs we may assign, X has a lower total cost than Y. For example, if one solution 'A' has 4 codivergences, 2 duplications, 3 losses and one host switch, if dominates another solution 'B' with 2 codivergences, 4 duplications, 3 losses and one host switch because we assert that codivergence has a lower cost than does duplication. On the other hand 'A' does not dominate a solution with 2 codivergences, 2 duplications, 2 losses and 1 host switch, because we cannot say which has the lower total cost. Clearly, if we do not know the event costs, then we must end up with a set of Pareto optimal maps, which we denote POpt. POpt contains all those maps that we cannot rule out as being optimal for some possible set of event costs.

Unfortunately this makes statistical testing of multiple maps in POpt rather difficult: TreeMap performs randomization testing based on user-input event counts rather than on the maps directly. That is, the user sets bounds on the maximum number of non- codivergence events (NCEs), or the minimum number of codivergence events (CEs), and tests the null hypothesis that a random P requires that many NCEs or permits that many CEs to reconcile it with H.

Virtually nothing is known about the *landscape* of the solution space of cophylogeny maps, which means that the design of intelligent heuristics to find (Pareto) optimal maps is somewhat crippled: while we can perform simple hill-climbing methods we don't know how well these methods are expected to perform. Also lacking are *in silico* studies akin to those in phylogenetics, where trees are "grown" and sequences "evolved" along them, these sequences then used as input for phylogenetic reconstruction to compare inference methods. In cophylogeny there are as yet no such studies, though the authors are attempting to address this shortfall.

The extent of concordance depends on the relative frequency of codivergence, duplication, host switching and loss. Many explanations for associated pathogen and host phylogenies can arise from the combination of these events, from which we must find the most likely, or most parsimonious. Several methods have been developed to determine the best explanation, and these are best implemented in conjunction with information about host and parasite population structure and geographical distribution, mode of pathogen transmission and sociality of hosts.

Genetic distances may be compared within and between host and pathogen datasets. If both host switching and codivergence are occurring frequently between H and P, then we expect some host switching events to take place after the last codivergence. In such a case we might expect the evolutionary distance between pathogens to be less than that of their hosts. If, on the other hand, codivergence dominates with little or no host switching, then we expect the similar speciation rates to be reflected in similar evolutionary distances. For example, speciation between louse species and seabirds has been found to be approximately synchronous [**24**]. In general, molecular divergence in the pathogen will be greater than the host if coevolution is dominated by duplications, lower if dominated by host switching and equal if codivergence is the primary mode.

Methodology

Recovering history

One of the principal motivations of cophylogenetic reconstruction is the recovery of ancient events: what happened, when it happened, and how often. From these inferences we can estimate model parameters to generate predictions. In nucleotide evolution, hidden mutations are masked by further mutation or eradicated by reversal. In host-pathogen coevolution, equivalent phenomena are switches to host lineages that subsequently go extinct, or attempted host switches that fail. For instance, the 1997 Hong Kong avian influenza epidemic was caused by a zoonotic event that did not lead to persistence of the emergent recombinant flu subtype. Cophylogenetic methods to determine the relative frequency of these 'hidden' events are currently under development [**14**].

What can be recovered?

Ronquist has confirmed that there are only four events that can be recovered, given the combination of host tree, H, pathogen tree, P, and known current associations, f, between them [**46**]. These events are codivergence, duplication, host switching and loss. Note that although extinction and missing the boat (Page's

"Sorting events") are different events, they cannot be recovered from the information given as above. Interestingly, although there has been a huge amount of work on what kinds of *phylogenetic* events can be recovered from molecular sequence data, the question of which cophylogenetic events are most likely to be recovered successfully remains unknown.

Matrices and maps

Various methods exist for the investigation of the underlying patterns of linked phylogenies, including matrices and cophylogeny maps, as outlined below.

Matrices

Not only can trees be coded as characters (and in particular binary trees as binary characters), but so can the associations between vertices in both phylogenies. If binary trees are used then the character matrix is composed of binary characters. As an example of coding two trees, consider H and P with trees $((A,B),(C,D))$ and $(1,(2,(3,4)))$ respectively where parasite 1 infects host A, 2 infects B, etc. That is, H has leaves A, B, C and D, (A,B) are siblings, (C,D) are siblings and the two pairs also form a sibling pair. In P, parasites 3 and 4 are siblings, and their next closest extant relatives are 2, followed by 1. We can code trees as columns of binary characters, where each column corresponds to a clade (which could be a single taxon), and a '1' is put in the column for each taxon in the clade. We code H in this way as

$$\begin{array}{lllllll} A & 1 & 0 & 0 & 0 & 1 & 0 \\ B & 0 & 1 & 0 & 0 & 1 & 0 \\ C & 0 & 0 & 1 & 0 & 0 & 1 \\ D & 0 & 0 & 0 & 1 & 0 & 1 \end{array} .$$

At the same time we code the tree P with each parasite using the same row as its host:

$$\begin{array}{lllllll} 1 & 1 & 0 & 0 & 0 & 0 & 0 \\ 2 & 0 & 1 & 0 & 0 & 0 & 1 \\ 3 & 0 & 0 & 1 & 0 & 1 & 1 \\ 4 & 0 & 0 & 0 & 1 & 1 & 1 \end{array} .$$

The optimization problem becomes one of finding a minimal coding of interior character states (0s and 1s) given the 'sequences' at the tips of H. This method has some apparent advantages: you can use existing tree-finding software, it is very easy to state and relatively easy to solve, and if you have more than one host for a given parasite then that can be encoded simply by duplicating the 'sequence' for that parasite. But these apparent features are the basis of its problems: there is no interpretation available for what constitutes a 'change of state' in cophylogenetic terms, such coding does not allow Pareto optimal solutions to be found. Columns in the 'sequence' derived from P, which would most likely correspond to nucleotide sites in a phylogenetic analysis, are treated as independent — which they are not. This is a fundamental problem for optimization methods such as Brooks' Parsimony Analysis (BPA) [**5, 6, 7**] which treat the characters as independent and then minimize the number of 'changes of state' of these pseudo-data.

Reconciliation analysis

The most intuitive method of recovering the most accurate explanation for the observed differences between two linked phylogenies is a map from one within the other [**36**]. Reconciliation analyses generate a map between host and parasite trees to calculate the cost of making the trees congruent given the parasite distribution. We avoid the use of the term 'onto', because not all the places in the host tree need have a parasite/pathogen on them. The number of feasible maps rises exponentially with the number of host and pathogen/parasite taxa [**10**], so even the application of the jungle [**8**] does not improve methodology as much as it might have been hoped. Nonetheless, TreeMap 2 correctly deals with the possibility of weakly incompatible host switches, that is, sets of host switches that in combination require the introduction of extra lineage sorting events. The first version of TreeMap (v1) would incorrectly discard such sets of host switches as impossible to reconcile. This correction means that for a given map with, say, k host switches, there may be up to $k!$ orderings of host switches to evaluate, so the advantage of global optimality comes with the expense of even more computational complexity.

Maximum Likelihood

Evaluation of the likelihood of cophylogenetic maps is essentially unsolved. It is a trivial task to choose a method (e.g., parsimony or likelihood) that gives a perfect score to a perfect match, but when the trees differ (not just their branch lengths, as in Huelsenbeck and Rannala's method [**27**], which required congruent trees), things are more complicated. One reasonable approach is to calculate a score for each feasible map Φ and then maximize this over all maps. The basic (usually implicit) assumption made by almost all cophylogeny methods is that parasite lineages, once diverged, are independent. One consequence of this is that the events that take place on each lineage are independent, and if each has a cost or score, then these can be summed to give a score for the complete map. The next question then is, what score can we assign these events? An overall likelihood can be calculated if we know the likelihood of each event, but no viable model exists for this: Huelsenbeck's method [**28**] prohibited more than one parasite/pathogen lineage on any one host lineage, so over-estimated the number of host switches required to reconcile trees. An event cost can be assigned to each type of event, equivalent to a fixed likelihood, and TreeMap permits this, though without prior knowledge as to what the event costs are, assignment of event costs usually leads to misinterpretation of results.

Both TreeMap and BPA hypothesize that the branching order of P and H are correct, which while being a reasonable working hypothesis can lead to over-interpretation of the results, if care is not taken to accommodate tree uncertainty by analyzing alternate phylogenies. An alternative approach by Johnson et al. assesses the degree to which the robustness in the underlying data affects confidence in the tree topologies [**32**]. Their method evaluates whether there is significant incongruence between the parasite/pathogen and host data, using a partition heterogeneity test, also known as an ILD test [**18, 48**]. If there is, the h/p pairs causing the incongruence are sequentially removed until congruence is achieved. This method is acknowledged to have its limitations, including the problem that occurs if incongruence is caused by an early host switch, a large set of taxa might be removed before the underlying congruent tree is estimated, then be put back in for tree reconciliation. This flaw may well lead to a bias in the estimation of which events led to the incongruence of the two trees, but as to how much, we do not yet know; nor is there a method to obviate this potential problem.

Event costs

It is assumed that pathogens and parasites tend to codiverge with their hosts unless other evolutionary events interfere with codivergence. Codivergence is thus considered a "default" action for coevolving species. Optimizing the total cost of a map from P into H, the other events must perforce be assigned a higher cost than that of codivergence. For computational reasons, we set $c = 0$, and all the other costs as positive, but we do not know how much more: the event costs of duplication (d), host switching (h), missing the boat (m), and extinction (e) are positive, but we cannot say whether $4c+m+e$ is more or less than $2c+2d+h$. Thus, we cannot rank these two solutions. Simply summing the number of non-codivergence events does not get around the problem (just as it does not when giving all characters the same weight in a Parsimony analysis); we still have a multi-objective optimization problem, and we have to include any solution that is not strictly worse than another.

Statistical testing

Randomization

As cophylogenetic reconstruction grows, statistical testing of results is lagging somewhat behind their acquisition. Various approaches have been used, from those which require both host and parasite tree to match [**27**], or contain matching subtrees [**32**] through to those which randomize the associations of the tip-most taxa in H and P. The consensus method now is randomization of the associate tree, but a few words must be said about the other options.

In only very few cases will host and parasite or pathogen tree be completely congruent, so that requirement is rather restrictive. The test is good for finding out whether the evolutionary rate of H and P are consistent with cophylogenetic evolution, but not for determining whether the number of codivergence events required to explain the degree of their similarity is "a lot" or only "a little". Given the complexity imposed by the set of Pareto optimal maps, we need to learn considerably more about the statistical aspects of this system before we can adequately test the significance of a match between P and H.

Randomizing H

Randomizing H answers the wrong question. The question answered is "How surprising is it that this particular host tree should match with the parasite tree, as compared with any other host tree?"

Randomizing f

Randomizing f is strongly affected by tree balance: consider the case where P and H are congruent and have a great many pairs of pendant sister taxa (cherries). Randomizing f effectively randomizes the tips of P (or H). Thus trees with more cherries will have deflated significance using this test. In the simplest possible case with 4 taxa, a back-of- the-envelope calculation shows $p = 1/3$ when H and P are fully balanced, and $p = 1/12$ for the unbalanced case. This is a much greater difference than would be expected by the frequency of observation of these tree shapes under the Yule model, which are in the ratio $1 : 2$ (balanced : unbalanced), not $3 : 12$.

Randomizing P

Randomizing P answers the *right* question. The question answered is "How surprising is it that this particular parasite tree should match with the host tree, as compared with any other parasite tree?"

Cherry-picking and preferential host switching

We call preferential host switching the phenomenon where host switching to closer hosts is more likely to be successful than would be switching to more distantly related hosts [**13**]. This phenomenon, when it occurs, leads to an increased agreement between the host and parasite / pathogen trees, and suggests to the unwary that more pure codivergence has taken place than actually has. This agreement is concentrated at the tips, and leads to the obvious "cherry picking" test, as implemented in TreeMap. There is significant cophylogenetic match between the phylogenies of primates and lentiviruses [**13**]. This pattern is frequently attributed to codivergence [**22, 35**]. However, codivergence is unlikely, because it would require roughly a 10000$\times$ higher rate of divergence than has been estimated [**13**].

A more biologically plausible explanation allows for preferential host switching of lentiviruses between primate species: switching between more closely related hosts is more likely to be successful. The simulation test by Charleston and Robertson [**13**] in which only preferential host switching was included showed that a significant level of apparent codivergence could be attributed to this kind of host switching, with no recourse to codivergence. Clearly the true situation is likely to include codivergence and host switching, but the 2002 study exemplifies that the combination of statistical analysis and biological interpretation is paramount to accurate explanation of cophylogenetic models.

Techniques used in cophylogenetic modeling

The basic cophylogenetic relationship is an asymmetric one, in which the host — independent — tree H grows according to a standard Yule model (i.e., Markov, applied to tree growth), and the parasite — dependent — tree P is grown on H according to a simple stochastic model. The simplest possible models have a fixed probability of codivergence at each host vertex (which may be 1), and may include parameters to mimic the birth-death process that most tree growth models use. Huelsenbeck's likelihood model did not incorporate these parameters, but did include the possibility of host switching. Given the range of cophylogenetic events that can possibly be recovered, even under "ideal" conditions [**46**], we can only expect to model events that give rise to codivergence, duplication, host switching, extinction and missing the boat; the last two of which lead to 'loss' events.

More complex models may include further interactions between host and pathogen (or among pathogens, as in Charleston and Galvani's host/pathogen model [**14**], but these are aimed at recovering more subtle biological phenomena, for example the effects of variability of the parameter R_0 on coevolutionary behavior.

Failure to diverge

Ideally cophylogenetic models would include a failure to diverge event, that is, when the host diverges, the pathogen continues on both new lineages without itself diverging. This shortfall has been the source of considerable concern for phylogeneticists attempting to uncover ancient relationships among phylogenies, and is unfortunately accompanied by a shortfall in methodology. Consequently, even if we were to incorporate such things in our models, we still lack the computational tools to recover them. In the canonical approach we would let the range of a single pathogen lineage be a subset of the extant host lineages, rather than just a single lineage: this would increase the computational complexity of the mapping problem by $O(2^{n \times n})$ where n is the number of host lineages. Given the already prohibitive computational complexity of cophylogeny mapping, this is not a feasible approach: clever heuristics need to be found.

Cophylogenetic modeling

The simplest model of cophylogeny stems from Fahrenholz' comment [**19**] that parasite phylogeny mimics host phylogeny; in the 'perfect' case, P is congruent with H and there is essentially no problem to solve. When we incorporate other events into our model we must adopt a mechanism of non-codivergence, i.e., failure to speciate or missing the boat, and independent events on P, that is, birth (duplication) and death (extinction) events. Page's 'reconciled tree' analysis [**38**] did not use a coevolutionary model per se, but was consistent with a model containing all the above events. While the omission of host switching from the above method means that its application is limited to cases in which lateral transfer is so rare that it can be ignored, this did curtail the computational complexity very well: solutions to the reconciled tree problem can be found in polynomial time.

Charleston's model incorporates the essential elements of codivergence and host switching, coupled with the normal Yule model for the generation of phylogenetic trees. It is the simplest possible model that accommodates all the events that can be recovered in a cophylogenetic analysis [**46**], and its parameters relate well to the event costs for cophylogenetic inference. By assumption H is the independent tree, thus it can be constructed according to the standard model first, and then P (or P_i) "grown" along it. At each vertex h in H, if there is a pathogen occupying it then it is tested for codivergence with its host. On the branches of H we treat the occupying pathogens p as though they are in a birth-death process; duplication and extinction (respectively) have a fixed rate, and we determine which events, if any, occur on each pathogen branch in the time interval for that host branch. Host switching occurs with a fixed probability, conditional on there being a duplication event: that is, if there is a duplication event then with a fixed probability one of the nascent parasite/pathogen lineages will attempt to switch to a new host. Thus all the processes of codivergence, duplication, extinction, host switching and missing the boat are included, and correspond directly with the event costs. A complicating factor is that while there must be two parameters giving rise to loss events (three, if we permit taxa to be imperfectly sampled), there is only one apparent 'event' from which to infer these parameters. Without assuming a convenient fossil record however, there is nothing one can do about this.

4. Discussion and challenges for future research

The field of cophylogenetic inference is in its infancy, yet its growth is being curtailed by the dearth of researchers involved. However, it is also a very broad area with great potential: it has a huge range of applications, from estimating rates of emergent diseases in humans, to estimating dates of extinction events, to inferring gene function. It also draws on many different disciplines — from mathematics, statistics, optimization and algorithm design, through evolutionary modeling and host/pathogen and host/parasite interactions. This rich field must be tapped of its resources if we are to continue to learn how ecology and evolution interact.

Cophylogenetic methods have proliferated over the last decade, but there remains considerable need for refinement and elaboration of these techniques. Future work is required to test the statistical significance of cophylogenetic matches and to develop maximum likelihood procedures. As cophylogenetic methods and comparative genomic studies advance, progress is anticipated in the evaluation of coevolution of pathogens and their hosts. New insights will be provided into the selective forces that affect host species and their variation.

Likelihood implementation is challenging, but is an essential next step in the statistical analysis and evaluation of cophylogenies. A viable likelihood model has yet to be developed. Huelsenbeck and colleagues' efforts paved the way [**27, 28**] but there has been little progress in ML methods in cophylogeny since. This is a non-trivial task and will require substantial dedication on the part of one or a few researchers, but it is an important task if we are to develop coherent, statistically sound analysis of cophylogeny. The first step will be to establish a likelihood formula that permits more than one pathogen per host and incorporates birth and death processes. Next will be to find an expression to take into account those invisible events which can never be traced: duplications in which all the descendants on one lineage perish, host switches that do not persist, etc. While challenging, we feel that this is a very fruitful goal.

Multiple methods may be used on the same data set. In some cases, they will all agree, such as on the extensive codivergence between Halipeurus lice on seabirds, increasing confidence in the results [**42**]. These results are further confirmed with seabird and louse biology. Seabirds of different species seldom interact, lice survive only briefly off the host, and louse transmission is limited to mating and brooding of chicks [**42**].

Cophylogenetic models can be most informatively interpreted in conjunction with information from other biological and biogeographical sources. Host transfer experiments are one approach to directly measuring the survival and transmission of pathogens that undergo host switches.

Currently the methods that allow reconstruction of all cophylogenetic events do not directly take into account the robustness of underlying data, although for example the original sequence data could be resampled (with or without replacement) to at least estimate the distribution of most likely trees, and reconciliation analysis performed for each replicate.

Requirements for successful cophylogenetic analysis

Accuracy of host and pathogen phylogenies is paramount. The more exhaustively sampled the host and pathogen clades the better, which is usually more feasible when sampling a closely related clade, such as a genus rather than a larger clade such as an order. Sorting events and duplications are more prone to be missed or misinterpreted when sampling is not exhaustive.

Reticulate phylogenies have not yet been implemented. Though there are very few cases of true reticulate evolution of parasite or host organisms, there are many more frequent instances of genetic recombination. A reticulate gene phylogeny can in principle be reconciled with the host (organismal) tree, but while the theory is at least drafted [**9**], the method is not available in any computer program. Likewise if the host phylogeny is reticulate, cophylogeny mapping is still theoretically possible, but at some expense: a branch of P whose beginning and end are mapped to locations prior and subsequent to a reticulate part of H may not have a uniquely determined pattern of associations with the host phylogeny. Reticulation in neither host nor parasite phylogeny is accommodated in current cophylogeny software.

Multi-host pathogens

Multi-host pathogens are those that can infect multiple hosts (Banks and Paterson 2004). Many indirectly transmitted pathogens fall into this category. The canonical mapping approach would have us permit a subset of the extant vertices and arcs in H as possible locations for each parasite, but that is computationally unfeasible: more sophisticated methods are needed. The best we can hope for in such circumstances, at present at least, is a heuristic or greedy algorithm.

How good can we expect reconstruction methods to be?

The efficacy with which we can expect to recover historical coevolutionary events is unknown. Remarkably, given the time and effort that has been spent on phylogenetic methods, no tests have been made on the frequency with which we can recover codivergence, duplication, host switch and loss events. We do not know at what frequency of host switching the codivergence events will be masked, nor the masking effects of duplication and loss. We do know that given perfect agreement between phylogenies, every cophylogeny method in use will infer a history of codivergence; also if the disagreement is not too great, and if we do not miss out any taxa, then we have a good chance of recovering the true history as one of the Pareto optimal maps. The overwhelming difficulty in cophylogenetic inference is currently the problem of interpretation: at the "end" of an analysis one is left with a set of Pareto optimal maps, the testing of whose quality is a subtle and computationally expensive exercise.

Sampling

To what extent does taxon sampling affect the efficacy of reconstruction? Taxon sampling has an enormous effect on the ability of cophylogenetic inference methods to recover the true history of associations as Jackson discovered [**30**]. As we have mentioned, exhaustive sampling of the host and pathogen clades aids the analyses, and is usually more feasible when sampling a closely related clade such as a genus, rather than a larger clade such as an order. Sorting events and duplications are more prone to be missed or misinterpreted when sampling is not exhaustive. Also in the case of ancient duplications, giving rise to paralogous sets of pathogens, poor sampling can imply much more incongruence than is truly present, manifesting as inflated numbers of host switches. However quantitative analyses of these effects remain to be performed; progress is hampered by the scarcity of tools.

Complexity

The number of solutions to the cophylogeny problem increases dramatically with the number of host and pathogen taxa (Charleston 2002). While the jungle is a useful tool enabling the complete elucidation of the solution space for this problem, it remains slow: intelligent heuristics must be devised in order to find at least some of the POpt solutions in a reasonable amount of time, rather than guaranteeing to find them all. Because TreeMap creates a bounded jungle, the search for POpt can be made more efficient by beginning with very tight bounds, e.g., restricting the solutions found to have at least 10 codivergence events, and then relaxing those bounds. If no solution can be found with 10 codivergences, then perhaps a jungle can be constructed with at least 8 codivergence events, and so on. Thus the complexity can be worked around to a certain extent, though the underlying issue remains.

Branch-and-bound search

Only very broad bounds exist for the cophylogeny mapping problem, so the exhaustive branch and bound search for POpt solutions in TreeMap is still prohibitively slow for more than a few taxa. Better theoretical bounds on the maximal numbers of events for a given instance of the cophylogeny problem will be a huge help, but none are forthcoming as yet. Given that optimal numbers of events are dependent on the tree shape (its topology to purists), and that the number of tree shapes increases exponentially with the number of leaves of a tree, it will not be possible to get tight bounds on the numbers of events without taking into account the tree shape in some algorithmic way such as traversing the host tree.

Pareto optimality and parameter finding

As Charleston pointed out [**10**] there are severe limitations to estimating parameter values in a coevolutionary system that yields very widely distributed behaviors for the same set of generating parameters. That is, if one set of parameters gives a very broad range of possible behaviors, then it becomes very difficult to infer a set of parameters from a single behavior. For example with a stochastic model in which prob.(codivergence)= 0.75, prob.(host switch)= 0.25, rate(duplication)= 1 = rate(extinction), we may find anything from 0% to 100% of the host vertices involved with actual codivergence events.

Incorporating epidemiology

Epidemiological dynamics underlie coevolutionary events between pathogens and their hosts. The key determinant of epidemiological dynamics and the likelihood that a zoonotic event will lead to a persistent host switching process is the basic reproductive number, R_0, which is defined as the number of secondary cases generated by an initial infection in a completely susceptible population [**1**]. Thus, an important next step is incorporating key determinants of epidemiological dynamics into cophylogenetic models.

Cophylogenetic techniques may provide a powerful approach for inferring the evolutionary history of clinically important loci in hosts and pathogens, such as receptor exploitation by pathogens and receptor evolution. Furthermore, integrating population genetics and epidemiology may have the potential to yield insight into the dynamic interplay between hosts and their pathogens and the epidemiological history of humans and of other species.

Acknowledgements

We sincerely thank Joe Felsenstein and another, anonymous, reviewer for their useful critiques, which significantly contributed toward improving this manuscript.

References

[1] R. Anderson and R. May, *Infectious Diseases of Humans: Dynamics and Control*, Oxford University Press, Oxford, 1991.

[2] R. Antia, R. R. Regoes, J. C. Koella, and C. T. Bergstrom, The role of evolution in the emergence of infectious diseases, *Nature* **426** (2003), 658–661.

[3] Banks and Paterson, in review, 2004.

[4] D. R. Brooks, Hennig's parasitological method: a proposed solution, *Syst. Zool.* **30** (1981), 229–249.

[5] D. R. Brooks, Macroevolutionary comparisons of host and parasite phylogenies, *Annu. Rev. Ecol. Syst.* **19** (1988), 235–259.

[6] D. R. Brooks and R. T. O'Grady, Crocodilians and their helminth parasites: macroevolutionary considerations, *Am. Zool.* **29** (1989), 873–883.

[7] D. R. Brooks, How to do BPA, really, *J. Biogeography* **28** (2001), 345–358.

[8] M. A. Charleston, Jungles: A new solution to the host/parasite phylogeny reconciliation problem, *Math. Biosci.* **149** (1998), 191–223.

[9] M. A. Charleston, Principles of cophylogeny maps, in (M. Lässig and A. Valleriani, ed.), *Biological Evolution and Statistical Physics*, Springer-Verlag, New York, 2002.

[10] M. A. Charleston, Recent advances in cophylogeny mapping, in (D. T. J. Littlewood, ed.), *Advances in Parasitology* **54**, Elsevier Academic Press, Amsterdam, 2003.

[11] M. A. Charleston and R. D. M. Page, TreeMap program, Macintosh program, available from "http://www.taxonomy.gla.ac.uk/software/treemap.html", 2002.

[12] M. A. Charleston and S. L. Perkins, Lizards, malaria, and jungles in the Caribbean, in (R. D. M. Page, ed.), *Tangled Trees: Phylogeny, Cospeciation, and Coevolution*, University of Chicago Press, Chicago, 2002, 65–92.

[13] M. A. Charleston and D. L. Robertson, Preferential host switching by primate lentiviruses can account for phylogenetic similarity with the primate phylogeny, 2002.

[14] M. A. Charleston and A. P. Galvani, Cophylogenetic modelling of hosts and pathogens, in preparation, 2005.

[15] Z. Chen, P. Zhou, D. D. Ho, N. R. Landau, and P. A. Marx, Genetically divergent strains of simian immunodeficiency virus use CCR5 as a coreceptor for entry, *J. Virol.* **71** (1997), 2705–2714.

[16] Z. Chen, D. Kwon, Z. Jin, S. Monard, P. Telfer, M. S. Jones, C. Y. Lu, R. F. Aguilar, D. D. Ho, and P. A. Marx, Natural infection of a homozygous delta24 CCR5 red-capped mangabey with an R2b-tropic simian immunodeficiency virus, *J. Exp. Med.* **188** (1998), 2057–2065.

[17] M. A. Clark, N. A. Moran, P. Baumann, and J. J. Wernegreen, Cospeciation between bacterial endosymbionts (Buchnera) and a recent radiation of aphids (Uroleucon) and pitfalls of testing for phylogenetic congruence, *Evolution* **54** (2000), 517–525.

[18] J. S. Farris, M. Källersjö, A. G. Kluge, and C. Bult, Testing significance of congruence, *Cladistics* **10** (1994), 315–320.

[19] H. Fahrenholz, Ectoparasiten und Abstammungslehre, *Zoologischer Anzeiger* **41** (1913), 371–374.

[20] A. P. Galvani and M. W. Slatkin, Evaluating plague and smallpox as historical selective pressures for the CCR5-?32 HIV-resistance allele, *Proc. Natl. Acad. Sci. USA* **100** (2003), 15276–15279.

[21] A. P. Galvani and J. P. Novembre, The evolutionary history of the CCR5-?32 HIV-resistance mutation, *Microbes and Infection*, in press.

[22] F. Gao, E. Bailes, D. L. Roberston, Y. Chen, C. M. Rodenburg, S. F. Michael, L. B. Cummins, L. O. Arthur, M. Peeters, G. Shaw, P. M. Sharp, and B. H. Hahn, Origin of HIV-1 in chimpanzee Pan troglodytes troglodytes, *Nature* **397** (1999), 436–441.

[23] M. S. Hafner and S. A. Nadler, Phylogenetic trees support the coevolution of parasites and their hosts, *Nature* **332** (1988), 258–259.

[24] M. S. Hafner and S. A. Nadler, Cospeciation in host-parasite assemblages: gophers and lice as a model system, *Syst. Zool.* **39** (1990), 192–204.

[25] B. H. Hahn, G. M. Shaw, K. M. De Cock, and P. M. Sharp, AIDS as a Zoonosis: Scientific and public health implications, *Science* **287** (2000), 607–614.

[26] E. Hoberg, D. R. Brooks, and D. Siegel-Causey, Host-parasite co-speciation: history, principles, and prospects, in (D. H. Clayton and J. Moore, eds.), *Host-Parasite Evolution: General Principles and Avian Models*, Oxford University Press, Oxford, 1997, 212–235.

[27] J. P. Huelsenbeck and B. Rannala, Phylogenetic methods come of age: testing hypotheses in an evolutionary context, *Science* **276** (1997), 227–232.

[28] J. P. Huelsenbeck, B. Rannala, and B. Larget, A Bayesian framework for the analysis of cospeciation, *Evolution* **54** (2000), 352–364.

[29] C. J. Humphries, J. M. Cox, and E. S. Nielsen, Nothofagus and its parasites: a cladistic approach to coevolution, in (A. R. Stone and D. L. Hawksworth, eds.), *Coevolution and Systematics*, Clarendon Press, Oxford, 1986, 77–91.

[30] A. Jackson, Doctoral dissertation, Oxford University, Oxford, 2004.

[31] A. Jackson and M. A. Charleston, A cophylogenetic perspective of RNA-virus evolution *Math. Biosci. Engineering*, **21**(1) (2004), 45–57.

[32] K. P. Johnson, D. M. Drown, and D. H. Clayton, A data based parsimony method of cophylogenetic analysis *Zool.*, scripta in press, 2001.

[33] S. Kossida, P. H. Harvey, P. de A. Zanotto, and M. A. Charleston, Lack of evidence for cospeciation between retroelements and their hosts, *JME* **50**(2) (2000), 194–201.

[34] J. Martin, E. Herniou, J. Cook, R. Waugh O'Neill, and M. Tristem, Interclass transmission and phyloetic host tracking in Murin Leukemia virus-related retroviruses, *J. Virol.* **73** (1999), 2442–2449.

[35] M. C. Muller, N. K. Saksena, E. Nerrienet, C. Chappey, V. M. Herve, J. P. Durand, P. Legal-Campodonico, M. C. Lang, J. P. Digoutte, and A. J. Georges, Simian immunodeficiency viruses from central and western Africa: evidence for a new species-specific lentivirus in tantalus monkeys, *J. Virol.* **67** (1993), 1227–1235.

[36] R. D. M. Page, Temporal congruence and cladistic analysis of biogeography and cospeciation, *Systematic Zoology* **39** (1990), 205–226.

[37] R. D. M. Page, Clocks, clades, and cospciation: comparing rates of evolution and timing of cospeciation events in host-parasite assemblages, *Systematic Zoology* **40** (1991), 188–198.

[38] R. D. M. Page, Parallel phylogenies: Reconstructing the history of host-parasite assemblages, *clad* **10** (1994), 155–173.

[39] R. D. M. Page and M. A. Charleston, Trees within trees: phylogeny and historical associations, *Trends Ecol. Evol.* **13** (1998).

[40] E. Palacios, L. Digilio, H. M. McClure, Z. Chen, P. A. Marx, M. A. Goldsmith, and R. M. Grant, Parallel evolution of CCR5-null phenotypes in humans and in a natural host of simian immunodeficiency virus, *Current Biology* **8** (1998), 943–946 and S1–S3.

[41] A. M. Paterson and R. Poulin, Have chondracanthid copepods cospeciated with their teleost hosts? *Syst. Parasitol.* **44** (1999), 79–85.

[42] A. M. Paterson and J. Banks, Analytical approaches to measuring cospeciation of host and parasites: through a glass, darkly, *Int. J. Parasitology* **31** (2001), 1012–1022.

[43] A. M. Paterson, R. L. Palma, and R. D. Gray, Drowning on arrival, missing the boat and x-events: How likely are sorting events? in (R. D. M. Page, ed.), *Tangled Trees: Phylogeny, Cospeciation, and Coevolution*, University of Chicago Press, Chicago, 2002, 287–309.

[44] A. S. Peek, R. A. Feldman, R. A. Lutz, and R. C. Vrijenhoek, Cospeciation of chemoautotrophic bacteria and deep sea clams, *Proc. Natl. Acad. Sci. USA* **95** (1998), 9962–9966.

[45] H. C. Proctor, Gallilichus jonesi sp. N. (Acari: Ascouracaridae): a new species of feather mite from the quills of the Australian brush-turkey (Aves: Megapodiidae), *Aust. J. Entomol.* **38** (1999), 77–84.

[46] F. Ronquist, Parsimony analysis of coevolving species associations, in (R. D. M. Page, ed.), *Tangled Trees: Phylogeny, Cospeciation, and Coevolution*, University of Chicago Press, Chicago, 2002, 22–64.

[47] P. M. Sharp, E. Bailes, F. Gao, B. E. Beer, V. M. Hirsch, and B. H. Hahn, Origins and evolution of AIDS viruses: estimating the time-scale, *Biochem. Soc. Trans.* **28** (2000), 275–282.

[48] D. L. Swofford, *PAUP*: Phylogenetic Analysis Using Parsimony*, Sinauer Associates, Inc. Publishers, Sunderland, 2001.

School of Information Technology, Madsen Building, University of Sydney, Sydney NSW 2006, Australia

E-mail address: `mcharleston@it.usyd.edu`

Department of Epidemiology and Public Health, Yale University School of Medicine, New Haven, CT 06520, USA

E-mail address: `alison.galvani@yale.edu`

DIMACS Series in Discrete Mathematics
and Theoretical Computer Science
Volume **71**, 2006

The influence of anti-viral drug therapy on the evolution of HIV-1 pathogens

Zhilan Feng and Libin Rong

ABSTRACT. An age-structured model is used to study the possible impact of drug treatment of infections with the human immunodeficiency virus type 1 (HIV-1) on evolution of the pathogen. Inappropriate drug therapy often leads to the development of drug-resistant mutants of the virus. Previous studies have shown that natural selection within a host favors viruses that maximize their fitness. By demonstrating how drug therapy may influence the within host viral fitness we show that while a higher treatment efficacy reduces the fitness of the drug-sensitive virus, it may provide a stronger force of selection for drug-resistant viruses allowing a wider range of resistant strains to invade.

1. Introduction

Reverse transcriptase inhibitors and protease inhibitors are the two major types of drugs that have been used as inhibitors of HIV-1 replication *in vivo*. Mathematical models of HIV-1 infection under the impact of drug treatments have been studied using ordinary and/or delay differential equations (see, for example [**4, 8, 11**]). Nelson et al. developed an age-structured model of HIV-1 infection (without drug treatments) and showed that this model is a generalization of ODE and DDE models mentioned above in the absence of treatments [**9**]. Such generalized models are considered to have greater flexibility that may better represent the underlying biology of an infection [**2**]. In [**9**] the local stability of both the infection-free and the infected steady states are shown for the case when the viral production rate has a special functional form (see Eq. (2.2)). These results are applied to a similar age-structured model in [**2**] to study the effect of life history parameters of HIV-1 on maximizing within host viral fitness under various assumptions on trade-offs between the virion production rate and other parameters. Drug treatments are again not included in this model.

In this chapter we generalize the model in [**9**] by incorporating the effect of two classes of anti-HIV drugs which help to reduce the HIV replication at two different stages of the cell infection. One class is the reverse transcriptase inhibitor which

Key words and phrases. HIV-1, Combination therapy, Drug-resistance, Optimal viral fitness, Age-structured model, Stability analysis.

This work is supported in part by NSF grant DMS-0314575 and by James S. McDonnell Foundation 21st Century Science Initiative.

blocks the translation of viral RNA into DNA so that CD4 cells cannot produce virion particles (the cells then become uninfected). The other class is the protease inhibitor which targets the HIV protease enzyme to prevent new copies of HIV from being made (so that some of the virion particles will remain non-infectious). Stability results are provided for a general form of the viral production rate, and the stability of the infection-free or the infected steady state is shown to depend on the reproductive ratio $\mathcal{R}$ being smaller or greater than 1. The formulation of this reproductive ratio also provides an appropriate measure for the within host viral fitness, which can be used to explore the optimal virion production rate for which $\mathcal{R}$ is maximized.

Recent clinical studies have suggested that prolonged treatment with a single anti-HIV drug may be responsible for the emergence of resistant virus [**3, 4, 5, 6, 10, 12**]. The impact of drug treatments on the dynamics of resistant stains of pathogens has been studied using age-independent mathematical models (see, for example, [**1, 4, 13**]). We show that if the viral production is linked to resistance, then higher treatment efficacy with antiretroviral agents (such as protease inhibitors) may lead to the establishment of multiple viral strains with a wider range of resistance levels.

This chapter is organized as follows: In Section 2, we formulate a mathematical model for HIV-1 infection which generalizes the age-structured model proposed in [**9**] by incorporating both types of drug treatments. Section 3 is devoted to the analysis of our model including the existence and stability of both the infection-free and the infection steady states. In Section 4, we derive a criterion for invasion by resistant strains and explore how drug treatments may affect the optimal viral fitness of resistant strains. Section 5 discusses the results.

2. Model formulation

In [**9**] the following age-structured model of HIV infection is proposed:

$$\begin{aligned}
&\frac{d}{dt}T(t) = s - dT - kVT, \\
&\frac{\partial}{\partial t}T^*(a,t) + \frac{\partial}{\partial a}T^*(a,t) = -\delta(a)T^*(a,t), \\
&\frac{d}{dt}V(t) = \int_0^\infty p(a)T^*(a,t)da - cV, \\
&T^*(0,t) = kVT,
\end{aligned} \tag{2.1}$$

with appropriate initial conditions. Here, $T(t)$ denotes the population of uninfected target T cells at time t, $T^*(a,t)$ denotes the density of infected T cells of infection age a (i.e. the time that has elapsed since an HIV virion has penetrated the cell) at time t, and $V(t)$ denotes the population of infectious virus at t. s is the recruitment rate of healthy T cells, d is the per capita death rate of uninfected cells, $\delta(a)$ is the age-dependent per capita death rate of infected cells, c is the clearance rate of an infectious virus, k is the rate at which an uninfected cell becomes infected by one infectious virus, and $p(a)$ is the virion production rate by an infected cell of age a.

One of the special forms of $p(a)$ considered in Model (2.1) is

$$p(a) = \begin{cases} p^* \left(1 - e^{-\theta(a-a_1)}\right) & \text{if } a \geq a_1, \\ 0 & \text{else} \end{cases} \tag{2.2}$$

where θ is a constant which determines how quickly $p(a)$ reaches the saturation level, p^*, and a_1 is the maximum age at which reverse transcription takes place.

We provide stability results in this chapter for an arbitrary function $p(a)$ which allows the possibility that an infected cell may start producing viruses before the age a_1 (i.e., in some infected cells the reverse transcription may occur earlier than a_1) and the possibility that $p(a)$ may not be a monotone function of a (e.g., it may have a peak production rate at some intermediate age). To account for the fact that an infected cell does not produce any virus before the reverse transcription has taken place we introduce a function $\beta(a)$ which describes the proportion of infected cells of age a that are not yet actively productive. Assume that $\beta(a) \in L^1[0,\infty)$ is a non-increasing function with the following properties:

$$0 \leq \beta(a) \leq 1, \quad \beta(0) = 1. \tag{2.3}$$

Then, $\beta(a)$ can be used to divide the class of infected cells, $T^*(a,t)$, into two subclasses, $T^*_{preRT}(a,t)$ and $T^*_{postRT}(a,t)$ which are defined by:

$$\begin{aligned} T^*_{preRT}(a,t) &= \beta(a)T^*(a,t), \\ T^*_{postRT}(a,t) &= (1-\beta(a))T^*(a,t). \end{aligned} \tag{2.4}$$

$T^*_{preRT}(a,t)$ represents the density of cells that have been "infected" by an HIV virion but reverse transcription has not been completed at infection age a, and an RT inhibitor could revert it back to uninfected class (because RT fails to occur) or reduce the probability that a preRT cell progresses to the postRT state. $T^*_{postRT}(a,t)$ represents the density of infected cells that have progressed to the postRT phase at infection age a, and the presence of a protease inhibitor could affect the rate at which new infectious virion particles are produced.

Let r_{rt} and r_p denote the efficacy of the treatment therapy with reverse transcriptase inhibitors and protease inhibitors, respectively ($0 \leq r_{rt},\ r_p \leq 1$), and let η denote the maximal (age-independent) per capita rate at which preRT cells of become uninfected. Then the rate at which preRT cells become uninfected is given by

$$\int_0^\infty r_{rt}\eta T^*_{preRT}(a,t)da,$$

and new infectious virion particles are produced at the rate

$$\int_0^\infty (1-r_p)p(a)T^*_{postRT}(a,t)da.$$

Incorporating these drug treatments in the equations for T, T^* and V in Model (2.1) we have:

$$\frac{d}{dt}T(t) = s - dT - kVT + \int_0^\infty r_{rt}\eta T^*_{preRT}(a,t)da,$$

$$\frac{\partial}{\partial t}T^*(a,t) + \frac{\partial}{\partial a}T^*(a,t) = -\delta(a)T^*(a,t) - r_{rt}\eta T^*_{preRT}(a,t),$$

$$\frac{d}{dt}V(t) = \int_0^\infty (1-r_p)p(a)T^*_{postRT}(a,t)da - cV.$$

Thus, using the relation (2.4) we modify Model (2.1) to get the following model:

$$\begin{aligned}
&\frac{d}{dt}T(t) = s - dT - kVT + \int_0^\infty r_{rt}\eta\beta(a)T^*(a,t)da,\\
&\frac{\partial}{\partial t}T^*(a,t) + \frac{\partial}{\partial a}T^*(a,t) = -\delta(a)T^*(a,t) - r_{rt}\eta\beta(a)T^*(a,t),\\
&\frac{d}{dt}V(t) = \int_0^\infty (1-r_p)(1-\beta(a))p(a)T^*(a,t)da - cV,\\
&T^*(0,t) = kVT,\\
&T(0) = T_0 > 0, \quad T^*(a,0) = T_0^*(a) \geq 0, \quad V(0) = V_0 > 0.
\end{aligned} \tag{2.5}$$

As mentioned earlier, we allow the virion production rate $p(a)$ to be an arbitrary function (e.g., it does not have to be a monotone function). $p(a)$ and $\delta(a)$ are assumed to be bounded.

System (2.5) can be reformulated as a system of Volterra integral equations. To simplify expressions we introduce the following notation:

$$\begin{aligned}
&K_0(a) = e^{-\int_0^a(\delta(\tau)+r_{rt}\eta\beta(\tau))d\tau}\\
&K_1(a) = r_{rt}\eta\beta(a)K_0(a)\\
&K_2(a) = (1-r_p)(1-\beta(a))p(a)K_0(a)\\
&\mathcal{K}_i = \textstyle\int_0^\infty K_i(a)da, \quad i = 0,1,2.
\end{aligned} \tag{2.6}$$

$K_0(a)$ is the probability of an infected cell remaining infected at age a, hereafter the age-specific survival probability of an infected cell. $K_1(a)$ gives the probability that an infected cell becomes noninfected at age a given that the cell has not died at age a. The probability that an infected cell has not died at age a is

$$e^{-\int_0^a \delta(\tau)d\tau}. \tag{2.7}$$

The probability that an infected cell is still in the preRT stage and has not been treated at age a by an RT inhibitor is $e^{-\int_0^a r_{rt}\eta\beta(\tau)d\tau}$ and hence the probability that the cell becomes noninfected at age a due to an RT inhibitor is

$$\frac{d}{dt}\left(1 - e^{-\int_0^a r_{rt}\eta\beta(\tau))d\tau}\right) = r_{rt}\eta\beta(a)e^{-\int_0^a r_{rt}\eta\beta(\tau))d\tau}. \tag{2.8}$$

$K_1(a)$ is the product of the two probabilities given by (2.7) and (2.8). $K_2(a)$ is a product of the the age-specific survival probability of an infected cell and the rate, $(1-r_p)(1-\beta(a))p(a)$, at which infectious virion particles are produced by an actively productive cell of age a. Thus, the integral of $K_2(a)$ over all ages, i.e.,

$$\mathcal{K}_2 = \int_0^\infty (1-r_p)(1-\beta(a))p(a)K_0(a)da$$

gives the total amount of infectious virion particles produced by one infected cell in its lifespan. This is an important quantity which will be used later to define the viral fitness.

For mathematical convenience we introduce the new variable, $B(t)$, to describe the rate at which an uninfected T cell becomes infected at time t,

$$B(t) = kV(t)T(t). \tag{2.9}$$

Integrating the T^* equation in System (2.5) along the characteristic lines, $t - a =$ constant, we get the following formula

$$T^*(a,t) = \begin{cases} B(t-a)K_0(a) & \textit{for } a < t, \\ T_0^*(a-t)\dfrac{K_0(a)}{K_0(a-t)} & \textit{for } a \geq t. \end{cases} \tag{2.10}$$

Substituting (2.10) into the T and V equations in (2.5):

$$\begin{aligned} \frac{d}{dt}T(t) &= s - dT - B(t) + \int_0^t K_1(a)B(t-a)da + \tilde{F}_1(t), \\ \frac{d}{dt}V(t) &= \int_0^t K_2(a)B(t-a)da - cV + \tilde{F}_2(t), \end{aligned} \tag{2.11}$$

where

$$\begin{aligned} \tilde{F}_1(t) &= \int_t^\infty r_{rt}\eta\beta(a)T_0^*(a-t)\frac{K_0(a)}{K_0(a-t)}da, \\ \tilde{F}_2(t) &= \int_t^\infty (1-r_p)(1-\beta(a))p(a)T_0^*(a-t)\frac{K_0(a)}{K_0(a-t)}da. \end{aligned} \tag{2.12}$$

Clearly, $\tilde{F}_1(t) \to 0$ as $t \to \infty$. Integrating the T equation in (2.11) and changing the order of integration:

$$\begin{aligned} T(t) &= T_0 e^{-dt} + \int_0^t e^{-d(t-u)}\Bigg[s - B(u) \\ &\quad + \int_0^u B(u-\tau)K_1(\tau)d\tau + \tilde{F}_1(u)\Bigg]du \\ &= \int_0^t \left[e^{-d(t-u)}(s - B(u)) + B(u)H_1(t-u)\right]du + F_1(t), \end{aligned} \tag{2.13}$$

where

$$H_1(t) = e^{-dt}\int_0^t e^{d\tau}K_1(\tau)d\tau,$$

$$(2.14)\qquad F_1(t) = T_0e^{-dt} + \int_0^t e^{-d(t-u)}\tilde{F}_1(u)du.$$

For the derivation of (2.13) we have used the following fact:

$$\begin{aligned}
&\int_0^t e^{-d(t-u)}\int_0^u B(u-\tau)K_1(\tau)d\tau du \\
&= \int_0^t e^{-d(t-u)}\int_0^u B(\alpha)K_1(u-\alpha)d\alpha du \\
&= \int_0^t B(\alpha)\int_\alpha^t e^{-d(t-u)}K_1(u-\alpha)dud\alpha \\
&= \int_0^t B(\alpha)e^{-d(t-\alpha)}\int_0^{(t-\alpha)} e^{d\sigma}K_1(\sigma)d\sigma d\alpha \\
&= \int_0^t B(\alpha)H_1(t-\alpha)d\alpha.
\end{aligned}$$

Similarly, by integrating the V equation in (2.11) we get

$$(2.15)\qquad \begin{aligned}
V(t) &= V_0e^{-ct} + \int_0^t e^{-c(t-u)}\left[\int_0^u B(u-\tau)K_2(\tau)d\tau + \tilde{F}_2(u)\right]du \\
&= \int_0^t B(u)H_2(t-u)du + F_2(t),
\end{aligned}$$

where

$$H_2(t) = e^{-ct}\int_0^t e^{c\tau}K_2(\tau)d\tau,$$

$$(2.16)\qquad F_2(t) = T_0e^{-ct} + \int_0^t e^{-c(t-u)}\tilde{F}_2(u)du.$$

Equations (2.13) and (2.15), with $B(t)$ replaced by $kV(t)T(t)$, form a system of Volterra integral equations which is equivalent to the original system (2.5). Hence, for the discussion of existence and uniqueness of the solutions we only need to consider the following system

$$(2.17)\qquad \begin{aligned}
T(t) &= \int_0^t \left[e^{-d(t-u)}(s - kV(u)T(u)) + kV(u)T(u)H_1(t-u)\right]du \\
&\quad + F_1(t), \\
V(t) &= \int_0^t kV(u)T(u)H_2(t-u)du + F_2(t),
\end{aligned}$$

where H_i and F_i ($i = 1, 2$) are given in (2.14) and (2.16).

3. Model analysis

In this section we provide analytic results on the existence of positive solutions as well as possible steady states and their stability.

3.1. Existence of positive solutions. Let $x(t) = (T(t), V(t))$. System (2.17) can be written in the form

$$x(t) = \int_0^t \kappa(t-u)g(x(u))du + f(t),$$

where $f(t) = (F_1(t), F_2(t))$ is a continuous function from $[0,\infty)$ to $[0,\infty)^2$, κ is the following 2×2 matrix with entries being locally integrable functions on $[0,\infty)$,

$$\kappa(t) = \begin{pmatrix} se^{-dt} & H_1(t) - e^{-dt} \\ 0 & H_2(t) \end{pmatrix},$$

and g is defined by,

$$g(x) = (1, kVT).$$

Obviously, $f \in C([0,\infty); \mathbf{R}^2)$, $g \in C(\mathbf{R}^2, \mathbf{R}^2)$, and $\kappa \in L^1_{loc}([0,\infty); \mathbf{R}^{2\times 2})$. Theorem 1.1 in Gripenberg *et al.* (1990), Section 12.1, now provides us with a continuous solution defined on a maximal interval such that the solution goes to infinity if this maximal interval is finite.

To see that all solutions will remain non-negative for positive initial data, we use the following system (see (2.9) and (2.11)) which is also equivalent to System (2.5):

$$\begin{aligned}
&\frac{d}{dt}T(t) = s - dT - B(t) + \int_0^t K_1(a)B(t-a)da + \tilde{F}_1(t), \\
(3.1) \qquad &\frac{d}{dt}V(t) = \int_0^t K_2(a)B(t-a)da - cV + \tilde{F}_2(t), \\
&B(t) = kV(t)T(t),
\end{aligned}$$

where $\tilde{F}_i$ is given in Eq. (2.12) and $\tilde{F}_i(t) > 0$, $\lim_{t\to\infty} \tilde{F}_i(t) = 0$ for $i = 1, 2$.

Suppose that there exists a $\bar{t} > 0$ such that $T(\bar{t}) = 0$ and $T(t), V(t) > 0$ for $0 \le t < \bar{t}$. Then $B(\bar{t}) = kV(\bar{t})T(\bar{t}) = 0$, $B(t) = kV(t)T(t) > 0$ for $0 \le t < \bar{t}$, and thus from the T equation in (3.1) we have

$$\frac{d}{dt}T(\bar{t}) = s + \int_0^{\bar{t}} K_1(a)B(\bar{t}-a)da + \tilde{F}_1(\bar{t}) > 0.$$

Hence, $T(t) \ge 0$ for all $t \ge 0$. Similarly we can show that $V(t) \ge 0$ and $B(t) \ge 0$ for all $t \ge 0$ and for all positive initial data.

3.2. Steady states and their stability. We use System (3.1) for our stability analysis. According to [**7**], any equilibrium of System (3.1), if it exists, must be a constant solution of the limiting system:

$$
\begin{aligned}
&\frac{d}{dt}T(t) = s - dT - B(t) + \int_0^\infty K_1(a)B(t-a)da,\\
(3.2)\qquad &\frac{d}{dt}V(t) = \int_0^\infty K_2(a)B(t-a)da - cV,\\
&B(t) = kV(t)T(t).
\end{aligned}
$$

System (3.2) has two constant solutions: the infection-free steady state

$$\bar{E} = (\bar{T}, \bar{V}, \bar{B}) = (\frac{s}{d}, 0, 0), \tag{3.3}$$

and the infected steady state

$$
\begin{aligned}
&E^\diamond = (T^\diamond, V^\diamond, B^\diamond) \qquad \text{where}\\
(3.4)\qquad &T^\diamond = \frac{c}{k\mathcal{K}_2}, \qquad V^\diamond = \frac{sk\mathcal{K}_2 - dc}{kc(1-\mathcal{K}_1)}, \quad B^\diamond = kT^\diamond V^\diamond
\end{aligned}
$$

with $\mathcal{K}_1$ and $\mathcal{K}_2$ given in (2.6). Notice that $\beta(a) = 0$ for $a \geq a_1$ and

$$
\begin{aligned}
\mathcal{K}_1 &< \int_0^\infty r_{rt}\eta\beta(a)e^{-\int_0^a r_{rt}\eta\beta(s)ds}da\\
&= -\int_0^\infty \frac{d}{da}\left(e^{-\int_0^a r_{rt}\eta\beta(s)ds}\right)da\\
&= 1 - e^{-\int_0^\infty r_{rt}\eta\beta(s)ds}\\
&\leq 1.
\end{aligned}
$$

Thus, $V^\diamond > 0$ if and only if $sk\mathcal{K}_2 - dc > 0$, or $\mathcal{R} > 1$, where

$$\mathcal{R} = \frac{sk\mathcal{K}_2}{dc}. \tag{3.5}$$

Clearly, the infected steady state (3.4) is feasible if and only if $\mathcal{R} > 1$. We can interpret $\mathcal{R}$ by noticing that s/d is the cell density in the absence of infection; k and c are the cell infection and viral clearance rates, respectively; and $\mathcal{K}_2$ gives the infectious virion particles produced by one infected cell during its entire life. Therefore, $\mathcal{R}$ is the reproductive ratio of the virus under the impact of drug treatments.

We now consider the stability of steady states. Let us first consider the infection-free steady state $\bar{E}$. The following result suggests that the population sizes of virus and infected cells will go to zero as $t \to \infty$ if the reproductive ratio is less than 1.

Result 1 *$\bar{E}$ is locally asymptotically stable if $\mathcal{R} < 1$, and it is unstable if $\mathcal{R} > 1$.*

This stability result can be verified as the follows. Using System (3.2) we get the characteristic equation at the steady state $\bar{E}$

$$\det\begin{bmatrix} -d-\lambda & -ks/d & \hat{K}_1(\lambda)\\ 0 & -c-\lambda & \hat{K}_2(\lambda)\\ 0 & ks/d & -1\end{bmatrix} = 0, \tag{3.6}$$

where λ is an eigenvalue and $\hat{K}_i(\lambda)$ denotes the Laplace transform of $K_i(a)$, i.e.,

$$\hat{K}_i(\lambda) = \int_0^\infty K_i(a)e^{-\lambda a}da, \quad i = 1, 2. \tag{3.7}$$

The characteristic equation (3.6) can be written as

$$(\lambda + d)\big[\lambda + c - \frac{sk}{d}\hat{K}_2(\lambda)\big] = 0. \tag{3.8}$$

One negative root of Eq. (3.8) is $\lambda = -d$ and all other roots are given by the equation

$$\lambda + c = \frac{sk}{d}\hat{K}_2(\lambda), \tag{3.9}$$

which can be rewritten as

$$\frac{\lambda}{c} + 1 = \mathcal{R}\frac{\hat{K}_2(\lambda)}{\mathcal{K}_2}. \tag{3.10}$$

Notice that $|\hat{K}_2(\lambda)| \leq \mathcal{K}_2$ for all complex roots λ with non-negative real part (i.e., $\Re\lambda \geq 0$). Hence, the modulus of the right-hand side of Eq. (3.10) is less than 1 provided that $\mathcal{R} < 1$. Since the modulus of the left hand side of Eq. (3.10) is always greater than 1 if $\Re\lambda \geq 0$, we conclude that all roots of (3.9) have negative real part if $\mathcal{R} < 1$. It follows that $\bar{E}$ is locally asymptotically stable when $\mathcal{R} < 1$.

For the case of $\mathcal{R} > 1$, let

$$f(\lambda) = \frac{\lambda}{c} + 1 - \mathcal{R}\frac{\hat{K}_2(\lambda)}{\mathcal{K}_2}. \tag{3.11}$$

Clearly, any real root of $f(\lambda) = 0$ is also a root of (3.9). Recognizing that

$$f(0) = 1 - \mathcal{R} < 0, \quad \lim_{\lambda\to\infty} f(\lambda) = \infty, \tag{3.12}$$

we know that $f(\lambda) = 0$ has at least one positive root $\lambda^* > 0$ which is a positive eigenvalue of the characteristic equation (3.8). This shows that the infection-free steady state is unstable when $\mathcal{R} > 1$.

Next, we consider the stability of the infected steady state $E^\diamond$. As noted earlier, this steady state exists if and only if $\mathcal{R} > 1$. The following result suggests that the virus population will be established if the reproductive ratio is greater than 1.

Result 2 *The infected steady state $E^\diamond$ is locally asymptotically stable if $\mathcal{R} > 1$.*

For the verification of Result 2 we now look at the characteristic equation at the steady state $E^\diamond$:

$$\det\begin{bmatrix} -d - kV^\diamond - \lambda & -kT^\diamond & \hat{K}_1(\lambda) \\ 0 & -c - \lambda & \hat{K}_2(\lambda) \\ kV^\diamond & kT^\diamond & -1 \end{bmatrix} = 0$$

or equivalently

$$\big[(\lambda + \frac{sk\mathcal{K}_2 - dc\mathcal{K}_1}{c(1-\mathcal{K}_1)}\big]\big[\lambda + c - c\frac{\hat{K}_2(\lambda)}{\mathcal{K}_2}\big] = \frac{sk\mathcal{K}_2 - dc}{c(1-\mathcal{K}_1)}\big[(\lambda + c)\hat{K}_1(\lambda) - c\frac{\hat{K}_2(\lambda)}{\mathcal{K}_2}\big].$$

Using the notation $\mathcal{R} = sk\mathcal{K}_2/dc$ we can rewrite the above equation as

$$\big[(1-\mathcal{K}_1)\lambda + d(\mathcal{R} - \mathcal{K}_1)\big]\big[\lambda + c - c\frac{\hat{K}_2(\lambda)}{\mathcal{K}_2}\big] = d(\mathcal{R} - 1)\big[(\lambda + c)\hat{K}_1(\lambda) - c\frac{\hat{K}_2(\lambda)}{\mathcal{K}_2}\big]$$

or as

$$\left(1+\frac{\lambda}{c}\right)\left(A(\lambda+d)+1-\hat{K}_1(\lambda)\right)=\frac{\hat{K}_2(\lambda)}{\mathcal{K}_2}A(\lambda+d), \tag{3.13}$$

where

$$A=\frac{1-\mathcal{K}_1}{d(\mathcal{R}-1)}. \tag{3.14}$$

The possibility of a negative real root of Eq. (3.13) can be excluded as follows. Suppose $\lambda \geq 0$. Then

$$\hat{K}_1(\lambda)\leq \hat{K}_1(0)=\mathcal{K}_1<1. \tag{3.15}$$

It follows that $A>0$ and

$$\left(1+\frac{\lambda}{c}\right)\left(A(\lambda+d)+1-\hat{K}_1(\lambda)\right)\geq A(\lambda+d).$$

From (3.13) we have

$$\frac{\hat{K}_2(\lambda)}{\mathcal{K}_2}>1. \tag{3.16}$$

However, since $\lambda \geq 0$, $\hat{K}_2(\lambda)\leq \hat{K}_2(0)=\mathcal{K}_2$, which contradicts with (3.16). Thus, Equation (3.13) has no non-negative real roots.

We can show that Eq. (3.2) or Eq. (3.13) has no complex roots λ with non-negative real part. Suppose not, then Eq. (3.2) has a root $\lambda = x_0+iy_0$ with $x_0\geq 0$, $y_0>0$. If $\mathcal{R}\to 1$, from (3.2) we have

$$(\lambda+d)\left(\lambda+c-c\frac{\hat{K}_2(\lambda)}{\mathcal{K}_2}\right)=0. \tag{3.17}$$

i) If $x_0>0$ then using a similar argument as in Result 1 we know that $\lambda=x_0+iy_0$ cannot be a root. ii) If $x_0=0$ then (3.17) has one negative root $-d$ and other roots that are determined by the equation

$$1+\frac{\lambda}{c}=\frac{\hat{K}_2(\lambda)}{\mathcal{K}_2}$$

or

$$1+\frac{y_0}{c}i=\frac{\int_0^\infty K_2(a)\cos(ya)da}{\mathcal{K}_2}-\frac{\int_0^\infty K_2(a)\sin(ya)da}{\mathcal{K}_2}i. \tag{3.18}$$

Comparison of the real parts of both sides yields that $\cos(ya)=1$. Thus, $\sin(ya)=0$ which implies that (3.18) does not hold. Therefore, (3.2) has no roots with non-negative real part when $\mathcal{R}\to 1$.

For general $\mathcal{R}$, by the continuous dependence of roots of the characteristic equation on $\mathcal{R}$ we know that the curve determined by the roots must cross the imaginary axis as $\mathcal{R}$ decreases and close to 1. That is, the characteristic equation (3.13) has a pure imaginary root, say, iy, with $y>0$. Replacing λ in (3.13) with iy we see that the modulus of the left-hand side of (3.13) satisfies

$$\begin{aligned}|LHS| &> |Ayi+Ad+1-\hat{K}_1(iy)|\\ &=\left|Ad+1-\int_0^\infty K_1(a)\cos(ya)da+i\left(Ay+\int_0^\infty K_1(a)\sin(ya)da\right)\right|.\end{aligned} \tag{3.19}$$

We claim that $\int_0^\infty K_1(a)\sin(ya)da \geq 0$. In fact, notice that

$$\int_0^\infty K_1(a)\sin(ya)da = \int_0^{a_1} K_1(a)\sin(ya)da. \tag{3.20}$$

Notice also that $K_1(0) = r_{rt}\eta$ and $K_1'(a) = r_{rt}\eta[\beta'(a)K_0(a) + \beta(a)K_0'(a)] \leq 0$ almost everywhere on $[0,\infty)$. Integrating $\int_0^{a_1} K_1(a)\sin(ya)da$ by parts,

$$\begin{aligned}\int_0^{a_1} K_1(a)\sin(ya)da &= \frac{r_{rt}\eta}{y} - \frac{1}{y}K_1(a_0)\cos(ya_1) + \frac{1}{y}\int_0^{a_1} K_1'(a)\cos(ya)da \\ &\geq \frac{r_{rt}\eta}{y} - \frac{1}{y}K_1(a_1)\cos(ya_1) + \frac{1}{y}\int_0^{a_1} K_1'(a)da \\ &= \frac{1}{y}K_1(a_1)(1-\cos(ya_1)) \\ &\geq 0.\end{aligned}$$

Thus $\int_0^\infty K_1(a)\sin(ya)da \geq 0$. We also observe that $1 - \int_0^\infty K_1(a)\cos(ya)da \geq 1 - \mathcal{K}_1 > 0$. It follows from (3.19) that

$$|LHS| > A|d+iy|. \tag{3.21}$$

On the other hand, the modulus of the right-hand side of (3.13) satisfies

$$|RHS| \leq A|d+iy|. \tag{3.22}$$

This leads to a contradiction.

We conclude that the characteristic equation (3.13) has no roots with nonnegative real part. Therefore, Result 2 holds.

It is obvious that the threshold condition $\mathcal{R} > 1$ and the magnitude of $\mathcal{R}$ play a key role in the maintenance of the virus. In the following section we use these results to demonstrate how drug efficacy may affect the invasion of drug resistant strains.

4. Influence of drug therapy on viral fitness

In this section we focus on the issue concerning the impact of drug treatments on evolution of pathogens. In particular we consider the development of mutant strains mediated by drug therapy.

Suppose that the drug-sensitive strain of HIV-1 infection is at the infected steady state $E^\diamond = (T^\diamond, V^\diamond, B^\diamond)$ (see (3.4)), and that a small number of drug resistant virus have been introduced into the population. We derive an invasion criterion for a resistant strain by using a heuristic argument as is done in [**2**]. Denote the reproductive ratio of the sensitive strain by $\mathcal{R}_s$ (which is the same $\mathcal{R}$ as defined in (3.5)). From results in Section 2 we know that $\mathcal{R}_s > 1$ and that the population size of uninfected cells is $T^\diamond = s/(d\mathcal{R}_s)$ (see (3.4)). Let $\tilde{K}_0(a)$ denote the age-specific survival probability of a T cell infected with the resistant strain (an equivalent quantity for the sensitive strain is given in (2.6)), $\tilde{r}_{rt}$ and $\tilde{r}_p$ denote the efficacy of the two types of drugs for the resistant strain, and $\tilde{p}(a)$ denote the virion production rate of the resistant strain. For ease of illustration we assume that all other parameters are the same for both strains. We derive the invasion criterion for the case when both types of drugs are included. This criterion will be applied to different scenarios of chemotherapy such as a single-drug therapy (e.g., $r_p > 0$ and $r_{rt} = 0$) or combination therapy (i.e., $r_p > 0$ and $r_{rt} > 0$).

Since $T^\diamond$ is the amount of available uninfected T cells, a typical resistant virus can infect $kT^\diamond/c$ cells in its lifespan. Each of these infected cells can produce a total of

$$N_r = \int_0^\infty (1-\tilde{r}_p)(1-\beta(a))\tilde{p}(a)\tilde{K}_0(a)da$$

virion particles during its whole life time (burst size). Thus the reproductive ratio of the resistant strain (when the sensitive strain is at its positive equilibrium), $\mathcal{R}_r^\diamond$, is

$$\mathcal{R}_r^\diamond = \frac{kT^\diamond}{c}\int_0^\infty (1-\tilde{r}_p)(1-\beta(a))\tilde{p}(a)\tilde{K}_0(a)da,$$

and the invasion criterion is $\mathcal{R}_r^\diamond > 1$. Substituting $s/(d\mathcal{R}_s)$ for $T^\diamond$ we obtain the condition for the resistant strain to invade the sensitive strain:

$$\mathcal{R}_r > \mathcal{R}_s \tag{4.1}$$

where the quantity

$$\mathcal{R}_r = \frac{s}{d}\frac{k}{c}\int_0^\infty (1-\tilde{r}_p)(1-\beta(a))\tilde{p}(a)\tilde{K}_0(a)da \tag{4.2}$$

actually represents the reproductive ratio of the resistant strain when the equilibrium density of uninfected cells is s/d (which is the value of T at the infection-free state). We use the quantity $\mathcal{R}_r$ as a measure of fitness of a resistant virus. Eq. (4.1) shows that natural selection within a host favors viruses that maximize its reproductive ratio. We can also define a relative viral fitness (cf. [**2**]) by dividing the reproductive ratio by the factor $(s/d)(k/c)$ in both $\mathcal{R}_s$ and $\mathcal{R}_r$ since they are assumed equal for both sensitive and resistant strains.

Next, for the calculation of the optimal reproductive ratio we consider the case when the viral production rates for both strains have the form given in (2.2). That is,

$$\tilde{p}(a) = \begin{cases} \tilde{p}^*\left(1-e^{-\theta(a-a_1)}\right) & \text{if } a \ge a_1, \\ 0 & \text{else} \end{cases} \tag{4.3}$$

where $\tilde{p}^*$ is the saturation level for the resistant strain. Accordingly, we choose $\beta(a)$ to be (see (2.3))

$$\beta(a) = \begin{cases} 1, & 0 \le a < a_1, \\ 0, & a \ge a_1. \end{cases} \tag{4.4}$$

The death rate of cells is assumed to be the same for both strains with form

$$\delta(a) = \begin{cases} \delta_0, & 0 \le a < a_1, \\ \delta_0 + \mu, & a \ge a_1, \end{cases} \tag{4.5}$$

where δ_0 and μ are positive constants with δ_0 representing a background death rate of cells and μ representing an extra death rate for actively reproductive cells.

Drug resistance is incorporated by assuming that the efficacy of chemotherapy for the resistant strain is lower than that for the sensitive strain by a factor σ, $0 \le \sigma \le 1$, i.e.,

$$\tilde{r}_{rt} = \sigma_{rt} r_{rt}, \quad \tilde{r}_p = \sigma_p r_p. \tag{4.6}$$

For ease of demonstration we assume in this chapter that $\sigma_{rt} = \sigma_p = \sigma$ which could be relaxed in later studies. $\sigma = 0$ corresponds to the completely resistant strain while $\sigma = 1$ corresponds to the completely sensitive strain. Other strains have an intermediate value $0 < \sigma < 1$. To incorporate the cost that resistant strains pay for the development of resistance we consider a trade-off between drug resistance and virion production rate $\tilde{p}(a)$. Various functional forms for the cost can be used. Herein, we consider two types of costs by which the saturation level p^* is reduced according to the following formulas

$$\text{Type I}: \quad \tilde{p}(a) = \sigma p^* \big(1 - e^{-\theta(a-a_1)}\big), \tag{4.7}$$

$$\text{Type II}: \quad \tilde{p}(a) = e^{-\phi(1/\sigma - 1)} p^* \left(1 - e^{-\theta(a-a_1)}\right), \tag{4.8}$$

where ϕ is a measure for the level of cost. We provide analytic results for Type I cost and illustrate that the qualitative properties of the two types of costs are similar. To make the calculation transparent we rewrite the reproductive ratio $\mathcal{R}_r$ using (4.4) – (4.7) as

(4.9)

$$\begin{aligned}\mathcal{R}_r &= \frac{s}{d}\frac{k}{c}\int_{a_1}^{\infty} (1 - \sigma r_p)\sigma p^* \big[1 - e^{-\theta(a-a_1)}\big] e^{-(\delta_0 + \sigma r_{rt}\eta)a_1} e^{-(\delta_0+\mu)(a-a_1)} da \\ &= \big(1 - \sigma r_p\big)\sigma e^{-(\delta_0+\sigma r_{rt}\eta)a_1} D,\end{aligned}$$

where

$$D = \frac{s}{d}\frac{k}{c}\int_0^{\infty} p^* \big(1 - e^{-\theta u}\big) e^{-(\delta_0+\mu)u} du$$

is a quantity that is independent of σ, r_{rt}, r_p, and a_1. Similarly we can rewrite $\mathcal{R}_s$ in the form

$$\mathcal{R}_s = (1 - r_p) e^{-(\delta_0 + r_{rt}\eta)a_1} D. \tag{4.10}$$

From Eqs (4.9) and (4.10) we get the following relationship between $\mathcal{R}_r$ and $\mathcal{R}_s$:

$$\mathcal{R}_r = \frac{\sigma(1 - \sigma r_p) e^{-r_{rt}\eta(1-\sigma)a_1}}{1 - r_p} \mathcal{R}_s. \tag{4.11}$$

Consider $\mathcal{R}_r = \mathcal{R}_r(\sigma)$ as a function of σ. A resistant strain with resistance σ can invade the sensitive strain if $\mathcal{R}_r(\sigma) > \mathcal{R}_s$, which (from the fact that $\mathcal{R}_r(1) = \mathcal{R}_s$) is possible only if $\frac{d\mathcal{R}_r(\sigma)}{d\sigma} < 0$ for some $\sigma \in (0, 1)$. Notice that

$$\frac{d\mathcal{R}_r}{d\sigma} \begin{cases} < 0 & \text{if} \quad \sigma_- < \sigma < \sigma_+ \\ = 0 & \text{if} \quad \sigma = \sigma_-, \sigma_+ \\ > 0 & \text{else,} \end{cases} \tag{4.12}$$

where

$$\sigma_{\pm} = \frac{2r_p + a_1 r_{rt}\eta \pm \sqrt{4r_p^2 + a_1^2 r_{rt}^2 \eta^2}}{2a_1 r_p r_{rt}\eta},$$

if $r_p > 0$ and $r_{rt} > 0$. Since

$$\sqrt{4r_p^2 + a_1^2 r_{rt}^2 \eta^2} = \sqrt{(2r_p + a_1 r_{rt}\eta)^2 - 4a_1 r_p r_{rt}\eta} \leq 2r_p + a_1 r_{rt}\eta$$

we know that $\sigma_{\pm} \geq 0$. Clearly $\mathcal{R}_r(\sigma) \leq \mathcal{R}_s$ for all $\sigma \in (0,1)$ if $\sigma_- \geq 1$, i.e., if r_p and r_{rt} satisfy

$$\frac{2r_p + a_1 r_{rt}\eta - \sqrt{4r_p^2 + a_1^2 r_{rt}^2 \eta^2}}{2a_1 r_p r_{rt}\eta} \geq 1. \tag{4.13}$$

Obviously the condition (4.13) is not easy to use to draw conclusions. Let us first derive some analytic understanding for a simpler case in which only a single-drug therapy with a protease inhibitor is considered, i.e., $r_p > 0$ and $r_{rt} = 0$. The case of combined therapy will be explored numerically.

Single-drug therapy. In this case, since $r_p > 0$ and $r_{rt} = 0$, Eq. (4.11) simplifies to

$$\mathcal{R}_r(\sigma) = \frac{\sigma(1 - \sigma r_p)}{1 - r_p}\mathcal{R}_s. \tag{4.14}$$

It is easy to check that in order to have $\mathcal{R}_r(\sigma) \geq \mathcal{R}_s$ for some $\sigma \in (0,1)$ it is necessary that $r_p > \frac{1}{2}$, in which case

$$\frac{d\mathcal{R}_r(\sigma)}{d\sigma} < 0, \quad \text{for} \quad \frac{1}{2r_p} < \sigma < 1.$$

The above inequalities suggest that there exists a maximum level of resistance, $\sigma_{\max}$, such that

$$\mathcal{R}_r(\sigma) > \mathcal{R}_s \quad \text{if and only if} \quad \sigma_{\max} < \sigma < 1. \tag{4.15}$$

(Remark: A higher resistance level corresponds to a smaller value of σ.) That is, strains with resistance $\sigma < \sigma_{\max}$ cannot invade. We can determine $\sigma_{\max}$ by noticing that it must be a solution of the equation $\mathcal{R}_r(\sigma) = \mathcal{R}_s$, from Eq. (4.14) we know that $\sigma_{\max}$ satisfies

$$\frac{\sigma(1 - \sigma r_p)}{1 - r_p} = 1. \tag{4.16}$$

Two solutions of (4.16) are

$$\sigma_1 = 1, \quad \sigma_2 = \frac{1 - r_p}{r_p}.$$

Since the condition $r_p > \frac{1}{2}$ guarantees that $\frac{1-r_p}{r_p} < 1$, we have

$$\sigma_{\max} = \sigma_2 = \frac{1 - r_p}{r_p} < 1, \quad \text{if} \quad r_p > \frac{1}{2},$$

and

$$\begin{cases} \mathcal{R}_r(\sigma) > \mathcal{R}_s, & \sigma_{\max} < \sigma < 1, \\ \mathcal{R}_r(\sigma) < \mathcal{R}_s, & \sigma < \sigma_{\max}, \\ \mathcal{R}(\sigma) = \mathcal{R}_s, & \sigma = 1, \sigma_{\max} \end{cases} \tag{4.17}$$

(see Figure 1). Clearly, if $r_p < \frac{1}{2}$ then $\sigma_{\max} > 1$ and hence $\mathcal{R}_r < \mathcal{R}_s$ for all σ. This indicates that when the drug efficacy is very low, the sensitive strain is favored. The intuitive reason for this is that if the cost of resistance is high, one would not

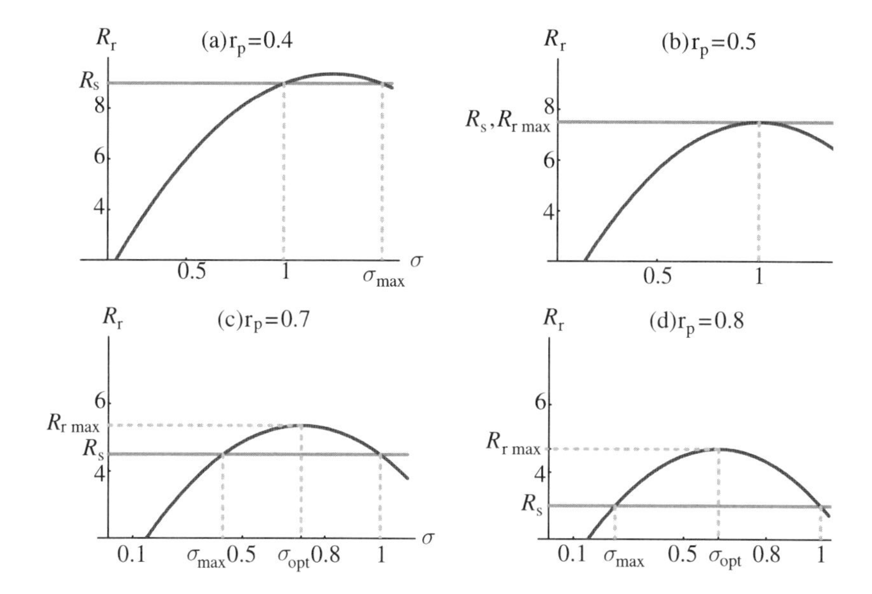

FIGURE 1. Plots of the reproductive ratio $\mathcal{R}_r$ vs. the resistant level $1/\sigma$ for different treatment efficacy r_p (r_{rt} is chosen to be 0). In (a) and (b) it is shown that $\mathcal{R}_r < \mathcal{R}_s$ for all $\sigma < 1$. Therefore, no resistant strains can invade. In (c) and (d) it is shown that resistant strains with resistance σ in $(\sigma_{\max}, 1)$ can invade. The optimal resistance is σ_{opt} at which $\mathcal{R}_r$ reaches its maximum $\mathcal{R}_{r\max}$.

expect resistance when there is little selection pressure from the drugs. Other non-resistant strains would outcompete it under these conditions. Resistant strains can only increase in frequency when the selection pressure (drug efficacy) is high.

We can also determine an optimal resistance, σ_{opt}, which maximizes the reproductive ratio. In fact, we can easily check that $\mathcal{R}_r(\sigma)$ has only one critical point in the interval $(\sigma_{\max}, 1)$, $\sigma = \frac{1}{2r_p}$, at which $\frac{d\mathcal{R}_r(\sigma)}{d\sigma} = 0$. Hence,

$$\sigma_{opt} = \frac{1}{2r_p} \tag{4.18}$$

(see Figure 1).

We summarize the following results for the case of a single-drug therapy. Recall that a resistant strain with resistance σ can invade the sensitive strain if and only if $\mathcal{R}_r(\sigma) > \mathcal{R}_s$.

1. There exists a threshold drug efficacy r_p^* ($r_p^* = 1/2$ for Type I cost) below which no resistant strains can invade (see Figure 1 (a) (b)). Analytically this is due to the fact that $\sigma_{\max} \geq 1$ when $r_p < r_p^*$. Hence $\mathcal{R}_r(\sigma) < \mathcal{R}_s$ for all $\sigma < 1$.
2. When the drug efficacy is above the threshold r_p^* there is a range of resistance levels for which the resistant strains are able to invade. This is because, analytically, $\sigma_{\max} < 1$ when $r_p > r_p^*$, and $\mathcal{R}_r(\sigma) > \mathcal{R}_s$ for all σ in $(\sigma_{\max}, 1)$.

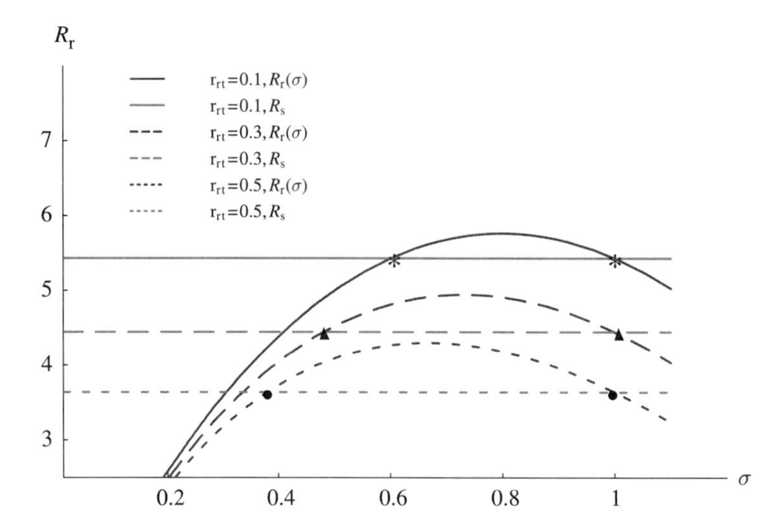

FIGURE 2. Plots of the reproductive ratio $\mathcal{R}_r$ vs. resistance σ for $r_{nt} = 0.1$ (solid), $r_{rt} = 0.3$ (long dashed), $r_{rt} = 0.5$ (short dashed). The value of r_p is fixed at $r_p = 0.6$ for which invasion is possible in the absence of the an RT drug (i.e., if $r_{rt} = 0$). For each given r_{rt}, the values of σ for which $\mathcal{R}_r(\sigma) > \mathcal{R}_s$ give the range for resistance invasion, which is the range between the two intersection points of the two corresponding curves.

3. When $\sigma_{\max} < 1$, the range of invasion strains, $(\sigma_{\max}, 1)$, increase with the drug efficacy r_p. The optimal resistance, σ_{opt}, decrease with the drug efficacy r_p (a more resistant strain corresponds to a smaller σ value, see Figure 1 (c) (d)). This increasing property is also clear from the formulas $\sigma_{\max} = (1 - r_p)/r_p$ and $\sigma_{opt} = 1/(2r_p)$.
4. As the drug efficacy increases, the optimal viral fitness, $\mathcal{R}_r(\sigma_{opt})$, decreases (see Figure 1 (c) (d)). We can also see this from the formula $\mathcal{R}_{r\max} = \mathcal{R}_r(\sigma_{opt}) = D/(4r_p)$ which is a decreasing function of r_p.

Combination therapy. We now consider the case of combination therapy, i.e., $r_p > 0$ and $r_{rt} > 0$. For simplicity we choose $\eta = 1$. From (4.9) and (4.10) we have

$$\mathcal{R}_r = \frac{\sigma(1 - \sigma r_p)e^{-r_{rt}(1-\sigma)a_1}}{1 - r_p}\mathcal{R}_s. \tag{4.19}$$

Again, consider $\mathcal{R}_r = \mathcal{R}_r(\sigma)$ as a function of σ. Then $\mathcal{R}_r(\sigma) > \mathcal{R}_s$ if and only if σ satisfies the inequality

$$\frac{\sigma(1 - \sigma r_p)e^{-r_{rt}(1-\sigma)a_1}}{1 - r_p} > 1. \tag{4.20}$$

To explore the role of r_{rt} we fix r_p (e.g., $r_p = 0.7$ in Figure 2). Eq. (4.20) cannot be solved analytically for σ. However, plots of $\mathcal{R}_r(\sigma)$ for different values of r_{rt} suggest that, as r_{rt} increases, the range for $\mathcal{R}_r(\sigma) > \mathcal{R}_s$ also increases (see Figure 2). Figure 3 illustrates the joined effect of r_{rt} and r_p on the reproductive ratios $\mathcal{R}_s$

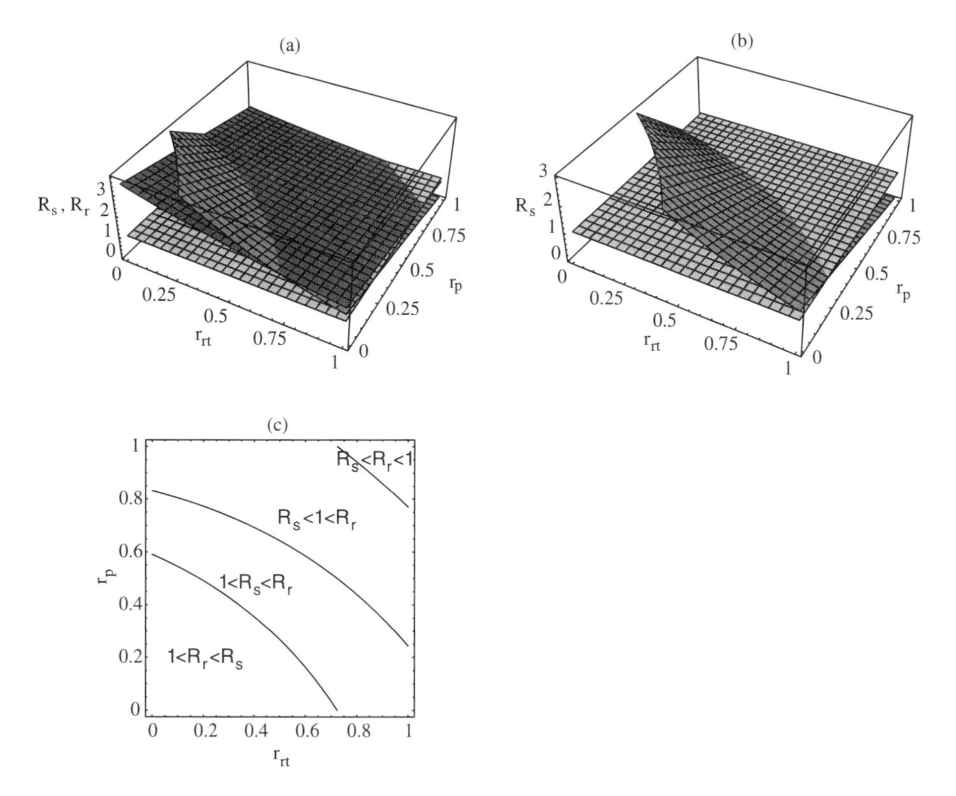

FIGURE 3. Plots of the reproductive ratios $\mathcal{R}_s$ and $\mathcal{R}_r$ as functions of r_{rt} and r_p. Three surfaces are plotted in Figure 3(a): $\mathcal{R}_r(r_{rt}, r_p)$ (the top surface near the origin), $\mathcal{R}_s(r_{rt}, r_p)$ (middle surface) and the constant 1 (the bottom surface). The intersection of the top two surfaces is the curve on which $\mathcal{R}_r = \mathcal{R}_s$. In Figure 3(b) two surfaces, $\mathcal{R}_s(r_{rt}, r_p)$ and the constant 1, are plotted to show the curve on which $\mathcal{R}_s = 1$. Figure 3(c) is a contour plot of the surfaces $\mathcal{R}_r(r_{rt}, r_p)$ and $\mathcal{R}_s(r_{rt}, r_p)$.

and $\mathcal{R}_r$. From the contour plot (see Figure 3(c)) we see that when the drug efficacy is low (the region in the lower-left corner in which $\mathcal{R}_s > \mathcal{R}_r > 1$) the resistant strain cannot invade. Neither strain can survive when the drug efficacy is high (the top-right region in which $\mathcal{R}_s < 1$ and $\mathcal{R}_r < 1$). In the middle region the invasion of resistant strains are possible as $\mathcal{R}_r > \mathcal{R}_s$.

Figure 4 shows that when Type II cost is used the qualitative property of the reproductive ratio $\mathcal{R}_r$ as a function of σ is very similar to those when Type I cost is used. For example, the function $\mathcal{R}_r(\sigma)$ admits a unique $\sigma_{\max}$ and a unique σ_{opt} for sufficiently small values of ϕ.

5. Discussion

We have formulated an age-structured model for HIV-1 infection with drug treatments to study the impact of chemotherapy on the emergence of resistant HIV-1 strains. We have exhibited the reproductive ratio of the drug sensitive strain $\mathcal{R}_s$,

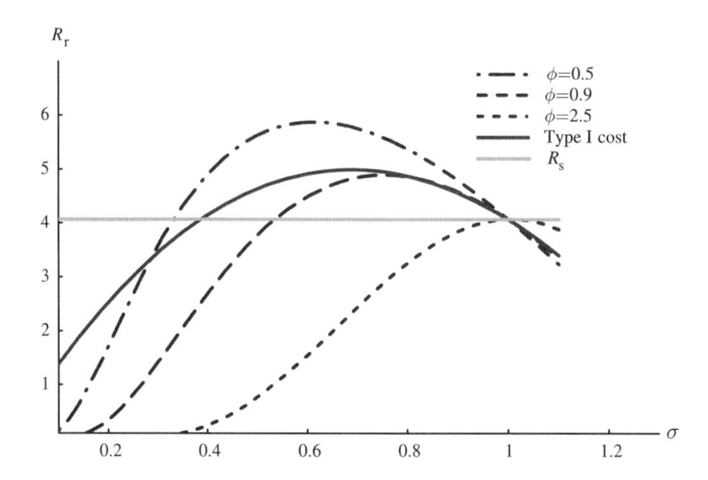

FIGURE 4. Plots of the reproductive ratio $\mathcal{R}_r$ vs. resistance σ when Type II cost is considered. The value of ϕ measures the level of cost. Invasion is possible for σ in the range between the two intersection points at which $\mathcal{R}_r = \mathcal{R}_s$. It also shows that invasion is impossible if the cost is too high (e.g., $\phi = 2.5$).

and demonstrated the asymptotical stability of the infection-free steady state $\bar{E}$ if $\mathcal{R}_s < 1$ and the infected steady state $E^\diamond$ if $\mathcal{R}_s > 1$. We have considered two types of drug treatments with reverse transcriptase inhibitors and protease inhibitors and the possible development of drug resistance. The cost for resistant strains is assumed to be a reduced viral production rate. We have calculated the reproductive ratio for the resistant strain $\mathcal{R}_r(\sigma)$ with resistance σ, and provided a criterion for the potential invasion of resistant strains, $\mathcal{R}_r(\sigma) > \mathcal{R}_s$, in an environment where the wild strain is already established. We argue that natural selection within a host favors viruses that maximize the reproductive ratio which is consistent with earlier findings (see, for example, [**2**]). Consequently, we show that natural selection should favor viral strains that have an intermediate level of resistance, and the optimal resistance (σ_{opt}) decreases with increasing drug efficacy (see Figure 1 and Figure 2). Mathematically, increasing the values of r_p and r_{rt} results in: (i) a reduction in the reproductive ratio $\mathcal{R}_s$ of the drug sensitive strain (see Figure 4) and hence a reduction in the equilibrium level of infection (see $T^\diamond$ and $V^\diamond$ in Eq. (3.4)); (ii) a decrease in the optimal viral fitness $\mathcal{R}_{r\max}$ of the resistant strain and a decrease in the optimal resistance σ_{opt} (see Figure 1 and Figure 3); and (iii) an increase in the range $\sigma_{\max} < \sigma < 1$ of resistance (that is, $\sigma_{\max}$ decreases with both r_p and r_{rt}, see Figure 1 and Figure 2). These are strains that has a high fitness (i.e., $\mathcal{R}_r(\sigma) > \mathcal{R}_s$) are hence are able to invade a host population. On the other hand, if r_p and r_{rt} are small such that $\sigma_{\max}$ is greater than 1, then $\mathcal{R}_r(\sigma) < \mathcal{R}_s$ for all resistance σ. These strains will not be maintained in a population.

Acknowledgments

We would like to thank M. Gilchrist and P. Nelson for very useful suggestions which improved this manuscript. We also would like to thank D. Coombs for helpful discussions.

References

[1] Z. Feng, J. Curtis, and D. Minchella, The influence of drug treatment on the maintenance of schistosome genetic diversity, *J. Math. Biol.* **43** (2001), 52–68.

[2] M. A. Gilchrist, D. Coombs, and A. S. Perelson, Optimizing within-host viral fitness: infected cell lifespan and virion production rate, *J. Theor. Biol.* **229** (2004), 281–288.

[3] D. D. Ho et al., Characterization of human immunodeficiency virus type 1 variants with increased resistance to a C2-symmetric protease inhibitor, *J. Virol.* **68** (1994), 2016–2020.

[4] D. Kirschner and G. F. Webb, Understanding drug resistance for montherapy treatment of HIV infection, *Bull. Math. Biol.* **59** (1997), 763–786.

[5] B. A. Larder and S. D. Kemp, Multiple mutations in HIV-1 reverse transcriptase confer high-level resistance to zidovudine (AZT), *Science* **246** (1989), 1155–1158.

[6] B. A. Larder, P. Kellam, and S. D. Kemp, Convergent combination therapy can select viable multidrug-resistant HIV-1 in vitro, *Nature* **365** (1993), 451–453.

[7] R. Miller, *Nonlinear integral equations*, W. A. Benjamin Inc., New York, 1971.

[8] P. Nelson and A. Perelson, Mathematical analysis of delay differential equations for HIV, *Math. Biosci.* **179** (2002), 73–94.

[9] P. Nelson, M. Gilchrist, D. Coombs, J. Hyman, and A. Perelson, An age-structured model of HIV infection that allows for variations in the production rate of viral particles and the death rate of productively infected cells, *Mathematical Biosciences and Engineering* **1** (2004), 267–288.

[10] M. A. Nowak, S. Bonhoeffer, G. M. Shaw, and R. M. May, Anti-viral drug treatment: dynamics of resistance infreevirus and infected cell populations, *J. Theor. Biol.* **184**(2) (1997), 203–217.

[11] A. Perelson and P. Nelson, Mathematical analysis of Hiv-1 dynamics in vivo, *SIAM Rev.* **41** (1999), 3–44.

[12] D. D. Richman, Zidovudine resistance of human immunodeficiency virus, *Rev. Infect. Dis.* **12** Suppl. 5 (1990), S507–510.

[13] D. Xu, J. Curtis, Z. Feng , and D. Minchella, On the role of shistosome mating structure in the maintenance of resistant strains, *Bull. Math. Biol.* in press, 2005.

Department of Mathematics, Purdue University, IN 47907, USA
E-mail address: zfeng@math.purdue.edu

E-mail address: rong@math.purdue.edu

DIMACS Series in Discrete Mathematics
and Theoretical Computer Science
Volume **71**, 2006

Do rhinoviruses follow the neutral theory? The role of cross-immunity in maintaining the diversity of the common cold

William J. Koppelman and Frederick R. Adler

ABSTRACT. Over 100 serotypes of rhinoviruses, one of the primary causes of the common cold, co-circulate in the human population. This high diversity makes it effectively impossible to develop a vaccine, even for those at risk of complications due to asthma or cystic fibrosis. Is the high mutation rate of these viruses sufficient to explain this diversity? We use parameters estimated from the literature to study whether immune interactions between different rhinovirus serotypes also play an important role in maintaining diversity. Our mathematical models indicate that high mutation rates alone may well be responsible for the observed levels of diversity. However, careful studies of a few communities have found that some serotypes persist for many years, in conflict with the predictions of the simplest models, hinting that there might be more to the story than is yet known.

1. Introduction

Diversity is usually regarded as an ecological question and is rarely considered in the realm of virology [**1**]. However, the immense diversity of rhinoviruses, the most frequent cause of the common cold, makes development of treatments and a vaccine virtually impossible [**15**]. We use mathematical and computer simulation models to attempt to understand the maintenance of this high diversity.

There are currently 102 serotypes of the human rhinovirus (HRV) that co-circulate in the human population, divided into type A (with 76 serotypes), type B (25 serotypes), and a single type which clusters with enteroviruses [**30**]. HRV is a genus in the family *Picornaviridae* and shares the characteristics of other genera within this family (e.g. enterovirus, poliovirus, foot and mouth disease virus, and hepatitis A virus). These characteristics include a non-enveloped icosahedral capsid and a single-stranded positive-sense RNA genome of approximately 7.2 kb. Viruses are spherical in shape with a diameter of 25–30 nm [**8**]. Within *Picornaviridae*, HRV is genetically most closely related to the enteroviruses [**21**]. The enteroviruses can tolerate a wider range of pH and therefore multiply mainly in the alimentary tract while HRV multiplies in the nasal epithelium [**31**]. HRV serotypes cause as

Key words and phrases. Biodiversity, Rhinovirus, Neutral theory, Cross-immunity.
First author was supported by IGERT grant NSF DGE-0217424.

many as 40% of upper respiratory illnesses, or "common colds" [**17**]. They have also been associated with bacterial sinusitis and otitis media [**8**], and appear in over 80% of acute asthma episodes in children [**14**].

Once a sufficient dose, as small as 1–30 particles, has deposited itself in the nasal passages it is transported back to the adenoid area [**9, 18**]. Then the virus attaches to a receptor on the surface of the nasal cells. HRV-A serotypes attach to the Intercellular Adhesion Molecule-1 (ICAM-1) receptor while serotypes in HRV-B attach to the low density lipoprotein receptor (LDLR) [**6**]. This attachment leads to the infection process. The incubation period is 8–12 hours with the peak of infection occurring in 36–72 hours [**15**].

The goal of this paper is to develop a model to explain the observed diversity of HRV using characteristics from its pathology and evolution. The maintenance of diversity may be explained by the high mutation rate of HRV, the cross-reactivity between serotypes, or immunodominance from previous infections.

RNA viruses typically have high rates of mutation and can be thought of as a quasi-species [**25**]. It has been shown that the mutation rate for HRV is 0.67 mutations per genome, or 10^{-4} per nucleotide per replication [**10**]. This high rate of mutation may lead to evolution of serotypes. In 1969 it was suggested that HRV-51 had undergone sufficient variation in two to four years to produce a new serotype [**32**]. A recent experiment found that HRV-17 could quickly evolve to escape the antisera used to identify serotypes, while 15 other serotypes showed little or no such evolution [**28**].

Cross-reactivity between serotypes led to the proposal that groups of HRV share common epitopes. In 1975, Cooney et al. [**7**] analyzed the published literature and in combination with their own data found evidence for a strong relationship between serotypes. They found that a serotype will elicit a heterotypic antibody response from an average of 3.75 other serotypes.

A rhinovirus in the human body presents a large number of epitopes to the immune system. The immune response focuses on only a few of the many potential epitopes, a process called immunodominance [**12**]. The sequence in which the host encounters antigenic variants influences the specificity of the immune response. This process was termed *original antigenic sin* [**3**]. For example, if the primary exposure in the host was with antigen **A** which elicited an immune response **a**, a secondary exposure to a related antigen $\mathbf{A}'$ could stimulate the same immune response **a**. A rapid response from the memory B or T cells may keep the antigen density below a threshold required for stimulation of naive B or T cells [**12**]. Although there may be a better response to a particular antigen, the immune system does not respond with it because the invasion has been regulated by existing memory cells.

A previous model by Pease [**29**] considered an evolutionary epidemic model in which a person would go from having immunity to a virus back to being susceptible after the virus had changed genetically, and hence immunologically. Andreasen et al. [**4**] developed an influenza model to consider partial cross-protection among strains. In the model, the immunity history of the host population was followed so that invading strains could be analyzed against it. This allowed for the existence of a multi-strain endemic equilibrium. Cross-immunity has been the focus of other papers as well [**5, 13**]. In particular, recent work has included cross-immunity, mutation, and demographic stochasticity in a study of the persistence and dynamics of a viral disease [**1**].

Other groups have developed a mathematical model of immunodominance in HIV-1 infections [**26, 27**]. These models focus on competition for stimulation by epitopes between immune cell lineages, concluding that the most efficient predator (immune cell lineage) reduces the prey (epitopes) down to a level where less efficient predators cannot survive. Models of immunological interactions between different dengue virus serotypes allow modelers to identify the immunological distance at which serotypes can stably coexist [**22**].

Hubbell [**20**] developed a neutral theory for species richness and relative species abundance. The theory is neutral because individual organisms do not differ in their demographic parameters, and there are no competitive interactions other than maintenance of constant population size. The theory proposes that only three parameters are needed to determine diversity: the fundamental biodiversity number (which includes metacommunity size and speciation rate), the probability of immigration, and the local community size.

Our aim is to estimate the relevant parameters for HRV, and to test what effects cross-immunity and immunodominance have on maintained diversity in comparison to the neutral case. As a test, we compare our model results with a detailed study of the small town of Tecumseh, Michigan [**24**], comparing the observed diversity and serotype frequency during the two time periods covered by the study, and checking whether the overlap in serotypes between the two periods is realistic.

2. The simulation

Rhinovirus serotypes are placed on a d-dimensional hyper-torus, with periodic boundaries of length 2. Each serotype is represented by a single hypercube with sides of length one, creating space for 2^d serotypes. The maximum distance between serotypes is $\sqrt{d}$, relative to the difference between serotypes. The difference between the most similar serotypes is roughly 10% nucleotide divergence [**28**], thus a dimension of $d = 10$ is consistent with the maximum divergence of 34 - 41% observed [**30**]. This dimension leaves room for $2^{10} = 1024$ possible serotypes, an order of magnitude more than have been observed.

Our model tracks the infectious and immune state of each individual in a population of size N. Hosts are born and die at equal and constant rates, set to give a mean lifespan of 70 years. Newborn hosts have no HRV immunity. Every time step (one day), infected hosts have a probability γ of clearing the infection. A value of $\gamma = 0.1$ is consistent with a mean infection duration of 10 days [**24**]. Upon recovery, these individuals gain immunity to the infecting serotype.

Susceptible individuals contract the illness through contact with a single pool of viruses, rather than direct contact with other individuals. Such contacts occur with a probability α. The pool of viruses is created by mixing a fraction $1 - m$ from infected individuals in the population and a fraction m of random viruses to represent immigration. We chose values of $m = 0$ and $m = 0.01$ to study the effects of immigration, assuming in the latter case that an individual re-enters the local area on about 1% of days, or four times per year. The value of α is adjusted until the disease prevalence matches a pre-determined target based on a frequency of approximately 0.5 HRV infections per person per year with a duration of 10 days [**17**]. If contact occurs, the potentially infecting strain is a mutated version of a strain chosen randomly from the pool. The mutation is in a random direction in the d dimensional space with a magnitude of μ times a standard log-normal distribution (with mean and variance equal to 1). The value of μ is chosen so that a new serotype is generated after about 50 infections [**32**], meaning that most mutations do not generate a new serotype. Given a mutation rate of 10^{-4} per nucleotide per replication [**10**], we thus assume 20 replications per infection to achieve the 10% sequence divergence between serotypes.

Successful infection occurs if the exposed individual does not have immunity from previous infections (Figure 1). Previous infections provide partial immunity out to a distance of x_i, with the probability of infection being

$$\text{Probability of infection} = \frac{\text{distance to nearest prior infection}}{x_i}.$$

The parameter x_i is varied in our simulations, but should take on values slightly larger than 1 to correspond to the observed degree of cross-reactivity. Immunity is determined by distance without explicit regard to serotype. Unsuccessful infections are included in the exposed individual's immune history when immunodominance is off, and are not included when it is on. In the latter case, we assume that the successful response was based entirely on the existing epitopes.

TABLE 1. Variables, parameters, and functions in the simulation

Parameters		
Symbol	Meaning	Value or values
N	Population size of hosts	1000 – 8000
p	Prevalence of rhinovirus infections	0.02 – 0.1
d	Dimension of genetic space	7 or 10
x_i	Cross-immunity distance	0.0 – 2.0
μ	Standard deviation of mutation distance	0.06
δ	Birth rate of hosts per day	$1/(70 \cdot 365)$
γ	Recovery rate of hosts per day	0.1
α	Contact rate of hosts with virus pool	Computed
m	Probability of immigration of infected host	0 or 0.01

3. Simulation results

How well do these parameter estimates (summarized in Table 1) predict the actual diversity of HRV? First, we compare our results with the neutral theory [**20**]. The fundamental biodiversity parameter θ is the product of the number $J = pN$ of infected individuals and twice the effective mutation rate ν, or

$$\theta = 2\nu pN = 2\nu J.$$

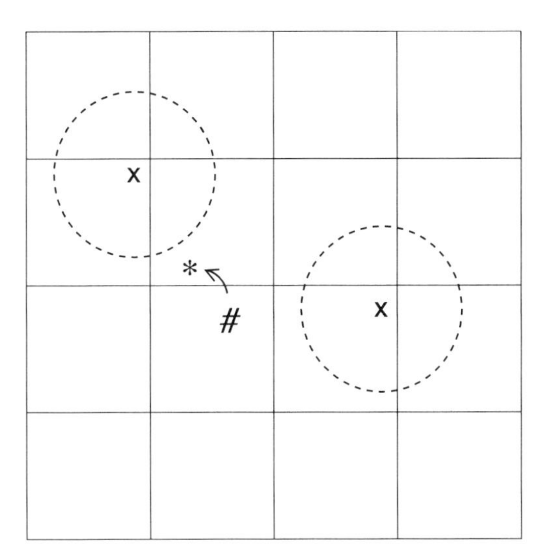

FIGURE 1. The structure of the simulation. An individual who started with the infection marked with # (displayed in a two-dimensional version of the serotype space) infects another individual by transmitting the mutated version indicated by a $*$. The new individual has previously had the two colds at locations marked with the x's, and has immunity (potentially only partial immunity) in the regions indicated by the dashed circles. The new infection lies outside these circles, and is thus successful. Because it has crossed one of the lines in the grid, it is designated as a new serotype.

The effective mutation rate ν includes the effects of both mutation and migration. The diversity D_v is then

$$\begin{aligned} D_v &= \sum_{j=1}^{J} \frac{\theta}{\theta + j - 1} \\ &\approx \theta \int_{x=0}^{J-1} \frac{1}{\theta + x} dx \\ &= \theta \ln\left(\frac{\theta + J - 1}{\theta}\right) \\ &= 2\nu pN \ln\left(\frac{2\nu pN + pN - 1}{2\nu pN}\right) \\ &\approx 2\nu pN \ln\left(\frac{1 + 2\nu}{2\nu}\right). \end{aligned}$$

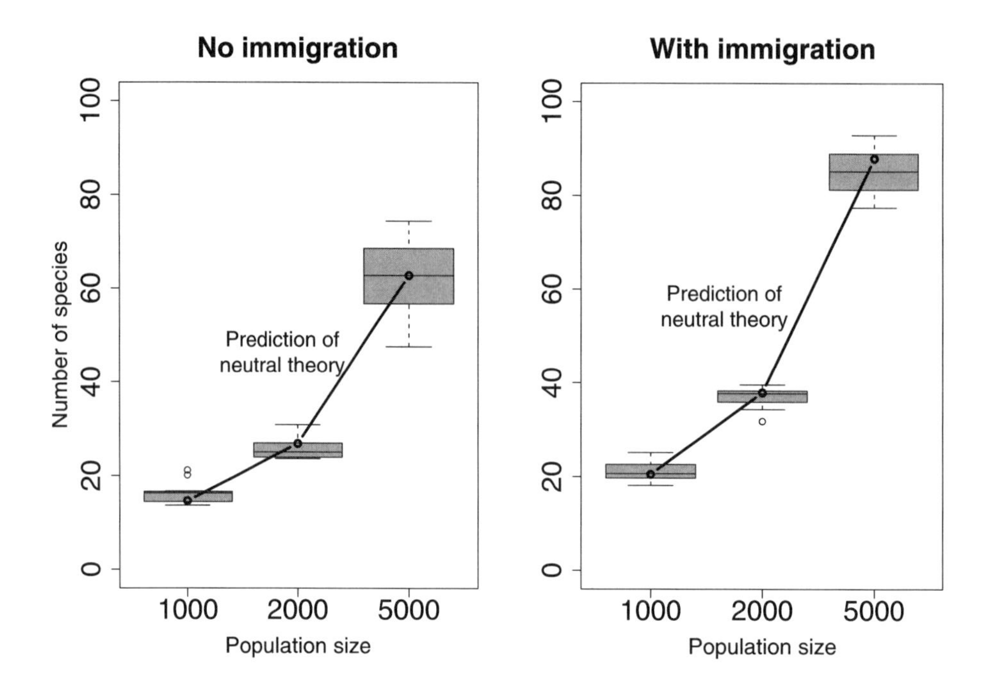

FIGURE 2. Comparison of neutral theory predictions with simulated diversity for the ranges of parameter values in Table 1 but with no cross-immunity ($x_i = 0$). To avoid disease die-out in the smaller populations, we used an unrealistically high prevalence of 10%. To match the actual number of infections carried by a person in their lifetime, we increased the birth and death rates by a factor of 7.0. **a.** With no immigration ($m = 0$). **b.** With 1% of infections from immigration ($m = 0.01$).

This model assumes that all new mutants form new serotypes, and our estimated dimension of $d = 10$ creates 1024 serotypes, a factor of 10 more than have been observed co-circulating in the human population.

We estimated a value of $\nu = 0.016$ from the simulations with $m = 0$ (fairly close to our target value of 0.02 based on requiring approximately 50 infections to create a new serotype). We used a value of $\nu = 0.016 + m$ in the cases with $m = 0.01$ because new migrants are essentially certain to be of a new serotype. In each case, the neutral theory predicts the overall diversity quite well in the absence of cross-immunity (Figure 2).

Cross-immunity generally increases diversity. For relatively large populations, the increase is maximized at an intermediate scale of cross-immunity ($x_i = 0.75$ to 1.0) roughly consistent with the scale of a serotype (Figure 3). Thus, the observed relatively limited cross-immunity between serotypes [**7**] may be at a level that maximizes the standing diversity of HRV. Contrary to our original hypothesis, immunodominance has no effect on the diversity maintained in the system (Figure 4).

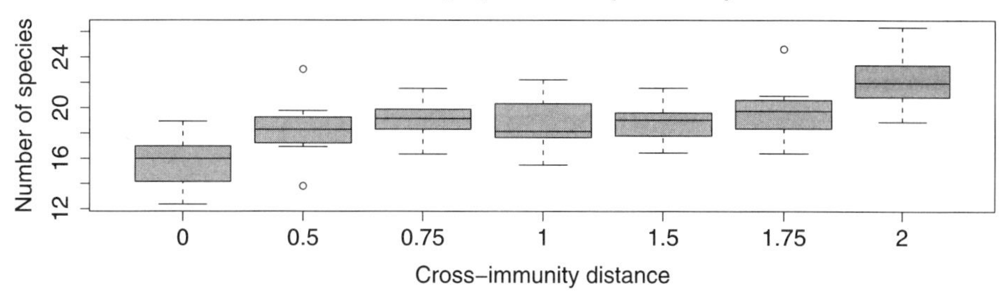

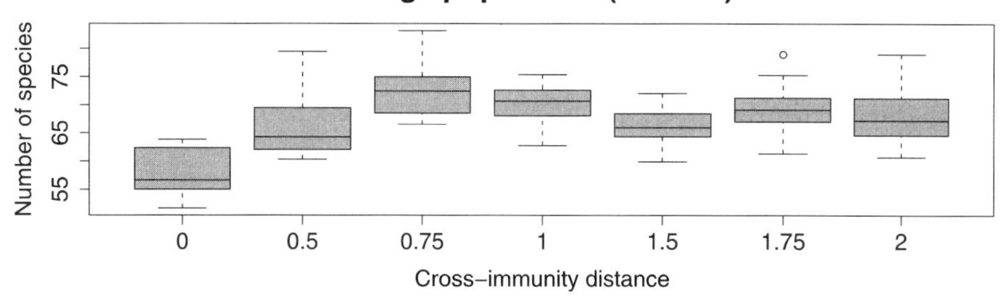

FIGURE 3. The effects of cross-immunity on diversity using the parameters in Figure 2 and $m = 0$. A cross-immunity distance of 2.0 corresponds to deriving partial cross-immunity to all cold serotypes from any infection. Bars which share the same label have serotype richnesses that are statistically indistinguishable (Tukey's honestly significant differences).

We then tested the model on a detailed data set from Tecumseh, Michigan, collected during the periods from 1966–1971 and 1976–1981 [**24**]. In this study, symptomatic individuals from the community were visited, and their infections sampled, identified and serotyped. There were 250 HRV samples taken during the first period and 194 during the latter. At that time, only 89 serotypes had been identified and fully 73 of these were identified at some point in the study. Our simulations were designed to match the size of this community (roughly 8000 people) with our other most reasonable guesses of parameter values. We randomly sampled infected individuals during two 6 year periods offset by 10 years from each other to match the design of the study.

The dominance-diversity plots match the data fairly closely, although the distributions with $m = 0.01$ have too many singletons (Figure 5). The observed diversity matches that seen in the Tecumseh study reasonably well, although the predicted diversities are too large when immigration is included at even a low level (Table 2).

However, the turnover during this time is far too large (the overlap in the last row of Table 2). In Tecumseh, most of the serotypes observed in the earlier samples reappeared in the later samples, in sharp contrast to our simulation results. The authors of the Tecumseh study argue that nearly all possible serotypes had been discovered [**24**]. There were 89 identifiable serotypes in 1987, and only 13

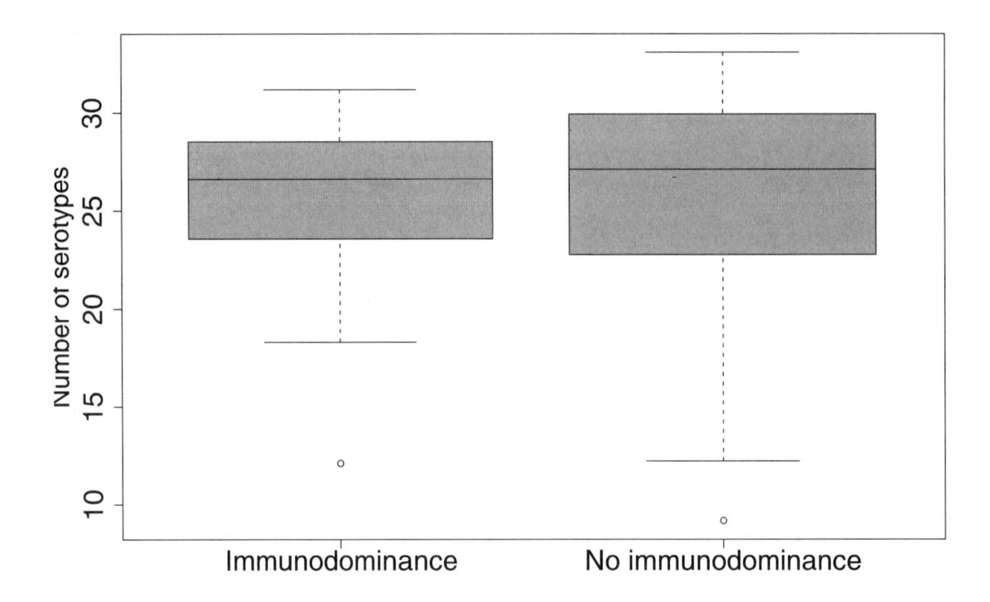

FIGURE 4. The lack of an effect of immunodominance using the parameters in Figure 2.

TABLE 2. Simulated and measured HRV serotype diversity and overlap: $d = 10$

	$x_i = 0$, $m = 0$	$x_i = 0$, $m = 0.01$	$x_i = 1$, $m = 0$	$x_i = 1$, $m = 0.01$	Measured
1966-71	45	88	73	115	62
1976-81	40	86	66	86	53
overlap	12	5	13	8	42

additional ones have been identified subsequently, at least 2 of which cross-react strongly with serotypes already identified [**23**], indicating that perhaps the numbers have saturated.

To test this hypothesis, we reran our simulation in seven dimensions, creating only 128 possible serotypes, much closer to the number of identified serotypes. Not surprisingly, we found a much larger proportion of overlap between the two sampling periods, more in line with the observed results (Table 3).

We used logistic regression [**19**] to test whether abundance in the earlier time period predicts presence in the later period. We find a highly significant effect ($p < 0.001$) with the actual data (after including the 16 absent serotypes with abundance zero), but no significant relationship in the samples from the simulations with any parameter values.

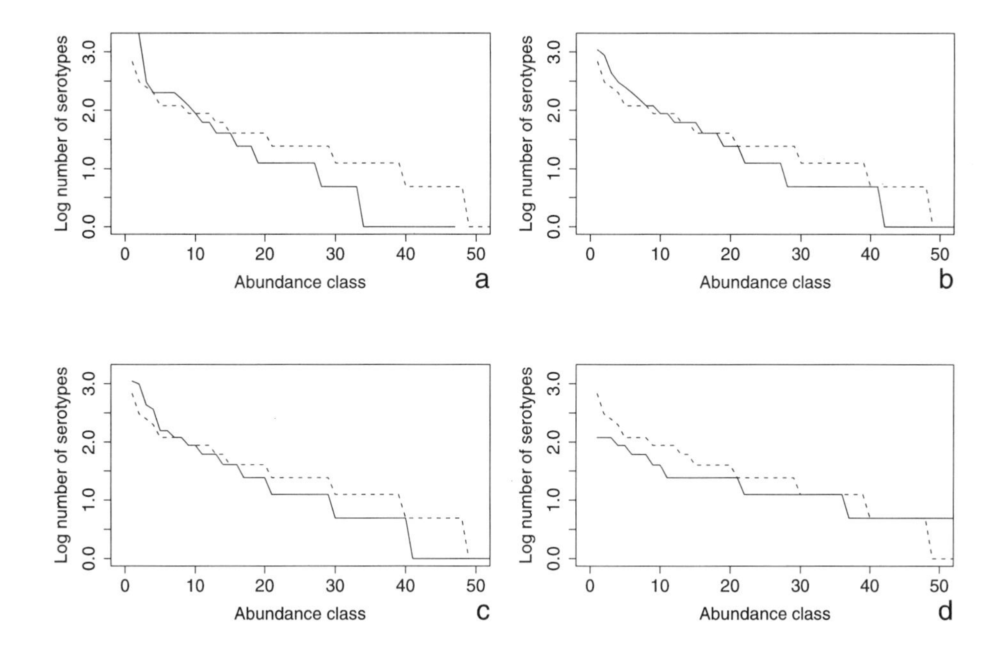

FIGURE 5. The predicted dominance-diversity curves for Tecumseh by running the simulation with $N = 8000$ and a more realistic prevalence of 10 days of infection per year (solid lines) compared with the actual data (dashed lines). The abundance rank of each serotype is based on the total number of individuals observed with that serotype. The Tecumseh study serotyped 250 individuals from 1966-1971, and we sampled the same number of individuals over the course of 6 years from the simulations. Parameter values are **a.** $x_i = 0$ and $m = 0$, **b.** $x_i = 0$ and $m = 0.01$, **c.** $x_i = 1$ and $m = 0$, and **d.** $x_i = 1$ and $m = 0.01$.

TABLE 3. Simulated and measured HRV serotype diversity and overlap: $d = 7$

	$x_i = 0$, $m = 0$	$x_i = 0$, $m = 0.01$	$x_i = 1$, $m = 0$	$x_i = 1$, $m = 0.01$	Measured
1966-71	29	69	56	70	62
1976-81	50	57	55	70	53
overlap	10	32	26	38	42

4. Discussion

We have found that a simple model with realistic parameters does a fairly good job of estimating the standing diversity of HRV. In general, the neutral theory well approximates the results of simulations, although our effective mutation rate ν differed slightly from that estimated directly from the underlying process. Inclusion of competition through cross-immunity increases diversity by as much as 25%. For larger population sizes (5000 individuals in this study), maximum diversity is maintained when cross-immunity acts at an intermediate scale which roughly matches scale seen in nature. Immunodominance, which describes the immune response to failed infections, has no detectable effect on diversity.

Using parameters based on the detailed Tecumseh study [**24**], we find a good quantitative fit to the number of serotypes present. However, we find that the turnover between serotypes is far too fast unless we restrict the number of possible serotypes. If only approximately 100 serotypes are viable, it would be extremely interesting to understand the constraints operating on the sites which determine serotype [**16**]. Even in this case, however, the simulations differ from the actual measurements in showing no relationship between abundance in the first census and presence in the later census. If the observed relationship were due to a much lower mutation rate than we assume, the observed diversity could not be maintained. Alternatively, the relationship could be due to heterogeneity among the serotypes, either in detectability or in transmission.

Our model deals with disease transmission in only the simplest way. We neglect age-dependence of prevalence and duration of infection [**17**], seasonality [**14**], and heterogeneity in contact rates. As noted above, we also neglect differences among serotypes in their overall parameters and in their unstudied effects on people of different ages or immune functions. We do not consider the fact that immunity may well last no more than a few years [**8**], and neglect the possible importance of recombination [**2**].

More detailed study of these models needs to address the observed phylogenetic relationships among coexisting serotypes [**30**]. Most simply, the models could use the two types A and B as roughly independent replicates of the experiment. The simulation could be made more explicitly genetic, rather than taking place in an abstract serotype space, and the results could then be presented explicitly as phylogenetic trees. Comparison with related studies of influenza, which shows a very different pattern of replacement of strains through antigenic drift and antigenic shift, could be particularly useful [**11**].

Given the wealth of genetic and epidemiological data available on this widespread and important class of viral infections, we believe that the common cold can provide new insights into the applicability of ecological theories, and that ecological theory can provide new insights into the transmission and evolution of the common cold.

Acknowledgments

Thanks to Lissy Coley and Jim Keener for useful comments on the Master's Thesis from which this chapter was developed, and to two anonymous reviewers for constructive comments.

References

[1] L. J. Abu-Raddad and N. M. Ferguson, The impact of cross-immunity, mutation and stochastic extinction on pathogen diversity, *Proc. Roy. Soc. Lond. B* **271** (2004), 2431–2438.

[2] V. I. Agol, Recombination and other genomic rearrangements in picornaviruses, *Seminars in Virology* **8** (1997), 77–84.
[3] D. E. Anderson, M. P. Carlos, L. Nguyen, and J. V. Torres, Overcoming original (antigenic) sin, *Clinical Immunology* **101** (2001), 152–157.
[4] V. Andreasen, J. Lin, and S. A. Levin, The dynamics of cocirculating influenza strains conferring partial cross-immunity, *J. Math. Biol.* **35** (1997), 825–842.
[5] R. Antia, M. A. Nowak, and R. M. Anderson, Antigenic variation and the within-host dynamics of parasites, *Proc. Natl. Acad. Sci. USA* **93** (1996), 985–989.
[6] S. Blomqvist, C. Savolainen, L. Raman, M. Roivainen, and T. Hovi, Human rhinovirus 87 and enterovirus 68 represent a unique serotype with rhinovirus and enterovirus features, *J. Clinical Microbiology* **40** (2002), 4218–4223.
[7] M. K. Cooney, J. A. Wise, G. E. Kenny, and J. P. Fox, Broad antigenic relationships among rhinovirus serotypes revealed by cross-immunization of rabbits with different serotypes, *J. Immunology* **114** (1975), 635–639.
[8] R. B. Couch, Rhinoviruses, in (B. N. Fields, D. M. Knipe, and P. M. Howley, eds.), *Fields Virology*, Lippincott-Raven Publishers, Philadelphia, 3rd edition, 1996, 713–734.
[9] R. G. Douglas Jr, Pathogenesis of rhinovirus common colds in human volunteers, *Annals of Otology, Rhinology, and Laryngology* **79** (1970), 563–571.
[10] J. W. Drake and J. J. Holland, Mutation rates among RNA viruses, *Proc. Natl. Acad. Sci. USA* **96** (1999), 13910–13913.
[11] N. M. Ferguson, A. P. Galvani, and R. M. Bush, Ecological and immunological determinants of influenza evolution, *Nature* **422** (2003), 428–433.
[12] S. A. Frank, *Immunology and Evolution of Infectious Disease*, Princeton University Press, Princeton, 2002, chapter 6.
[13] J. R. Gog and B. T. Grenfell, Dynamics and selection of many-strain pathogens, *Proc. Natl. Acad. Sci. USA* **99** (2002), 17209–17214.
[14] S. B. Greenberg, Respiratory consequences of rhinovirus infection, *Archives of Internal Medicine* **163** (2003), 278–284.
[15] J. M. Gwaltney Jr, J. O. Hendley, G. Simon, and W. S. Jordan Jr, Rhinovirus infections in an industrial population. II. Characteristics of illness and antibody response, *J. American Medical Association* **202** (1967), 494–500.
[16] D. T. Haydon and M. E. Woolhouse, Immune avoidance strategies in RNA viruses: fitness continuums arising from trade-offs between immunogenicity and antigenic variability, *J. Theor. Biol.* **193** (1998), 601–612.
[17] T. Heikkinen and A. Jarvinen, The common cold, *Lancet* **361** (2003), 51–59.
[18] J. O. Hendley, W. P. Edmondson Jr, and J. M. Gwaltney Jr, Relation between naturally acquired immunity and infectivity of two rhinoviruses in volunteers, *J. Infect. Dis.* **125** (1972), 243–248.
[19] D. W. Hosmer and S. Lemeshow, *Applied Logistic Regression*, Wiley-Interscience, New York, 2000.
[20] S. P. Hubbell, *The Unified Neutral Theory of Biodiversity and Biogeography*, Princeton University Press, Princeton, 2001.
[21] A. L. Hughes, Phylogeny of the Picornaviridae and differential evolutionary divergence of picornavirus proteins, *Infection, Genetics and Evolution* **4** (2004), 143–152.
[22] I. Kawaguchi, A. Sasaki, and M. Boots, Why are dengue virus serotypes so distantly related? Enhancement and limiting serotype similarity between dengue virus strains, *Proc. R. Soc. Lond. B* **270** (2003), 2241–2247.
[23] R. M. Ledford, N. R. Patel, T. M. Demenczuk, A. Watanyar, T. Herbertz, M. S. Collett, and D. C. Pevear, VP1 sequencing of all human rhinovirus serotypes: Insights into genus phylogeny and susceptibility to antiviral capsid-binding compounds, *J. Virol.* (2004), 3663–3674.
[24] A. S. Monto, E. R. Bryan, and S. Ohmit, Rhinovirus infections in Tecumseh, Michigan: frequency of illness and number of serotypes, *J. Infect. Dis.* **156** (1987), 43–49.
[25] M. A. Nowak, What is a quasispecies? *Trends in Ecology and Evolution* **7** (1992), 118–121.
[26] M. A. Nowak, R. M. May, R. E. Phillips, S. Rowland-Jones, D. G. Lalloo, S. McAdam, P. Klenerman, B. Köppe, K. Sigmund, C. R. M. Bangham, and A. J. McMichael, Antigenic oscillations and shifting immunodominance in HIV-1 infections, *Nature* **375** (1995), 606–611.

[27] M. A. Nowak, R. M. May, and K. Sigmund, Immune responses against multiple epitopes, *J. Theor. Biol.* **175** (1995), 325–353.

[28] L. J. Patterson and V. V. Hamparian, Hyper-antigenic variation occurs with human rhinovirus type 17, *J. Virol.* **71** (1997), 1370–1374.

[29] C. M. Pease, An evolutionary epidemiological mechanism, with applications to type A influenza, *Theor. Pop. Biol.* **31** (1987), 422–452.

[30] C. Savolainen, S. Blomqvist, M. N. Mulders, and T. Hovi, Genetic clustering of all 102 human rhinovirus prototype strains: serotype 87 is close to human enterovirus 70, *J. General Virology* **83** (2002), 333–340.

[31] C. Savolainen, P. Laine, M. N. Mulders, and T. Hovi, Sequence analysis of human rhinoviruses in the RNA-dependent RNA polymerase coding region reveals large within-species variation, *J. General Virology* **85** (2004), 2271–2277.

[32] E. J. Stott and M. Walker, Antigenic variation among strains of rhinovirus type 51, *Nature* **224** (1969), 1311–1312.

Department of Mathematics, University of Utah, 155 South 1400 East, Salt Lake City, UT 84112-0900

E-mail address: `william@math.utah.edu`

Department of Mathematics and Department of Biology, University of Utah, 155 South 1400 East, Salt Lake City, UT 84112-0900

E-mail address: `adler@math.utah.edu`

DIMACS Series in Discrete Mathematics
and Theoretical Computer Science
Volume **71**, 2006

Drug Resistance in Acute Viral Infections: Rhinovirus as a Case Study

Alun L. Lloyd and Dominik Wodarz

Abstract. The emergence and spread of drug resistant virus variants reflects both within-host and between-host processes. We develop an epidemiological model that can be used to address the spread of resistance at the population level, and a virus dynamics model that can be used to study the dynamics of virus over the time course of an individual's infection. The dynamics depend in an important way on the competition between drug sensitive and drug resistant virus strains. A key observation is that the strength of competition between strains is strongly modulated by the degree of cross-immunity that infection with one strain confers against infection with the other. At the within-host level, we see that an efficient immune response can reduce the likelihood of the emergence of resistant virus. Consequently, resistance poses more of a problem for chronic infections in which there is significant immune impairment than for acute infections. These findings are discussed in the setting of rhinovirus infections, which are an important cause of infection in humans and for which novel antiviral drugs are being developed.

1. Introduction

Viruses are responsible for a large number of infectious disease cases each year. Many viruses are associated with severe disease and high levels of mortality: examples include the human immunodeficiency virus (HIV) and pandemic strains of influenza such as the 1918 Spanish flu. Even when the case fatality ratio is not so high, the sheer number of cases can lead to a large number of disease-related deaths. Many other viruses, on the other hand, are associated with much less severe disease, but their high levels of morbidity lead to their having a major economic impact (see [**19**]). As an example, rhinovirus infections— responsible for roughly a half of 'common cold' cases— are one of the leading reasons for people to visit their physician in the US (see [**16, 32**]).

Until recently, there were only a few drugs that could be used to combat viral infections. This situation has changed with the development of a large number of antiviral drugs, for instance in the settings of HIV, hepatitis B and influenza. The potential benefit of these drugs is enormous, both in terms of improving the

---

Key words and phrases. Competition model, Within-host, Between-host, Rhinovirus, Drug resistance.

prognosis for infected individuals and for the potential reduction of transmission of the infection at the population level (see [**8, 52**]).

The use of an effective drug treatment imposes a strong selection pressure on a virus (or any other infectious disease agent). Variants of the virus that are less sensitive to the action of the drug have a considerable replicative advantage over more sensitive variants. As a consequence, their relative frequency will increase. The emergence of viral variants that are resistant to one or more drugs threatens to curb the benefits that might be gained from the use of antivirals: concern over the emergence of resistant forms of HIV and influenza virus has long been expressed and indeed such forms have often been found in the wake of the introduction of drugs against these infections (see [**15, 31, 36, 46, 48**]).

Many viruses have the potential to undergo rapid genetic changes. For instance, in the case of HIV, the reverse transcription of viral RNA to DNA during the virus's replication cycle lacks a proof reading mechanism and so is error-prone, leading to the generation of many mutants. Since, in many instances, drug resistance can be conferred by small genetic changes, such as even a single point mutation, there is ample opportunity for the acquisition of resistance. (In the case of bacteria, genetic material can be acquired from other bacteria or the wider environment, offering them yet more opportunities to acquire resistance.) The acquisition of resistance, however, often comes at a cost: the resistant type is typically less fit (i.e. has a lower replicative ability) in the absence of the drug than the sensitive virus from which it evolved (see, for example, [**4, 26**]).

This combination of potentially rapid genetic change and the strong selection pressure imposed by drug treatment provides an ideal environment within which drug resistant variants can emerge and spread. Any attempt to deploy drug treatment on a wide scale should, therefore, take resistance, and its management, into account [**37**]. In particular, issues surrounding resistance are an important consideration when new drugs are brought into clinical use.

In order to fully understand drug resistance, two distinct issues must be considered. The emergence of a novel drug resistant viral variant typically occurs within an infected individual over the time course of their infection. The population level impact of this new resistant variant can only be assessed by understanding its spread from individual to individual in the population. Mathematical modeling provides an ideal framework within which such questions can be addressed, as witnessed by the growing literature in this area (see, for example, [**6, 7, 9, 10, 11, 21, 24, 25, 37, 38, 39, 47, 50**]).

In this chapter we shall address both of these issues, using modeling approaches to examine the likelihood of emergence and spread of resistance. We shall focus on the setting of human rhinovirus infections, for which an effective drug, pleconaril, has only recently been developed (see [**32, 33, 49**]). We shall see that the epidemiology and natural history of rhinovirus infections plays a crucial role in terms of resistance. Consequently, we shall compare and contrast behavior seen here with what occurs in other settings such as HIV and influenza.

This chapter is organized as follows. After discussing pertinent details of rhinovirus epidemiology, we shall introduce the basic epidemiological framework that we employ. After discussing the spread of resistant virus at the epidemiological level, we turn to the dynamics of resistant and sensitive virus within an infected individual.

2. Rhinovirus Epidemiology

Roughly 100 serotypes of human rhinoviruses have been described (see [**16, 29**]). Epidemiological surveys typically indicate that a large number of these co-circulate within a given community at any given time, with little indication of clear geographic or temporal patterns of their distribution (see [**22, 28, 30, 42, 41**]). Not all serotypes are equally transmissible in all settings, but the impact of any differences in transmissibility on population-level patterns of serotype prevalence is not clear.

An interesting observation is that new serotypes do not appear to be emerging over time (see [**23, 42**]). This is in marked contrast to influenza, where the appearance of new strains, which over a period of time replace the previously existing strains, is well documented (see [**12, 20**]).

Infection with a given rhinovirus typically leads to the development of cold symptoms following a short (1-2 day) incubation period (see [**16, 30**]). Symptomatic individuals have been shown to shed large amounts of virus, suggesting that the duration of infectiousness typically echoes the exhibition of symptoms. About 25% of infections appear to be subclinical, with few symptoms being exhibited: such individuals are presumably much less infectious.

Recovery from infection leads to long-lived, if not lifelong, immunity against the infecting serotype (see [**14, 27, 35, 51**]). Cross-immunity between serotypes appears to be low since individuals repeatedly acquire rhinovirus infections, but the infection rate declines with age (adults typically experience roughly 2-4 colds per year, and children between 4 and 6, see [**29**]). The higher infection rate amongst children is consistent with this picture of immunity: repeated infection leads to immunity against an ever larger repertoire of serotypes over time.

The antiviral agent pleconaril interrupts the viral replication cycle by blocking the attachment of virus to its cellular target. Clinical trials of the drug have shown that its use can reduce both the duration and severity of cold symptoms (see [**32, 33, 49**]). Given the observed relationship between disease symptoms and infectiousness, this will presumably also reduce the transmissibility of infection. An interesting *in vitro* observation is that pleconaril is not effective against all rhinovirus serotypes (see [**44**]): there are what might be termed 'naturally resistant' serotypes.

The limited clinical trials that have studied the use of pleconaril have yet to yield much information on the emergence of resistance. In a small fraction of treated individuals, virus recovered after the end of infection exhibited somewhat decreased drug susceptibility (see [**33**]). The clinical significance of these virus variants is not, however, clear from these studies: neither the level to which these variants grew nor the timing of their appearance were ascertained. Additional information has been gleaned from *in vitro* cell culture studies, from which drug resistant mutants have also been recovered. Such mutants often show severely reduced growth (see [**34, 45**]).

An important observation in the rhinovirus setting, therefore, is that there are two distinct types of resistant variants. Naturally resistant variants existed and circulated long before the introduction of treatment, whereas novel resistant variants would only appear in the wake of the introduction of treatment. Notice that these novel resistant variants are likely to be serotypically similar or identical to existing forms of the virus. The form of our model in the 'novel variant' setting echoes existing models for the population-level spread of resistance in which it is typically assumed that individuals are either infected with resistant or sensitive variants, but not both. The natural resistance setting, however, has received much less attention in the literature.

Clinical trials involving pleconaril have focused on therapeutic use of the drug, where already infected patients are treated. Beyond the setting of clinical trials, its availability would likely be restricted to prescription usage, requiring a visit to a physician. In the early stages of its deployment, therefore, it would be reasonable to assume that it use would not be extremely widespread. The possibility of its use as a prophylactic has been considered as a means of reducing the potential for secondary infections. In this study we restrict attention to therapeutic use of pleconaril.

3. The Epidemiological Model

Our population-level model is based on the well-known multi-strain model of Castillo-Chavez et al. [**13**]. Their model is built upon the standard SIR (susceptible/infective/recovered) framework (see, for example, [**1**]), but incorporates two distinct strains of infection. (A further extension to additional strains is given by Andreasen et al. [**2**]). We extend this model to include treatment so that it can be used to describe the transmission dynamics of drug sensitive and resistant virus (see [**40**]).

The infections are assumed to be non-fatal, with recovery from a particular strain leading to life-long immunity against reinfection with that strain. An important feature of the model is that recovery from one infection confers partial protection against infection with another strain. One way to model this cross-immunity is by reducing the susceptibility of an individual to infection by other strains (see [**2, 13**]). The model must keep track of the infection histories of individuals: for combinatorial reasons, this rapidly becomes difficult as the number of strains in the model increases (see [**2**]). For this reason, we restrict attention to a two strain setting.

The fractions of the population that are susceptible, infective and recovered are denoted by S, I and R, respectively. We subdivide the S and I classes according to infection history (superscripts) and the current strain with which an individual is infected (subscripts). So, for example, S denotes the fraction of individuals that have never experienced infection. S^1 denotes individuals that have recovered from strain one but are still susceptible to strain two. In a similar way, I_2 denotes individuals who are infected with strain two and who have not experienced strain one, while I_2^1 denotes individuals that are infected by strain two but who have experienced strain one in their past. Notice that the total fraction of individuals that are infected with strain two equals $I_2 + I_2^1$.

All of the infection classes, and the possible transitions between them, are illustrated in Figure 1. Notice that for simplicity we neglect any latent period between infection and the start of infectiousness, assume that individuals can only be infected with one strain at a time and that individuals who recover from one strain are immediately susceptible to the other strain (if they have not previously encountered that strain).

Making standard assumptions (for instance that the population is well-mixed and that recovery probabilities are independent of the time since infection) Castillo-Chavez et al. obtain the following set of equations for the epidemiological model

$$\dot{S} = \mu - \beta_1 S(I_1 + I_1^2) - \beta_2 S(I_2 + I_2^1) - \mu S \tag{3.1}$$

$$\dot{I}_1 = \beta_1 S(I_1 + I_1^2) - (\mu + \gamma) I_1 \tag{3.2}$$

$$\dot{I}_2 = \beta_2 S(I_2 + I_2^1) - (\mu + \gamma) I_2 \tag{3.3}$$

$$\dot{S}^1 = \gamma I_1 - \sigma \beta_2 S^1 (I_2 + I_2^1) - \mu S^1 \tag{3.4}$$

$$\dot{S}^2 = \gamma I_2 - \sigma \beta_1 S^2 (I_1 + I_1^2) - \mu S^2 \tag{3.5}$$

$$\dot{I}_2^1 = \sigma \beta_2 S^1 (I_2 + I_2^1) - (\mu + \gamma) I_2^1 \tag{3.6}$$

$$\dot{I}_1^2 = \sigma \beta_1 S^2 (I_1 + I_1^2) - (\mu + \gamma) I_1^2 \tag{3.7}$$

$$\dot{R} = \gamma I_1^2 + \gamma I_1^2 - \mu R. \tag{3.8}$$

Here, the parameters β_1 and β_2 are the transmission parameters of the strains, γ is the rate of recovery (assumed to be the same for both strains) and μ is the birth and death rate of the population. We assume the birth and death rates to be equal, corresponding to the assumption of a constant population size.

The remaining parameter, σ, is an inverse measure of cross immunity: under the mass-action (perfect mixing) assumption, the rate at which naive individuals become infected by strain one is given by the product $\beta_1 S(I_1 + I_1^2)$. In contrast, the rate at which individuals that have already experienced infection with strain two become infected with strain one is $\sigma\beta_1 S^1(I_1 + I_1^2)$. If σ equals zero, recovery from strain two leads to complete protection against strain one. If σ equals one, recovery from strain two offers no protection against strain one. (In this case, the model of [**13**] reduces to an earlier model that was employed by [**18**].)

Mutation between strains can be considered by allowing a small fraction of infections, ν, with one strain to lead to infection with the other strain. The parameter ν attempts to capture the complex process by which a strain might mutate over the course of an individual's infection. A more realistic description of this process requires the development of a within-host model, as we outline towards the end of this chapter. This mutation term is most likely to be important when one of the strains is either absent or present at a low frequency.

4. Behavior of the Epidemiological Model

4.1. Behavior in the Absence of Treatment. As is often the case with epidemiological models of this sort, the behavior of the system depends on the basic reproductive numbers of the two strains [**13, 18**]. R_0^1 quantifies the transmissibility of strain one: in a completely susceptible population, the introduction of one individual infected with strain one would lead to an average number of $R_0^1 = \beta_1/(\gamma+\mu)$ secondary infections of type 1. Similarly, $R_0^2 = \beta_2/(\gamma + \mu)$. A necessary (but not sufficient) condition for the establishment of a given strain is that its R_0 value is greater than one.

The two strains compete for susceptible hosts, with the level of competition dependent on the level of cross-immunity (see [**2, 13**]). If there were complete cross-immunity, individuals that became infected with one strain could never become infected with the other strain. In this case, the strain with the larger R_0 value would out-compete the other strain, driving it to extinction [**13**]: this is an example

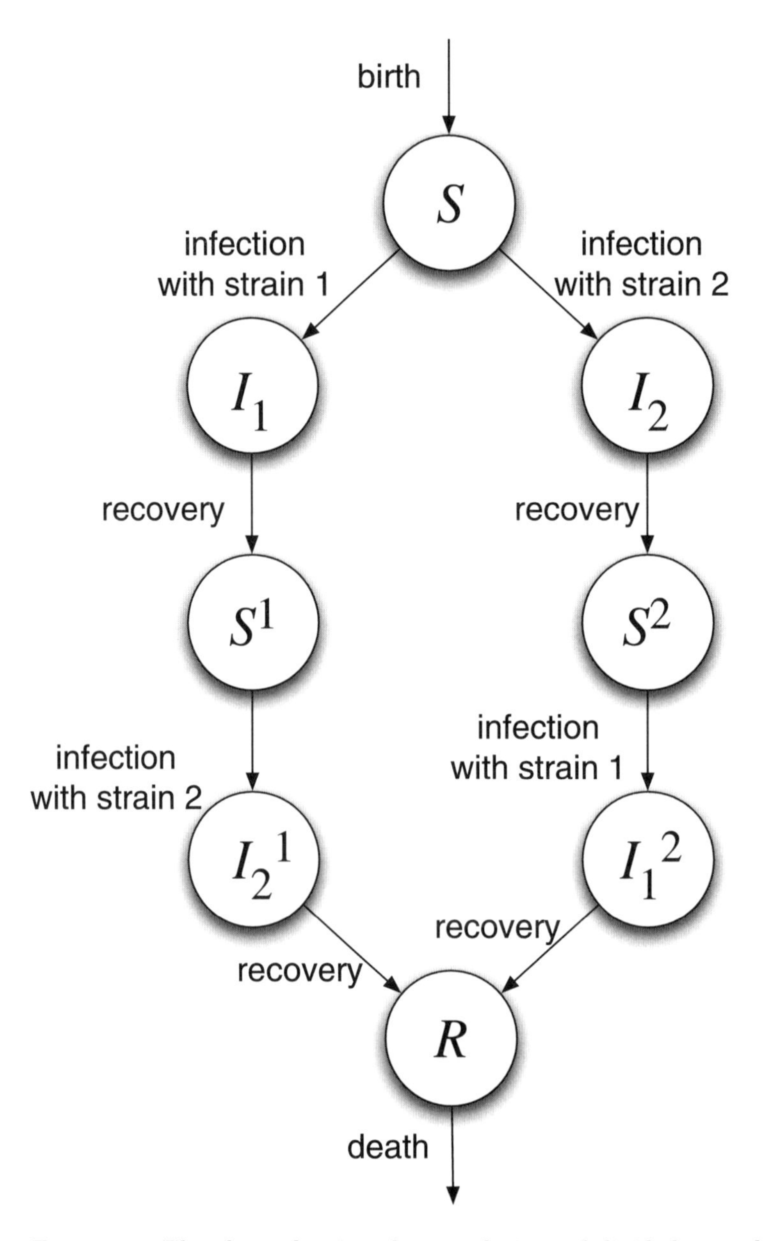

FIGURE 1. Flowchart showing the population subdivided according to infection status and the possible transitions between classes. The model assumes a constant disease-independent background mortality rate of μ. For reasons of clarity, this figure does not depict deaths from classes other than the recovered class.

of competitive exclusion. At the other extreme, if there were no cross-immunity, competition between strains would be very weak: the only effect of strain one on

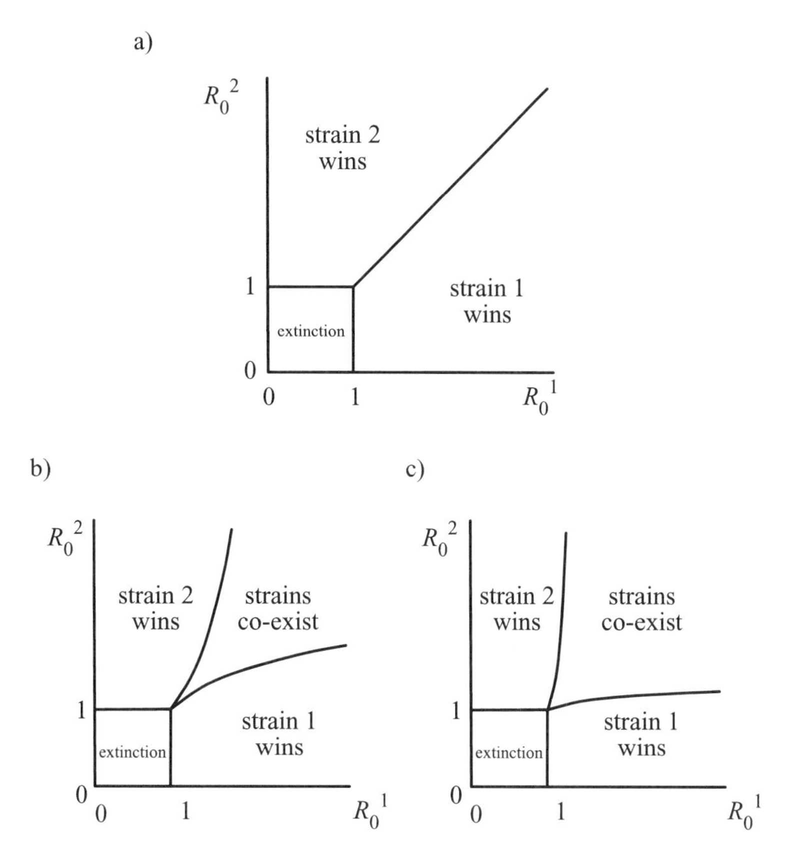

FIGURE 2. Schematic figure illustrating the outcome of the competition between strains in the two strain epidemiological model, equations (3.1-3.8). The three panels depict three different strengths of cross-immunity: (a) complete cross-immunity ($\sigma = 0$), (b) partial cross-immunity, (c) no cross-immunity ($\sigma = 1$). The curves separate different regions of the (R_0^1, R_0^2) plane according to the long-term (stable equilibrium) behavior seen. (The exact boundary curves that appear in this figure can be calculated exactly using the analytic results of [**13**] and [**18**], as outlined in the text.)

strain two arises because coinfection is assumed not to occur (see [**18**]). If the duration of infection is short then this effect would be very small. The dependence of the equilibrium (long-term) behavior of this model, in the absence of drug treatment, on the two values R_0^1 and R_0^2 is illustrated in Figure 2. Analytic expressions for the curves that separate the different regions in Figure 2 can be found in [**13**] in the general case and in [**18**] in the $\sigma = 1$ case.

The inclusion of mutation within this model prevents the less fit strain from going extinct when it would otherwise do so. It will be continually produced by mutation from the other strain and so will be present at the low equilibrium level that results from the balance between mutation (producing the strain) and selection (which reduces the strain's relative frequency).

4.2. Inclusion of Treatment. Treatment is included within the model by allowing for treated classes of individuals. We imagine that a certain fraction of infections, ϵ, are treated and that the effect of treatment is to increase the rate at which treated individuals recover from γ to γ_T. This corresponds to a reduction in the average duration of infection. We assume that strain one is drug sensitive and that strain two is completely drug resistant.

In this extended model, the class T_1 denotes infectious individuals whose first infection is with strain 1 and are being treated. The class T_1^2 denotes infectious individuals whose second infection is with strain 1 and are being treated. The classes I_1 and I_1^2 now refer to untreated infectious individuals. (Since treatment has no effect on infections with strain 2, there is no need to make a similar distinction for those individuals.) The total fraction of individuals that are infected with strain 1 is given by $I_1 + T_1 + I_1^2 + T_1^2$. Equation (3.2) (the previous equation for $\dot{I}_1$) is replaced by the following pair of equations

$$\dot{I}_1 = (1-\epsilon)\beta_1 S(I_1 + I_1^2 + T_1 + T_1^2) - (\mu + \gamma)I_1 \tag{4.1}$$

$$\dot{T}_1 = \epsilon\beta_1 S(I_1 + I_1^2 + T_1 + T_1^2) - (\mu + \gamma_T)T_1. \tag{4.2}$$

A similar pair of equations replaces the previous $\dot{I}_1^2$ equation.

As mentioned earlier, an important feature of many resistant variants is that the acquisition of resistance comes at some fitness cost. In terms of the model parameters, this means that $\beta_2 < \beta_1$. An important realization in the rhinovirus setting is that resistance is not always associated with a fitness cost: some serotypes are naturally resistant to pleconaril. Given that these strains have been circulating in the human population for some time, and there is no evidence that these strains are present at lower prevalence than drug sensitive strains, we argue that their transmissibility must be comparable to other, drug sensitive, serotypes.

4.3. Behavior of the Model in the Presence of Treatment. The imposition of treatment reduces the basic reproductive number of strain 1 by reducing the average duration of infections with that strain. Treatment, therefore, shifts the competitive balance in favor of strain 2. Since the strength of competition is modulated by the degree of cross-immunity between strains, the impact of treatment depends on whether we are considering the setting of natural resistance (two serotypically distinct strains with low or no cost of resistance) or the setting of novel resistant strains (two serotypically identical strains with a significant cost of resistance).

When cross-immunity between strains is not too high, as in the 'natural resistance' setting, the imposition of treatment can do little to alter the prevalence of drug sensitive and drug resistant strains. The model predicts that the use of antiviral drugs is unlikely to lead to a major change in the relative frequencies of sensitive and resistant strains, provided that the fraction of individuals treated is not too high. This situation is illustrated in Figure 3a, in which we assume that there is no transmission cost to having resistance. If a large fraction of individuals is treated (a situation which is unlikely to hold if people have to visit a physician in order to obtain a drug for what is in most cases a mild infection), then the use of the drug can have a more significant impact.

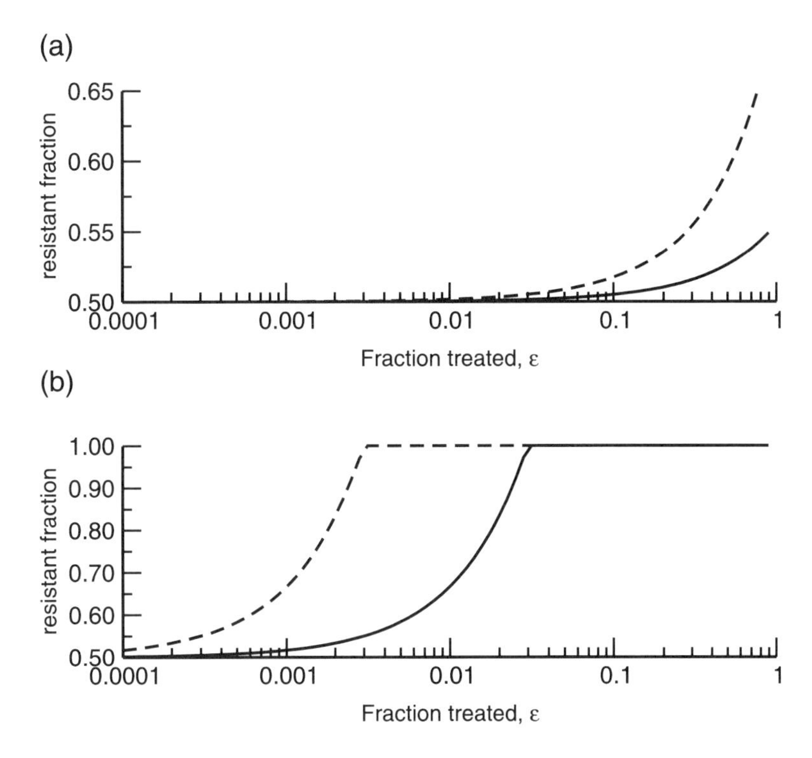

FIGURE 3. Dependence of the equilibrium fraction of infections that are due to the resistant strain on the fraction of infections that are treated (ϵ). Panel (a) depicts the cases of no cross-immunity ($\sigma = 1$, solid curve) and partial cross-immunity ($\sigma = 0.1$, broken curve). The impact of treatment is small or modest in these cases. Panel (b) illustrates strong cross-immunity ($\sigma = 0.001$, solid curve and $\sigma = 0.0001$, broken curve), and shows that treatment can have a major impact on the relative prevalence of resistant infections if cross-immunity is strong. In the limiting case of complete cross-immunity ($\sigma = 0$), all infections, at equilibrium, are with the resistant strain whenever ϵ is positive. In this figure, we assume that both strains have an R_0 value of six: since the basic reproductive number of rhinovirus infections has yet to be quantified, we take this value for illustrative purposes. We assume that untreated infections last, on average, six days while treated infections last five days. The average lifespan of the population is taken to be 70 years. All curves were generated by numerical integration of the equations governing the system.

If there is strong cross-immunity between strains (i.e. when the competition between strains is strong), the imposition of treatment can have a major impact (Figure 3b). This scenario, in which the resistant strain is serotypically similar to the sensitive strain, is most likely to arise when the resistant strain is a novel mutant derived from an existing drug sensitive strain. The acquisition of resistance in this setting, however, is typically associated with a fitness cost. This reduces the competition benefit that the resistant strain experiences due to the imposition of treatment. Consequently Figure 3b provides an overly pessimistic estimate of the increase of the resistant strain since the parameter values chosen did not include any cost of resistance. The drug-resistant strain will only increase its relative frequency if the reduction in R_0 that it experiences due to the cost of resistance is lower than the reduction in R_0 that the sensitive strain experiences due to the use of treatment. In this case, we would expect the drug resistant strain to largely replace the sensitive strain because of the strong competition. (The sensitive strain will not completely disappear, assuming that back mutation of the resistant strain can generate the sensitive strain.) A previously sensitive serotype would be largely replaced by a resistant serotype.

In order for this second scenario to occur, a novel resistant mutant must have appeared over the course of an individual's infection. For us to consider the likelihood of this event, we must turn to a within-host model that can describe the time course of infection in an individual. Before doing so, we remark that this discussion has only examined long-term (equilibrium) dynamics. More details of the short-term dynamics, in particular a discussion of the time-scale on which resistance spreads, can be found in [**40**].

5. The Within-Host Model

The time course of infection within an individual is modeled by describing the interaction between the virus, the cells that it infects (the so-called 'target cells') and an immune response that develops against the infection. Our model is based on the basic model of virus dynamics that has received much attention in the setting of HIV infection (early examples include [**24, 43**]). As a consequence, we can draw upon the literature that models the emergence of drug resistant HIV strains (see, for example, [**10, 11, 24, 47**]). There are, however, important differences that arise because rhinovirus infections are acute infections that are resolved by the immune system, as opposed to chronic HIV infections in which the immune system experiences significant impairment.

Two processes must occur in order for a novel drug resistant viral strain to be generated within a patient and then transmitted to another individual. First, the resistant virus strain must be created by mutation: this event occurs over the course of the virus's transmission and replication cycle. Second, the mutant virus must replicate to a sufficiently high level within the infected individual in order to allow the possibility of transmission to occur. We shall consider these two processes separately, and since the second process is in some ways analogous to the population-level process already studied, our discussion will first focus on this issue.

The basic model of virus dynamics for a single strain can be written as

$$\dot{x} = \lambda - dx - \beta xy \tag{5.1}$$

$$\dot{y} = \beta xy - ay - pyz \tag{5.2}$$

$$\dot{z} = cyz - bz. \tag{5.3}$$

Here, x denotes the number of uninfected (target) cells, y the number of infected cells and z measures the strength of the immune response. We make standard assumptions regarding uninfected and infected cell dynamics. Free virus particles are assumed to undergo a rapid turnover, so their numbers closely echo those of the infected cell populations (in the model, we make a quasi-steady state assumption between free virus and infected cells). The rate at which target cells become infected is assumed to be described by the mass action term βxy, i.e. is proportional to both the numbers of target cells and infected cells, with constant of proportionality β. Uninfected cells are produced at a constant rate λ and die at a constant per-capita rate d. Infected cells die at rate ay (so that their lifespan is $1/a$) in the absence of an immune response. The immune response that develops against the infection further removes infected cells at rate pyz. The presence of infected cells stimulates the immune response: for simplicity and definiteness, we take this stimulation to be described by the bilinear term cyz. (For more detailed discussions of alternative forms of this immune response, see [**17**] or [**53**].) Finally, the immune response (in the absence of stimulation) is reduced at rate bz.

The behavior of this model is determined by the basic reproductive number of the within-host system. During the initial stages of infection, when the target cell population is close to its infection free equilibrium level and the immune response is weak, the average number of secondary infected cells that result from the introduction of a single infected cell is given by $R_0 = \lambda\beta/da$. When R_0 is less than one, the level of virus within an individual can never increase. If R_0 is greater than one, the level of virus can increase and an infection can be established: the patient's viral load will reach an equilibrium level over time. It is important to keep in mind that this within-host measure of the virus's replicative ability is different to the population-level (between-host) R_0 discussed above. (The two quantities are linked, however: if the virus is unable to replicate effectively within an individual, then an infected individual would be very unlikely to be able to transmit infection.)

An important observation is that the immune response cannot completely clear the infection within this model. This reflects the predator-prey dynamics between the immune response and the infection: if the level of infection falls, the level of the immune response falls which in turn allows the level of infection to rise. Eventually, an equilibrium is established between the two. The level of this equilibrium, however, is higher or lower depending on how quickly the immune response can kill infected cells.

Since we are interested in the dynamics of an acute infection, our primary interest is in the short term behavior of the within-host model in response to the introduction of infection. It is difficult to perform much mathematical analysis on these dynamics, but it turns out that some insight can be gained by understanding the equilibrium behavior of the model. An equilibrium approach is also informative when comparing the behavior of the rhinovirus system with that previously described in the setting of HIV infection. Our discussion will initially focus on equilibrium dynamics before later returning to the short term dynamics of acute infection.

5.1. The Two Strain Within-Host Model and Drug Treatment. The within-host model can be extended to consider the interaction between two virus strains and the impact of treatment on the strains (see [**54**]). We account for wild-type (drug sensitive) strain and a drug resistant strain, and the numbers of cells infected with these strains are written as y_w and y_r. Furthermore, we assume that these strains will be equally well recognized by the immune response: this assumption corresponds to the belief that a novel resistant strain will most likely be serotypically identical to the strain from which it evolved. We also assume that the lifespan of infected cells (in the absence of an immune response) is the same for cells that are infected with either strain.

The infection parameter β is allowed to differ between the two strains, as expressed by the two parameters β_w and β_r. As mentioned before, the acquisition of resistance usually comes at a fitness cost, so we have that $\beta_r < \beta_w$. The imposition of treatment reduces the replication of virus: we have that $\beta'_w < \beta_w$, where β'_w denotes the treatment value of the infection parameter. In order for the mutant strain to be considered resistant, we require that β'_r is greater than β'_w: the 'resistant' strain is better able to replicate during treatment than the wild-type. Notice that the replication rate of the resistant strain may be affected by treatment, but the operational definition of resistance is that treatment overturns the competitive advantage that the wild-type enjoys (due to the cost of resistance) in the absence of treatment (see [**10, 11**]).

From these infection parameters, we can write down the basic reproductive numbers of the strains in the presence and absence of treatment. We denote these quantities by $R_0^{(\mathrm{w})}$, $R_0^{(\mathrm{w})'}$, $R_0^{(\mathrm{r})}$ and $R_0^{(\mathrm{r})'}$, where the superscripts denote wild-type and resistant and the primes denote values during treatment. Since we have assumed that the infected cell lifespan (and indeed the rate at which the immune response clears infected cells) is the same for both strains, we see that the basic reproductive numbers are obtained by multiplying the corresponding infection parameters by λ/da.

Mutation can lead to the generation of resistant strains in patients that were only initially infected with drug sensitive virus. To model this possibility, we assume that infection of cells by the sensitive strain can lead to the appearance of a resistant strain. The per infection mutation probability (usually assumed to be small) is written as μ. Since we are interested in the emergence and spread of resistant virus, we ignore back mutation from the resistant to the sensitive strain.

The two strain model, in the presence of mutation from wild-type to resistant virus, can be written as

$$\dot{x} = \lambda - dx - \beta_w x y_w - \beta_r x y_r \tag{5.4}$$

$$\dot{y}_w = (1-\mu)\beta_w x y_w - a y_w - p y_w z \tag{5.5}$$

$$\dot{y}_r = \beta_r x y_r + \mu \beta_w x y_w - a y_r - p y_r z \tag{5.6}$$

$$\dot{z} = c(y_w + y_r)z - bz \tag{5.7}$$

in the absence of treatment. During treatment, the same model holds except that the β parameters are replaced by the β'.

5.2. Equilibrium Behavior of the Within-Host Model. As for the between-host model, the outcome of the within-host model depends on the competition between strains. Our first step towards understanding the within-host dynamics involves investigating the competition between sensitive and resistant strains. In this section we shall assume that resistant strains are already present at an early stage of infection: discussion of the appearance of resistant strains will be deferred until later.

An important difference between the within-host and between-host models is that target cells only experience infection with one strain or the other: the within-host model is in some ways similar to the between-host model with complete cross immunity ($\sigma = 0$). Competition between strains, therefore, is strong. In the absence of mutation and treatment, the wild-type strain out-competes the resistant strain, driving it to extinction. Assuming that $R_0^{(\mathrm{w})}$ is greater than one, an equilibrium level of infection is approached in the long term. As in the between-host model, mutation allows for the maintenance of the resistant strain at a low level, reflecting the within-host balance between mutation and selection.

When treatment is applied, the sensitive strain loses its competitive advantage and so the frequency of the resistant strain may increase. The presence of an effective immune response, however, can temper this increase. Assuming that the target cell level quickly rebounds to its pre-treatment level following the start of therapy and that the level of the immune response does not fall appreciably over this time, the following condition for the growth of resistant virus can be derived (see [**54**])

$$R_0^{(\mathrm{r})'} - 1 - (p/a)z^* > 0. \tag{5.8}$$

Here z^* is the equilibrium level of the immune response that developed in response to the growth of the sensitive virus. As would be expected, since the immune response effectively increases the death rate of infected cells, the established immune response makes it more difficult for the resistant strain to grow (its R_0 must lie well above one).

In the limit of a strong enough immune response, it can be shown that this condition reduces to

$$R_0^{(\mathrm{r})'} > R_0^{(\mathrm{w})}. \tag{5.9}$$

In order for resistant virus to grow during treatment, its basic reproductive number must be greater than that of the sensitive virus in the absence of treatment. This is very unlikely to be the case in reality: the cost of resistance means that the resistant virus should have a lower replicative ability than the wild-type.

If the immune response is weaker, this condition is relaxed somewhat. The degree to which this condition is relaxed depends on the relative importance of the immune response, compared to the natural death of infected cells, in the turnover of infected cells. It has been argued (see [**54**]) that if the immune response is efficient, i.e. is largely responsible for the removal of infected cells, then condition (5.9) provides a good approximation.

The level of resistant virus can be kept in check by two distinct mechanisms. In the absence of therapy, resistant virus is out-competed by sensitive virus. In the presence of therapy, resistant virus can be kept at low values by the immune response that was developed in response to the sensitive virus. This second effect assumes that the immune response does not quickly fade upon treatment (and the subsequent decline in the level of sensitive virus). Over time, the immune response will decline, relaxing the condition for the growth of resistant virus. The model, therefore, predicts that the timing of the rise of resistant virus will depend on the time taken for the immune response to fade away.

If the immune response is absent (or is seriously impaired), the level of resistant virus will increase in the presence of therapy provided that $R_0^{(\mathrm{r})'}$ is greater than one. (Recall that the the definition of resistance guarantees that $R_0^{(\mathrm{r})'} > R_0^{(\mathrm{w})'}$, that is to say that the resistant type has a competitive advantage over the wild-type during therapy.) This is precisely the situation that has been described for HIV infection (see [**10, 11**]). An interesting observation in this case is that the equilibrium level of resistant virus that is established upon treatment will, in general, not be much different to the equilibrium level of sensitive virus that was established before treatment (see [**11**] for more details).

5.3. Acute Infection Dynamics. Over the time course of an acute infection, the immune response develops in response to the growth of the virus. This observation is significant in light of the preceding discussion: the competitive advantage of resistant virus during therapy is modulated by the strength of the immune response that has developed. If therapy is started early during the time course of infection —before a strong immune response has developed— the resistant virus will find it easier to grow than if therapy is started later in the time course of infection— by which time a strong immune response will have been mounted.

This surprising behavior is illustrated by the numerical simulations presented in Figure 4, which show the levels of sensitive and resistant virus over the time course of infection when treatment is either started early or late. We remark that it is difficult to obtain a precise analytic understanding of this behavior since the levels of the immune response and target cells are changing, often rapidly, in this acute phase. General properties of the behavior, such as the immune response's ability to limit the growth of the resistant virus, can be gained in terms of the preceding equilibrium discussion.

It should be pointed out that our model does not provide an entirely satisfactory description of acute infection dynamics. As discussed above, if we take the simple description of the immune response provided by equation (5.3), immunity is unable to clear the virus. (A more complex description of the immune response— in particular one whose expansion is not proportional to the instantaneous level of the virus— could allow for elimination of the virus. See [**3**] for an example.) The immune response and target cell depletion leads to a low level of virus (or infected cells) following the primary infection, but this level is non-zero. Since the model is deterministic, the virus can persist at any low level and later re-emerge. In our simulations, we assume that the virus has been cleared if it is reduced below some threshold level.

5.4. Generation of Resistant Virus Strains. The preceding discussion only considered the competition dynamics between wild-type virus and pre-existing resistant virus, focusing on the question of whether resistant virus could ever reach levels at which transmission to another individual would be possible. We now turn to the question of the likelihood of mutation leading to the generation of such resistant virus in the first place.

We assume that there is a fixed probability that any viral replication cycle will lead to a cell becoming infected with the novel resistant mutant. The chance of generating a resistant mutant, therefore, is proportional to the number of replication events that occur. We can estimate the relative probabilities of a resistant mutant being generated before the start of therapy and during therapy by simply counting the number of times that wild-type virus replicates before and after the start of therapy. This question of whether resistant virus is more likely to evolve before or during treatment has been studied in detail for HIV infection (see [**11**]).

Figure 5 illustrates the scaled number of replication events that occur before and after the start of therapy for a wild-type infection, assuming a range of treatment efficacies (defined as $1-R_0^{(\mathrm{w})'}/R_0^{(\mathrm{w})}$). The number of replication events before

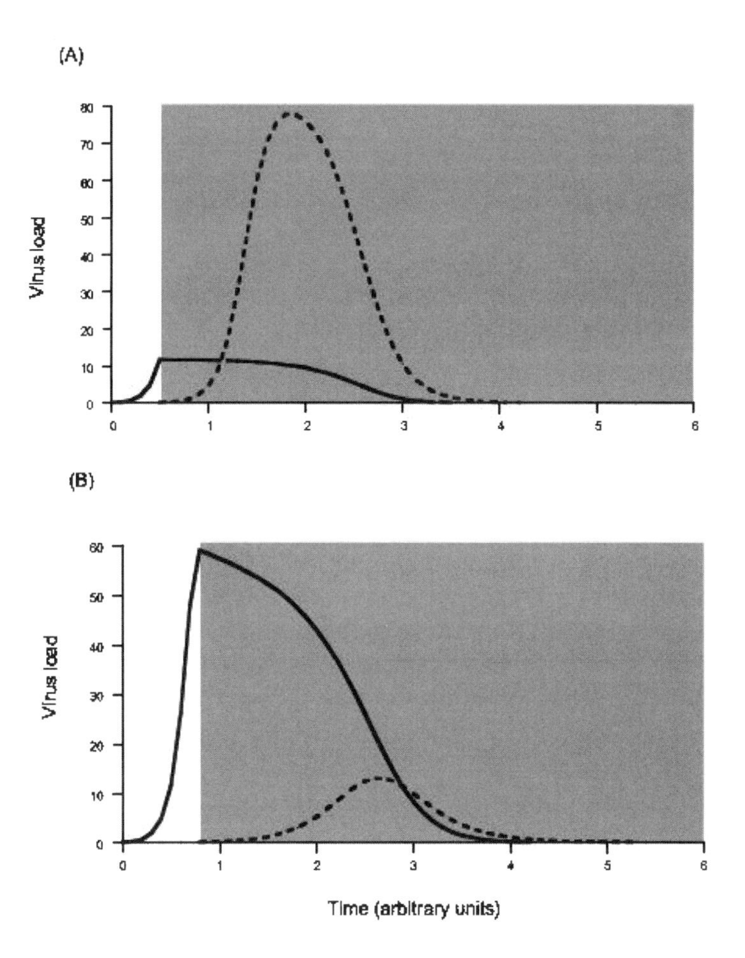

FIGURE 4. The time at which therapy is started affects the level to which resistant virus (dashed lines) can grow over the course of a treated infection. (Solid lines denote the level of sensitive, wild-type, virus.) In panel (a), therapy is started early, before the immune response has fully developed. Consequently, drug resistant virus can grow to a high level. In panel (b), therapy is started later and so a well developed immune response hinders the growth of resistant virus. In both cases it is assumed that resistant virus exists before the start of treatment. An illustrative set of parameter values was chosen as follows: $\lambda = 10$, $d = 0.1$, $a = 0.2$, $p = 1$, $c = 0.05$, $b = 0.01$, $\beta_w = 0.1$, $\beta_r = 0.09$, $\beta'_w = 0.0021$, $\beta'_r = 0.09$. The shaded area denotes the time over which treatment is deployed.

therapy is, as should be expected, a monotonic function of the time at which treatment is started.

Notice that we can easily produce a second curve that shows the number of infection events that occur *after* a given time point. This second curve is shown on

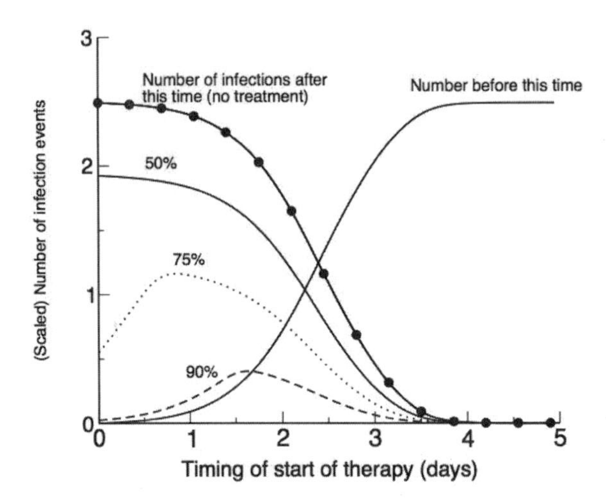

FIGURE 5. Number of infection events occurring before and during therapy for a wild-type infection, as a function of the time at which therapy is started. The solid curve depicts the number of infection events that occur before a given time point in the absence of therapy. The solid curve with symbols depicts the complementary quantity, namely the number of infection events that occur after the given time point. The remaining curves show the number of infection events that occur after the start of therapy, as a function of the time at which therapy is started, assuming that treatment is either 50, 75 or 90 percent effective. All curves were obtained by numerical simulation of the within-host model using the parameter values $\lambda = 0.25$, $d = 0.1$, $\beta_w = 1.0$, $a = 0.25$, $c = 4.0$, $p = 4.0$ and $b = 0.1$.

Figure 5 as the curve with symbols superimposed. In the limit of zero treatment efficacy, this second curve gives the number of infection events that occur after the onset of therapy. The corresponding curves for moderately effective therapies follow a similar shape, although with lower levels since the drug now diminishes wild-type replication.

If the drug efficacy is high, wild-type replication is efficiently curtailed following the start of therapy. In this case, the number of replication events during therapy echoes the viral load achieved before the start of therapy. Consequently, as can be seen in Figure 5, the curves for high drug efficacy essentially mimic the plot of viral load against time seen in a wild-type infection. The generation of resistant mutants in this case appears to be most likely if treatment is started near the peak level of viraemia.

6. Discussion

The key point that emerges from these models is the crucial role played by competition between strains, either for susceptible hosts on the epidemiological level or for target cells on the immunological level. In turn, we see that this competition depends on the details of the epidemiology and virology/immunology of the infection. At both the within and between host levels, competition is modulated by some factor. At the epidemiological level, competition is modulated by the degree of cross-immunity conferred by infection. In the rhinovirus context, this gives rise to two quite different scenarios when naturally resistant serotypes and novel resistant variants are considered. At the immunological level, competition is modulated by the immune response that is generated against the wild-type infection. Consequently, the effectiveness of the immune response and whether the infectious agent leads to acute or chronic infection has a major impact on the emergence of resistance.

Despite gaining these insights, much work remains to be done here. We have yet to provide a complete description of the emergence and rise of resistance at the within-host level: the work described here considers these problems as two separate issues. These viewpoints need to be integrated more closely. This will likely require the deployment of a stochastic modeling framework. This approach has been described in the setting of HIV infection (see [**47**]), but it leads to considerable additional complexity. Furthermore, as discussed earlier, the simple immune response that we employed in the model is in many ways unsatisfactory for describing an acute infection.

The most appropriate parameter values for this system are unclear at both the epidemiological and within-host levels. In the absence of these parameter values, we view the model predictions presented here as giving qualitative, rather than quantitative, insights into the dynamics of the system. Epidemiological parameters could be elucidated from more detailed transmission experiments. These could easily be accommodated within the framework of future clinical trials that assess the efficacy of drug treatment with pleconaril (or indeed similar drugs that are developed). Within-host parameters could, in analogy with the development of HIV models, be measured if trials investigated the viral load seen in patients over the time course of their infection. Furthermore, the use of well-designed clinical trials (such as family-based transmission studies) could yield important information regarding both the impact of treatment on reducing secondary transmissions and the likelihood of the generation and transmission of resistance.

Both the epidemiological and within-host models ignore heterogeneity within the population. A key heterogeneity is likely to be the strength of individuals' immune responses: these differ across a population, with certain groups (such as the immunocompromised or elderly) having significantly weakened immune responses. Resistance is much more likely to emerge in immunocompromised individuals, so from the viewpoint of managing resistance we might be less inclined to treat infection in this subgroup. But these are precisely the people who are likely to benefit most from treatment. This raises an interesting conflict between an individual's benefit of treatment and the population-level costs of treatment (i.e. the emergence of resistance). Mathematical frameworks such as the one presented here provide an ideal way to study such conflicts in more detail. (See [**5**] for a game-theoretic approach to a somewhat related individual/population conflict in the context of a mass-vaccination policy.)

Although this study has presented both between- and within-host models, the two were considered in isolation. An ambitious goal is to combine the two within a single, multi-scale, framework. Clearly, this will provide a much more realistic depiction of the dynamics of resistance since there is a close connection between the emergence of resistance within a patient and its spread at a population level. Such multi-scale models, however, pose significant challenges, particularly in terms of how we might gain general understanding of general patterns of behavior.

References

[1] R. Anderson and R. May, *Infectious Diseases of Humans*, Oxford University Press, Oxford, 1991.

[2] V. Andreasen, J. Lin, and S. A. Levin, The dynamics of cocirculating influenza strains conferring partial cross-immunity, *J. Math. Biol.* **35** (1997), 825–842.

[3] R. Antia, C. T. Bergstrom, S. S. Pilyugin, S. M. Kaech, and R. Ahmed, Models of CD8+ responses: 1. What is the antigen-independent proliferation program, *J. Theor. Biol.* **221** (2003), 585–98.

[4] N. Back, M. Nijhuis, W. Keulen, C. Boucher, B. Oude Essink, A. van Kuilenburg, A. van Gennip, and B. Berkhout, Reduced replication of 3TC-resistant HIV-1 variants in primary cells due to a processivity defect of the reverse transcriptase enzyme, *EMBO J.* **15** (1996), 4040–9.

[5] C. T. Bauch, A. P. Galvani, and D. J. Earn, Group interest versus self-interest in smallpox vaccination policy, *Proc. Natl. Acad. Sci. USA* **100** (2003), 10564–7.

[6] S. M. Blower T. C. Porco, and G. Darby, Predicting and preventing the emergence of antiviral drug resistance in HSV-2, *Nat. Med.* **4** (1998), 673–8.

[7] S. M. Blower and J. L. Gerberding, Understanding, predicting and controlling the emergence of drug-resistant tuberculosis: a theoretical framework, *J. Mol. Med.* **76** (1998), 624–36.

[8] S. M. Blower, H. B. Gershengorn, and R. M. Grant, A tale of two futures: HIV and antiretroviral therapy in San Francisco, *Science* **287** (2000), 650–654.

[9] S. M. Blower, A. N. Aschenbach, H. B. Gershengorn, and J. O. Kahn, Predicting the unpredictable: transmission of drug-resistant HIV, *Nat. Med.* **7** (2001), 1016–20.

[10] S. Bonhoeffer, R. M. May, G. M. Shaw, and M. A. Nowak, Virus dynamics and drug therapy, *Proc. Natl. Acad. Sci. USA* **94** (1997), 6971–6.

[11] S. Bonhoeffer and M. A. Nowak, Pre-existence and emergence of drug resistance in HIV-1 infection, *Proc. R. Soc. Lond. B* **264** (1997), 631–7.

[12] R. M. Bush, C. A. Bender, K. Subbarao, N. J. Cox, and W. M. Fitch, Predicting the evolution of human influenza A, *Science* **286** (1999), 1921–1925.

[13] C. Castillo-Chavez, H. W. Hethcote, V. Andreasen, S. A. Levin, and W. M. Liu, Epidemiological models with age-structure, proportionate mixing, and cross-immunity, *J. Math. Biol.* **27** (1989), 233–258.

[14] T. R. Cate, R. B. Couch, and K. M. Johnson, Studies with rhinoviruses in volunteers: Production of illness, effect of naturally acquired antibody, and demonstration of a protective effect not associated with serum antibody, *J. Clin. Invest.* **43** (1964), 56–67.

[15] J. M. Coffin, HIV population dynamics in vivo: implications for genetic variation, pathogenesis, and therapy, *Science* **267** (1995), 483–9.

[16] R. B. Couch, Rhinoviruses, in (D. M. Knipe and P. M. Howley, eds.), *Fields' Virology*, fourth ed., vol. 1, Lippincott Williams and Wilkins, Philadelphia, 2001, 777–797.

[17] R. J. De Boer and A. S. Perelson, Target cell limited and immune control models of HIV infection: a comparison, *J. Theor. Biol.* **190** (1998), 201–14.

[18] K. Dietz, Epidemiological interference of virus populations, *J. Math. Biol.* **8** (1979), 291–300.

[19] A. M. Fendrick, A. S. Monto, B. Nightengale, and M. Sarnes, The economic burden of non-influenza-related viral respiratory tract infection in the United States, *Arch. Intern. Med.* **163** (2003), 487–94.

[20] N. M. Ferguson, A. P. Galvani, and R. M. Bush, Ecological and immunological determinants of influenza evolution, *Nature* **422** (2003), 428–33.

[21] N. M. Ferguson, S. Mallett, H. Jackson, N. Roberts, and P. Ward, A population-dynamic model for evaluating the potential spread of drug-resistant influenza virus infections during community-based use of antivirals, *J. Antimicrob. Chemother.* **51** (2003), 977–90.

[22] J. P. Fox, M. K. Cooney, and C. E. Hall, The Seattle virus watch. V. Epidemiologic observations of rhinovirus infections, 1965-1969, in families with young children., *Am. J. Epidemiol.* **101** (1975), 122–43.

[23] J. P. Fox, M. K. Cooney, C. E. Hall, and H. M. Foy, Rhinoviruses in Seattle families, 1975-1979., *Am. J. Epidemiol.* **122** (1985), 830–46.

[24] S. Frost and A. McLean, Quasispecies dynamics and the emergence of drug resistance during zidovudine therapy of HIV infection, *AIDS* **8** (1994), 323–32.

[25] H. B. Gershengorn and S. M. Blower, Impact of antivirals and emergence of drug resistance: HSV-2 epidemic control, *AIDS* **14** (2000), 133–42.

[26] J. Goudsmit, A. De Ronde, D. Ho, and A. Perelson, Human immunodeficiency virus fitness in vivo: calculations based on a single zidovudine resistance mutation at codon 215 of reverse transcriptase, *J. Virol.* **70** (1996), 5662–4.

[27] J. M. Gwaltney, Jr., J. O. Hendley, G. Simon, and W. S. Jordan, Jr., Rhinovirus infections in an industrial population: II. Characteristics of illness and antibody response, *JAMA* **202** (1967), 494–500.

[28] J. M. Gwaltney, Jr., J. O. Hendley, G. Simon, and W. S. Jordan, Jr., Rhinovirus infections in an industrial population. 3. Number and prevalence of serotypes, *Am. J. Epidemiol.* **87** (1968), 158–66.

[29] J. M. Gwaltney, Jr., Rhinoviruses, in (A. S. Evans and R. A. Kaslow, eds.), *Viral Infections of Humans. Epidemiology and Control*, fourth ed., Plenum, New York, 1997, 815–838.

[30] D. Hamre, A. P. Connelly, Jr., and J. J. Procknow, Virologic studies of acute respiratory disease in young adults. IV. Virus isolations during four years of surveillance, *Am. J. Epidemiol.* **83** (1966), 238–49.

[31] F. G. Hayden and A. J. Hay, Emergence and transmission of influenza A viruses resistant to amantadine and rimantadine, *Curr. Top. Microbiol. Immunol.* **176** (1992), 119–30.

[32] F. G. Hayden, H. A. Hassman, T. Coats, R. Menendez, and T. Bock, Pleconaril treatment shortens duration of picornavirus respiratory illness in adults (abstract), Proceedings of the 39th InterScience Conference on Antimicrobial Agents and Chemotherapy, San Francisco, CA, 1999.

[33] F. G. Hayden, D. T. Herrington, T. L. Coats, K. Kim, E. C. Cooper, S. A. Villano, S. Liu, S. Hudson, D. C. Pevear, M. Collett, M. McKinlay, and the Pleconaril Respiratory Infection Study Group, Efficacy and safety of oral pleconaril for treatment of colds due to picornaviruses in adults: results of 2 double-blind, randomized, placebo-controlled trials, *Clin. Infect. Dis.* **36** (2003), 1523–32.

[34] B. A. Heinze, R. R. Rueckert, D. A. Shepard, F. J. Dutko, M. A. McKinlay, M. Fancher, M. G. Rossman, J. Badger, and T. J. Smith, Genetic and molecular analysis of spontaneous mutants of human rhinovirus 14 that are resistant to an antiviral compound, *J. Virol.* **63** (1989), 2476–2485.

[35] J. O. Hendley, J. M. Gwaltney, Jr., and W. S. Jordan, Jr., Rhinovirus infections in an industrial population. IV. Infections within families of employees during two fall peaks of respiratory illness, *Am. J. Epidemiol.* **89** (1969), 184–96.

[36] B. A. Larder and S. D. Kemp, Multiple mutations in HIV-1 reverse transcriptase confer high-level resistance to zidovudine (AZT), *Science* **246** (1989), 1155–8.

[37] B. R. Levin, R. Antia, E. Berliner, P. Bloland, S. Bonhoeffer, M. Cohen, T. DeRouin, P. I. Fields, H. Jafari, D. Jernigan, M. Lipsitch, J. E. McGowan Jr, P. Mead, M. Nowak, T. Porco, P. Sykora, L. Simonsen, J. Spitznagel, R. Tauxe, and F. Tenover, Resistance to antimicrobial chemotherapy: a prescription for research and action, *Am. J. Med. Sci.* **315** (1998), 87–94.

[38] M. Lipsitch, T. Bacon, J. Leary, R. Antia, and B. Levin, Effects of antiviral usage on transmission dynamics of herpes simplex virus type 1 and on antiviral resistance: predictions of mathematical models, *Antimicrob. Agents Chemother.* **44** (2000), 2824–2835.

[39] M. Lipsitch, C. T. Bergstrom, and B. R. Levin, The epidemiology of antibiotic resistance in hospitals: Paradoxes and prescriptions, *Proc. Natl. Acad. Sci. USA* **97** (2000), 1938–1943.

[40] A. L. Lloyd and D. W. Wodarz, Potential for spread of antiviral drug resistance in an acute, multi-strain rhinovirus setting, Submitted (2005).

[41] A. S. Monto and K. M. Johnson, A community study of respiratory infections in the tropics. II. The spread of six rhinovirus isolates within the community, *Am. J. Epidemiol.* **88** (1968), 55–68.

[42] A. S. Monto, E. R. Bryan, and S. Ohmit, Rhinovirus infections in Tecumseh, Michigan: frequency of illness and number of serotypes, *J. Infect. Dis.* **156** (1987), 43–9.

[43] M. Nowak and C. Bangham, Population dynamics of immune responses to persistent viruses, *Science* **272** (1996), 72–9.

[44] D. C. Pevear, Antiviral therapy for picornavirus infections: pleconaril, *Abstracts of the 15th Annual Clinical Virology Symposium*, Clearwater, FL, 1999.

[45] D. C. Pevear, T. M. Tull, M. E. Seipel, and J. M. Groarke, Activity of pleconaril against enteroviruses, *Antimicrob. Agents Chemother.* **43** (1999), 2109–2115.

[46] D. Pillay, S. Taylor, and D. D. Richman, Incidence and impact of resistance against approved antiretroviral drugs, *Rev. Med. Virol.* **10** (2000), 231–53.

[47] R. M. Ribeiro and S. Bonhoeffer, Production of resistant HIV mutants during antiretroviral therapy, *Proc. Natl. Acad. Sci. USA* **97** (2000), 7681–6.

[48] D. D. Richman, Drug resistance in viruses, *Trends Microbiol.* **2** (1994), 401–7.

[49] H. A. Rotbart, Treatment of picornavirus infections, *Antiviral Res.* **53** (2002), 83–98.

[50] N. I. Stilianakis, A. S. Perelson, and F. G. Hayden, Emergence of drug resistance during an influenza epidemic: Insights from a mathematical model, *J. Inf. Dis.* **177** (1998), 863–873.

[51] D. Taylor-Robinson, Studies on some viruses (rhinoviruses) isolated from common colds, *Arch. Ges. Virusforch.* **13** (1963), 281–93.

[52] J. X. Velasco-Hernandez, H. B. Gershengorn, and S. M. Blower, Could widespread use of combination antiretroviral therapy eradicate HIV epidemics? *Lancet Infect. Dis.* **2** (2002), 487–93.

[53] D. Wodarz, Helper-dependent vs. helper-independent CTL responses in HIV infection: implications for drug therapy and resistance, *J. Theor. Biol.* **213** (2001), 447–459.

[54] D. Wodarz and A. L. Lloyd, Immune responses and the emergence of drug-resistant virus strains in vivo, *Proc. R. Soc. Lond. B* **271** (2004), 1101–1109.

Biomathematics Graduate Program and Department of Mathematics, North Carolina State University, Raleigh, NC 27695

E-mail address: alun_lloyd@ncsu.edu

Department of Ecology and Evolutionary Biology, 321 Steinhaus Hall, University of California, Irvine, CA 92697

E-mail address: dwodarz@uci dot edu

DIMACS Series in Discrete Mathematics
and Theoretical Computer Science
Volume **71**, 2006

Dynamics and Control of Antibiotic Resistance in Structured Metapopulations

David L. Smith, Maciej F. Boni, and Ramanan Laxminarayan

ABSTRACT. The evolution of resistance to antimicrobial drugs is a major public health concern. Mathematical models for the spread of resistance have played an important role as a conceptual tool for understanding how and why resistance emerges and spreads. Here, we present a new, general mathematical model for the spread of resistance within a population that accounts for several biologically plausible effects of antimicrobial drug use. Except for the evolution of *de novo* resistance, the model is mathematically identical to Lotka-Volterra competition. The simple model is extended to include the spread of resistance among several patches, and the evolution of multi-drug resistance. The models are used to illustrate some simple ideas about the spatial spread and spatial control of resistance and the evolution of multi-drug resistance.

1. Introduction

General concern about the evolution of resistance to antimicrobial drugs is growing because the frequency of resistant infections has increased [**27, 35**]. During the 1980s and 90s chloroquine resistance was responsible for a global rise in malaria mortality [**28, 35**]. At the same time, vancomycin-resistant enterococci (VRE) spread epidemically among hospitalized patient populations [**9, 27, 29**]. The spread of VRE, called "superbugs" because they were naturally resistant or had acquired resistance to all approved antimicrobials, was accompanied by widespread fear that methicillin-resistant *Staphylococcus aureus* (MRSA) would acquire vancomycin resistance genes from VRE to create another, more virulent superbug, vancomycin-and-methicillin-resistant *S. aureus* (VRSA), an event that occurred in 2003 [**11**]. VRE and the fear of VRSA led to the alarmist speculation that we might be entering a "post-antimicrobial era".

Meanwhile, the rate that new antimicrobials drugs are being approved has declined [**34**]. Although two new drugs were approved for gram-positive bacterial infections (including enterococci and *S. aureus*) in 1999 and 2000, and a handful of new antimicrobial agents are in the pipeline, these new agents do not represent novel mechanisms of action. For now, most malaria, enterococcal, and *S. aureus*

Key words and phrases. Antimicrobial resistance, Metapopulations, Multi-drug resistance, Spatial spread.

Second author is supported by NIH grant GM28016 to Marcus W. Feldman.

infections are treatable, but the high frequency of resistance will inevitably lead to some treatment failure when patients take the wrong antimicrobial agent. However, excess morbidity and mortality caused by resistant infections and the slow response of the drug industry to a superbug suggest our efforts to keep up with evolution of resistance through pharmaceutical innovation may be futile and that part of the solution is better management of existing antimicrobials. The underlying cause of the evolution of resistance is the use of antimicrobial drugs – drug pressure creates a niche for resistant bacteria in a population that is being treated [**8**].

The evolution of resistance, defined broadly as the change in the frequency of resistance, involves the emergence of novel resistance and subsequent spread – two different problems with their own conceptual problems. The evolutionary origins of novel resistance are related to mutation and within-host selection [**15**]. Spread is related to human-to-human transmission, the movement of humans, inter- and intra-specific microbial competition, and other aspects of microbial ecology and host immunity. Emergence and spread are complex phenomena that occur within bacterial populations, within and among humans. Such phenomena are difficult to understand and analyze without the use of mathematical models [**10, 22, 26, 36**]. Indeed, VRE and anti-malarial resistance have initiated a new era of research on the population dynamics of drug resistance [**2, 3, 4, 5, 7, 15, 17, 18, 24, 25, 30, 31**]

Once resistance has emerged, public health responses should begin to shift focus from preventing emergence to reducing transmission and limiting spread. The measures to prevent the origins of new resistant pathogens and the spread of existing resistance types may not always be the same. Novel or *de novo* resistance, defined as the emergence of resistance by mutation within a population that was sensitive, encompasses one basic process for eukaryotes, but it can happen in two or more ways in prokaryotes [**20**]. The dominant view of the origins of antimicrobial resistance is that antimicrobials select for pre-existing mutations. High-level resistance may require several mutations, and since each one is rare, the process probably occurs step by step in partially resistant pathogens. Thus, the persistence of these partially resistant pathogens in a population plays a role in the origins of resistance [**16**]. This perspective is sufficient for *Plasmodium* and other eukaryotic pathogens, but bacteria are different [**20**]. In some cases, high-level resistance has emerged in bacteria when genes were acquired from other bacteria [**11**]. Novel resistance from the horizontal transmission of high-level resistance genes requires exposure to the genes as well as selection. Thus, the emergence of high-level resistance in one bacteria species is related to the prevalence of resistance in other species or microbial communities [**30**].

The spread of resistance from host to host is responsible for the majority of resistant infections. Resistant strains can spread in a population, just as their sensitive relatives do, once they have evolved, as long as they are not inhibited from infecting a host by the presence of another genotype. Thus, the underlying ecological model for the evolution of resistance is intra-specific competition [**2, 7**]. The spread of resistance can also play a role that is similar to mutation. As a result of drug chemotherapy, an infection will clear or recrudesce as an infection that remains drug-sensitive if no resistant mutant is present. Exposure to resistant strains just before or during drug treatment can "seed" a resistant infection that will respond as if it were pre-existing mutant. Thus, horizontal transmission of resistant pathogens is as important to understand as within-host selection. Here, we develop simple models that capture some of these effects.

The amount of antimicrobials used in a population is important, but it is equally important to understand where those antimicrobials are used and how humans move around. Our main focus is on the spread of resistance in structured populations, such as hospitals or networks of rural towns, building on the efforts of others to understand the spread and persistence of resistance to antimicrobial drugs in bacteria [**31**]. We have developed models that focus on the spread of resistance, or epidemiological models [**20**]. Possible candidate organisms for applying these models include *Plasmodium falciparum*, gram positive bacteria that are leading causes of hospital-acquired infections, enterococci, and *S. aureus*. Despite the enormous differences between these different pathogens, there are important similarities in their epidemiology. The immune responses are relatively weak, persistence times are relatively long, and asymptomatic infections are far more common than symptomatic infections. Thus, the pathogen dynamics can be usefully understood with SIS models. One important difference is that the use of antimicrobials to treat VRE and MRSA tends to cause more "collateral damage" by selecting for resistance in non-target bacteria. This is also true for some anti-malarials, but many anti-malarial drugs are used only for infections with plasmodia.

These models are intended to illustrate some general principles that affect the emergence and spread of resistance. We begin by formulating a new, general model for the spread of resistance in well-mixed populations. Next, we extend the model to a spatial context and focus on the spread of resistance among populations. Finally, we consider resistance to two drugs in space, and illustrate some interesting phenomena. The results should be interpreted with circumspection since antimicrobial policy will also be affected by many other concerns.

2. Well-mixed populations

We begin with a model for the spread of resistance in a well-mixed population, and derive equations for the spread of resistance. We assume that individuals are either uninfected, infected by drug-sensitive pathogens, or infected by drug-resistant pathogens. This model posits the strongest form intra-specific competition: no individuals are simultaneously infected with both drug-sensitive and drug-resistant "strains". Let U denote the proportion of patients who are uninfected, W the proportion who are infected by a drug-sensitive strain (wild-type), and X the proportion who are colonized by a drug-resistant strain. We assume the population is constant, so $U = 1 - X - W$.

We assume that populations are locally well-mixed. Let β denote the contact parameter for directly transmitted pathogens and vectorial capacity for vector transmitted pathogens. Let λ denote the rate at which infections clear. With no further assumptions, the dynamics are described by the following coupled differential equations:

$$\begin{aligned} \dot{W} &= \beta W U - \lambda W \\ \dot{X} &= \beta X U - \lambda X. \end{aligned} \tag{2.1}$$

The superdot denotes the derivative with respect to time.

Equations (2.1), are only a common starting point for understanding the evolution of resistance. Total prevalence of the pathogen, $P = W + X$ is described by a simple equation:

$$\dot{P} = \beta PU - \lambda P \tag{2.2}$$

The basic reproductive number for the pathogen is $R_0 = \beta/\lambda$, and if $R_0 > 1$, prevalence approaches the equilibrium $1 - 1/R_0$. There is no selection for or against resistance in Eqs. 2.1 because drug-sensitive and drug-resistant pathogens are competitively equivalent.

Intra-specific competition is modified by antimicrobial drug use and by antimicrobial resistance. Let ρ denote the rate that people are prescribed antimicrobials, and let ξ denote the fraction of the population under chemo-prophylaxis, defined as having concentrations of the drug that favor drug-resistant strains over drug-sensitive ones (see Appendix 1). Here, treatment and prophylaxis are not directly related to infection status. The assumption reflects the fact that most people who carry VRE, MRSA, or malaria infections are asymptomatic. Moreover, VRE and MRSA carriers are often treated for other infections, and selection for VRE or MRSA is a collateral effect. In areas where malaria is hyperendemic, malaria is often presumed to be the underlying cause of fever, so anti-malarials are often taken without respect to infection status. For malaria, new treatment programs being contemplated deliver anti-malarials to pregnant women and children, without respect to their infection status [**32, 33**]. For these reasons, we have not made the prescription rate a direct function of infection status.

Antimicrobial use and antimicrobial resistance fundamentally change intra-specific competition in one or more of the following ways:

(1) **Treatment and Clearance of Drug Sensitive Pathogens:** Infections clear in those who are undergoing chemo-prophylaxis. Total clearance rates increase to $(\lambda + \nu\xi)W$, where $\nu\xi$ is always less than or equal to ρ, the antimicrobial prescription rate, since if all infections are cleared instantaneously upon starting treatment, then $\nu\xi = \rho$. (The constraint helps to avoid errors in Eqs. 3 when considered in isolation from Eqs. 14).

(2) **Chemo-prophylaxis of Susceptibles:** An obvious effect of taking antimicrobials is that uninfected individuals are protected from infection by drug-sensitive strains. The incidence of infection with drug-sensitive strains is lowered by chemo-prophylaxis to $\beta WU(1 - \xi)$

(3) **Chemo-prophylaxis of Drug-Sensitives:** A secondary effect may be that those who are colonized by drug-sensitive strains do not transmit as efficiently while they are chemo-prophylaxed: shedding from prophylaxed individuals occurs at the rate $\zeta\xi$. Shedding from susceptibles occurs at the rate $(1-\xi+\zeta\xi)W$ shed. Combined with chemo-prophylaxis of susceptibles, the incidence of infection with drug-sensitive strains is lowered further to $\beta(1 - \xi)(1 - \xi + \zeta\xi)WU$. If susceptibles don't shed at all, then incidence is $\beta(1 - \xi)^2 WU$.

(4) **Biological Cost of Resistance – Clearance:** A biological cost of resistance is often incorporated into these models by assuming a higher rate of spontaneous clearance for resistant pathogens, $(\phi + \lambda)X$.

(5) **Biological Cost of Resistance – Super-infection:** A biological cost of resistance may allow drug-sensitive strains to displace drug-resistant ones. This allows individuals to convert directly from resistant to sensitive without ever clearing an infection. We assume that this only occurs when

neither host is chemo-prophylaxed. We further assume that the resident has an inherent advantage; the probability of conversion, per contact, is $q \leq 1$. Thus, resistant hosts become sensitive at the rate $q(1-\xi)(1-\xi+\zeta\xi)\beta XW$.

(6) **Transfer of Resistance Factors** *or* **Super-infection under Chemo-prophylaxis:** Drug-sensitive bacteria may acquire resistance factors from drug-resistant ones by gene transfer, for example, on a plasmid. We assume that this transfer is most likely if the host is being treated with antimicrobial drugs. Acquisition of resistance factors would be indistinguishable from super-infection during or just before chemo-prophylaxis that allows drug-resistant pathogens to displace drug-sensitive ones. We assume that this occurs through ordinary contact, and that the probability of conversion by either mechanism is r. Thus, people change status from sensitive to resistant at the rate $r\xi\beta XW$.

(7) **Novel Resistance:** The use of drugs may favor the evolution of resistance within a host, either through mutation or through the inter-specific transfer of high-level resistance factors. People change status from infected with drug-sensitive to drug-resistant pathogens at the rate $c\xi W$.

This list may not include all the advantages or disadvantages of resistance, but it describes several plausible mechanisms. In sum, antimicrobial drug use provides an advantage to resistant strains by reducing transmission and increasing clearance rates of drug-sensitive pathogens. This is countered by a biological cost of resistance that leads to more rapid clearance of resistant bacteria. Finally, drug use can change the rate that people change status without becoming uncolonized through super-infection, from infected with resistant to sensitive, or vice versa.

TABLE 1. State variables and parameters for Eqs. 2.3 and Eqs. 3.1.

Symbol	Description
U	Proportion of the population that is uninfected
W	Proportion that is infected with sensitive pathogens
X	Proportion that is infected with resistant pathogens
β	Contact parameter
ξ_a	Proportion of the population that is prophylaxed by antibiotic a
ν	The proportion of prophylaxed populations that clear infections
λ	The rate that infections are naturally cleared
ϕ	Cost of resistance, higher clearance
q, r	Probability of displacement by superinfection
ζ	Reduced transmission by prophylaxed susceptibles
c	Probability that treatment leads to *de novo* resistance
$1/\sigma_i$	Average time spent in i^{th} population
$\psi_{i,j}$	Proportion of immigrants to i^{th} population that come from the jth population

Incorporating all these effects gives the more complicated equations describing local competition under some specific level of antimicrobial drug use:

$$
\begin{aligned}
\dot{W} &= \beta(1-\xi)(1-\xi+\zeta\xi)WU + \left(q(1-\xi)(1-\xi+\zeta\xi) - r\xi\right)\beta XW \\
&\quad - (\lambda+\nu\xi)W - c\xi W \\
\dot{X} &= \beta XU - \left(q(1-\xi)(1-\xi+\zeta\xi) - r\xi\right)\beta XW - (\lambda+\phi)X + c\xi W.
\end{aligned}
\tag{2.3}
$$

A slightly more formal derivation of these equations is provided in Appendix 1.

Importantly, in the new environment the basic reproductive numbers for each type change, depending on the amount of antimicrobial drug that is used.

$$
\begin{aligned}
\mathcal{R}_{0,w} &= \frac{\beta(1-\xi)(1-\xi+\zeta\xi)}{\lambda+\nu\xi} \\
\mathcal{R}_{0,x} &= \frac{\beta}{\lambda+\phi}
\end{aligned}
\tag{2.4}
$$

We denote the basic reproductive number in the new environment with a different script and a subscript for each strain, to distinguish it from the basic reproductive number of the wild-type in an untreated population R_0

One could consider the susceptible population as a resource, and treat the dynamics as an example of "resource-based competition." The competitive dynamics would be entirely determined by the basic reproductive number for each strain in a particular treatment environment. $\mathcal{R}_{0,w}$ and $\mathcal{R}_{0,x}$. The pathogen with the highest $\mathcal{R}_{0,s}$ would also have the highest carrying capacity and would draw the susceptible population below the other strain's threshold level and prevent it from invading.

Super-infection makes this sort of analysis incorrect because coexistence is determined by the ability of each strain to invade the other at carrying capacity, and each strain is able to use the other pathogen as a resource. For example, the resistant strain can invade all susceptibles as well as sensitives, $U + r\xi W$, while persistence times, the average waiting time until a strain is either cleared or replaced, are shorter, approximately $1/(\lambda+\phi+q(1-\xi)(1-\xi+\zeta\xi)W)$. Similarly, the resource for sensitive strains invading a population is $U + q(1-\xi)^2 X$, and persistence times are approximately $1/(\lambda+\nu\xi+r\xi X)$. Thus, it is no longer possible to say which pathogen will be dominant by simply finding the one with the highest $\mathcal{R}_{0,s}$.

Goldilocksian Coexistence. The dynamics of this system are transparent once Eqs 2.3 are rewritten in the following way:

$$
\begin{aligned}
\dot{W} &= r_w W\left[1 - (W + \alpha_w X)/K_w\right] - c\xi W \\
\dot{X} &= r_x X\left[1 - (X + \alpha_x W)/K_x\right] + c\xi W
\end{aligned}
\tag{2.5}
$$

In Eqs 2.5, the parameters are recast as maximum growth rates, carrying capacities and competition coefficients (Table 2). In other words, ignoring the background evolution of novel resistance (i.e. assuming $c = 0$), the underlying dynamics are mathematically equivalent to the well-understood Lotka-Volterra competition equations.

Ignoring the evolution of novel resistance (i.e. $c = 0$) antimicrobial drug use reduces the prevalence of sensitive pathogens from $1 - 1/R_0$ to K_w, if resistance never appears. If $K_x - \alpha_x K_w > 0$, resistance will eventually be able to invade once it appears. If resistance is already present at a low frequency, the frequency will not begin to increase until antimicrobials reduce the prevalence of the sensitive

TABLE 2. The coefficients when Eqs. 2.3 are rewritten as Lotka-Volterra competition equations.

$i =$	w	x
r_i	$\beta(1-\xi)(1-\xi+\zeta\xi)-\nu\xi-\lambda$	$\beta-\phi-\lambda$
K_i	$1-1/\mathcal{R}_{0,w}$	$1-1/\mathcal{R}_{0,x}$
α_i	$1-q+\frac{r\xi}{(1-\xi)(1-\xi+\zeta\xi)}$	$1+q(1-\xi)(1-\xi+\zeta\xi)-r\xi$

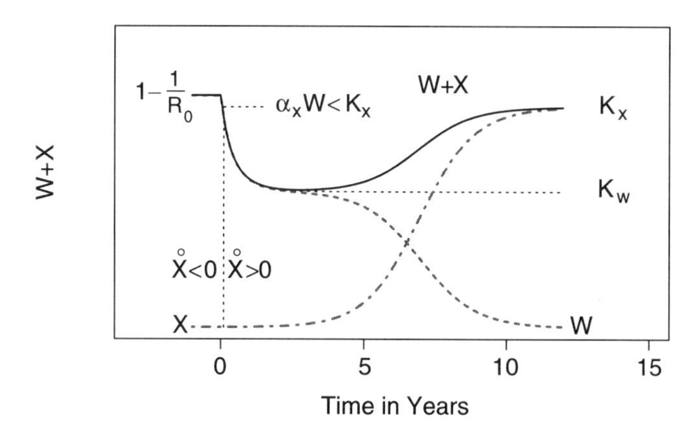

FIGURE 1. The anatomy of an epidemic of resistance. Before antimicrobials, the prevalence of a sensitive phenotype is assumed to be at equilibrium, $1 - 1/R_0$. Once drug use starts, at time $t = 0$, the prevalence of sensitive bacteria declines and would eventually reach a new equilibrium, K_w, if resistance never invaded (the dotted line shows W without competition from X). Once the prevalence of sensitive bacteria decline below a threshold ($W < K_x/\alpha_x$), the prevalence of resistance (dot-dash, X) begins to increase (i.e. $\dot{X} > 0$) and approaches a new equilibrium, K_x . The frequency of resistance is initially rare, but in this case, it eventually goes to fixation and drives the sensitive bacteria extinct (dashed lines show W with competition). Total prevalence ($W + X$, solid dark line) drops when the antimicrobial is initially introduced, but eventually rebounds. The parameters and initial conditions are the following: $q = r = \lambda = 1/500$ days, $\phi = \lambda/5$, $\xi = 3\%$, $\zeta = 0$, $\nu = 0.1$, $R_0 = 5$, $W(-365) = K_w$, and $X(-1) = 0.001$ or 0.

phenotype to $W < K_x/\alpha_x$. If $K_w - \alpha_w K_x < 0$, the resistant types will increase to a new equilibrium K_x and the sensitive phenotypes will be eliminated (Figure 1).

Coexistence requires a Goldilocksian balance:

$$\alpha_x < K_x/K_w < 1/\alpha_w \tag{2.6}$$

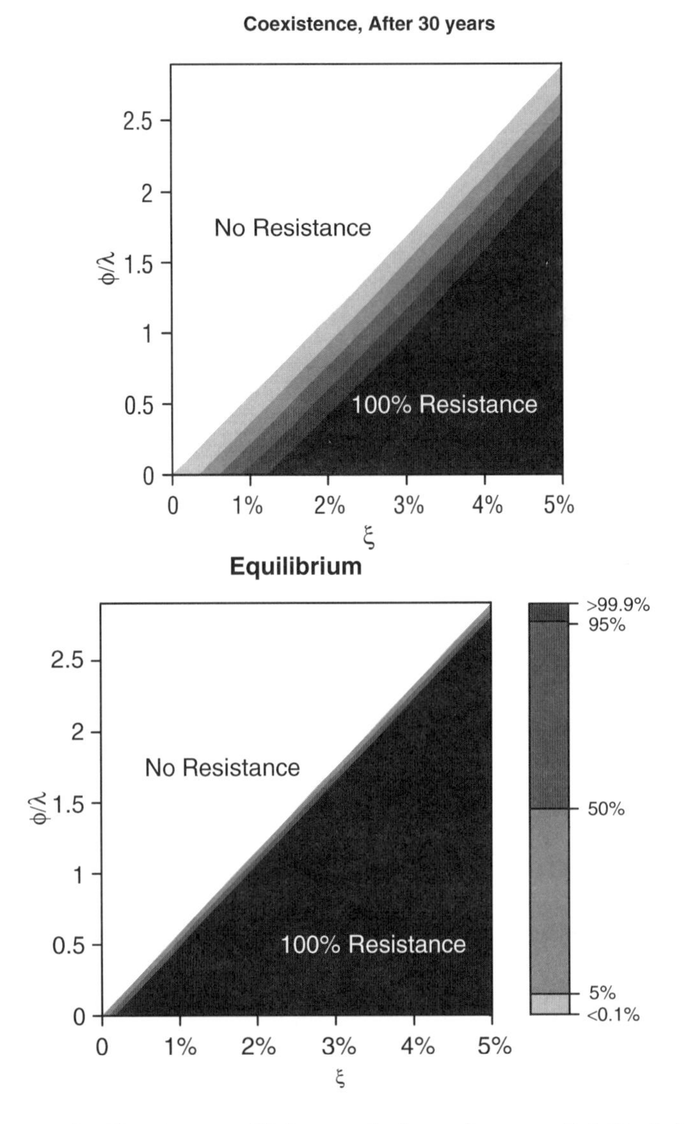

FIGURE 2. Coexistence (light gray) depends on a Goldilocksian balance between the rate of antimicrobial drug use and the cost of resistance, but when generation times ($1/\lambda$) are long, two strains can coexist for a very long time. The colors indicate the frequency of resistance–the darker, the higher frequency of resistance (see the key). Here, we've plotted the frequency of resistance after 30 years, and the equilibrium.

It is possible to translate this into a formula related to the proportion of the population under chemo-prophylaxis, ξ, but the resulting expressions are complicated. Crudely summarized, if selection pressure is too strong, sensitive phenotypes will be eliminated. If selection pressure is too weak, resistance will remain absent. Coexistence does not occur unless selection pressure is just right (Figure 2).

In natural populations, resistance has rarely become fixed. One explanation for coexistence is that the biological cost of resistance is extremely high. Some circumstantial evidence undermines this hypothesis, for a very high biological cost would be relatively easy to measure. A high biological cost of resistance has been reported relatively rarely [**1**]. One possible reason for a low biological cost of resistance is that compensatory mutations can arise that minimize the biological cost of resistance [**21**].

An alternative hypothesis is that coexistence is a transient phenomenon–resistance has had insufficient time to become fixed (Figure 2b). This sort of transient resistance also requires that the rate of antimicrobial use be delicately balanced, although the constraints are not quite as Goldilocksian (Silverlocksian?).

Another mechanism that could also explain coexistence is that prescription rates adjust to the frequency of resistance, for example, patients switch to another drug or avoid treatment when resistance becomes very frequent, coexistence would be more robust. An alternative explanation, explored below, is population heterogeneity.

3. Resistance in Structured Populations

The rate of antimicrobial drug use and local transmission can vary, with important implications for the dynamics and control of resistance. To understand epidemics in structured populations, we extend the previous model to link several locally well-mixed populations. Let subscript i denote the ith population, and let $1/\sigma_i$ denote the average length of stay in the ith population. We have assumed that the size of each local population remains constant, so every individual who leaves one population is replaced by an arrival from elsewhere. Let $\psi_{i,j}$ denote the proportion of all immigrants (sensitive or resistant) to population i that come from population j. The migration fractions are implicitly related to the relative population sizes and the migration rates between each pair of populations (see Appendix 2).

We assume that transmission rates (β_i) and the proportion chemo-prophylaxed (ξ_i) can vary from place but other parameters are fixed, no matter where a person resides at the time. The local dynamics are described by the following:

$$
\begin{aligned}
\dot{W}_i =& \beta_i(1-\xi_i)^2 W_i U_i + \left(q(1-\xi_i)^2 - r\xi_i\right)\beta_i X_i W_i \\
& - \nu\xi_i W_i - \lambda W_i - c\xi_i W_i - \sigma_i(W_i - \sum_j \psi_{i,j} W_j)) \\
\dot{X}_i =& \beta_i X_i U_i - \left(q(1-\xi_i)^2 - r\xi_i\right)\beta_i X_i W_i \\
& - (\phi+\lambda)X_i + c\xi_i W_i - \sigma_i(X_i - \sum_j \psi_{i,j} X_j).
\end{aligned}
\tag{3.1}
$$

Spatial Coexistence. Coexistence between sensitive and resistant phenotypes is relatively easy when migration rates are very low and antimicrobial use is heterogeneous. For example, consider a two-patch model where no antimicrobials are used in patch one, but antimicrobial use in patch two is high enough to fix resistance. Coexistence is trivial if the patches remain separated. With high migration rates, the population behaves as if well-mixed, with respect to coexistence. The amount of migration required to undermine this spatial coexistence is surprisingly small in the two-patch model (Figure 3).

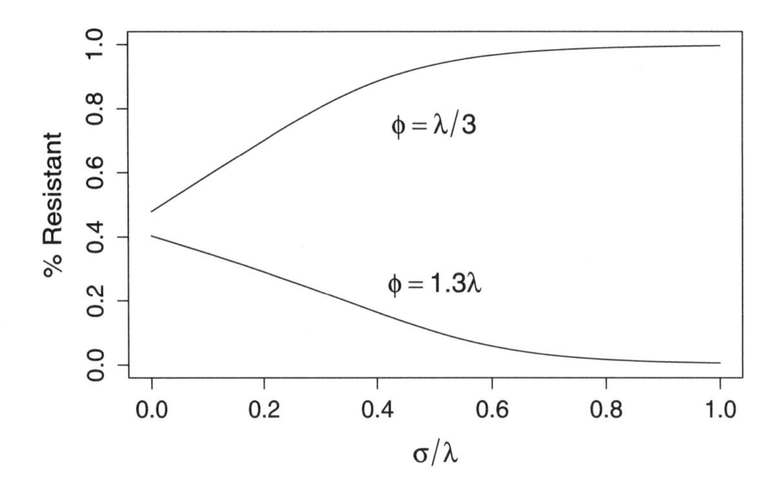

FIGURE 3. Spatial heterogeneity in antimicrobial use promotes coexistence when migration rates are low. Here, the equilibria are plotted as a function of the migration rate. For these parameters, the population is approximately well-mixed when migration occurs on approximately the same time scale as clearance.

An alternative explanation is that the human population is composed of many subgroups that vary in the amount of time spent in the prophylaxed location. For example, the elderly population spend more time, on average, than the non-elderly in hospitals and long-term care facilities [**31**]. Those who are frequently hospitalized play a role in the spread of antimicrobial resistant hospital-acquired infections that is analogous to those who are most sexually active in spreading sexually-transmitted diseases.

The lessons learned from hospital-acquired infections may be played out in structured populations where the sub-populations have a spatial relationship. A simple illustration of the principle is the frequency of resistance on an array (Figure 4). To keep the point as simple as possible, we allow two patches to be treated, but we vary the distance separating the treated patches. When the two treated patches are close together, individuals who have acquired resistance in one patch are more likely to enter the other, where they continue to transmit. Thus, the closer two patches are to one another, the more they amplify each other. A similar phenomenon happens at the edge, where we assume that no individuals leave at the edges so individuals are more likely to return to the treated patch. To put it simply, the prevalence of antimicrobial resistance in one subpopulation is affected by the rate of antimicrobial use in surrounding populations [**31**].

The spread of resistance in one relatively simple spatial network was described as a part of a study in Tanzania – the prevalence of resistance in two treated areas, and in surrounding areas provide some evidence that these principles are at work in real populations [**12**]. Enzi, an untreated town between two treated towns had a higher frequency of resistance than the treated towns, or than any of the untreated

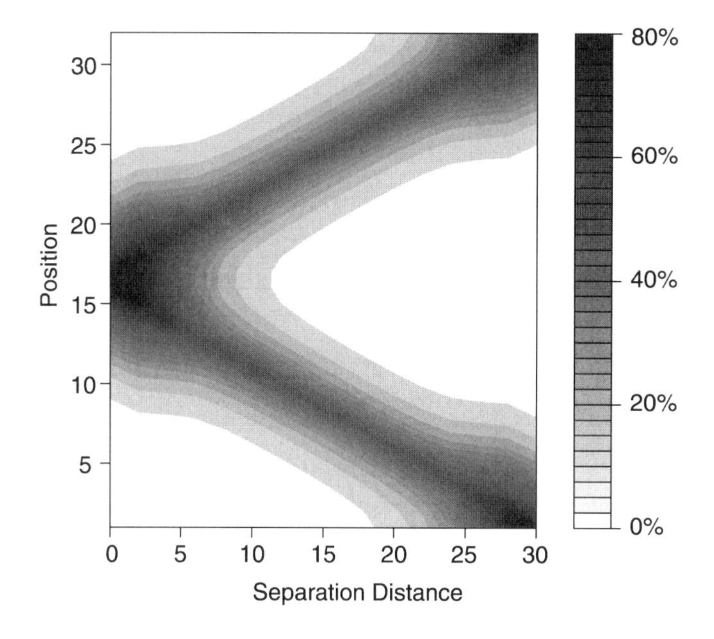

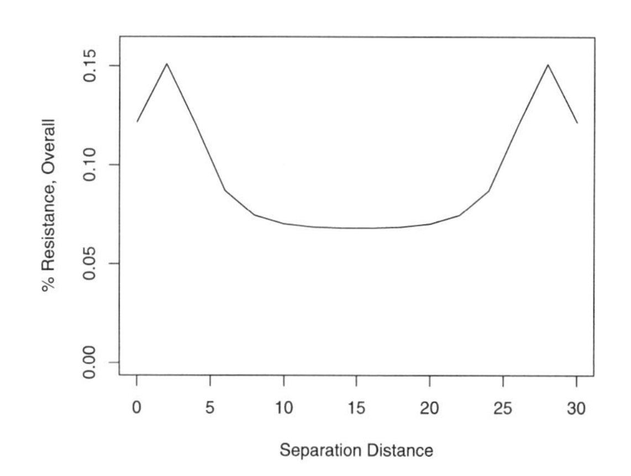

FIGURE 4. The frequency of resistance is higher when some hosts spend more time in populations where antimicrobials are heavily used. In the top graph, we show the frequency of resistance simulated on an array where two treated patches surrounded by untreated patches were separated by the indicated distance (vertical transects). When the patches are close to one another, hosts infected with resistant phenotypes are more likely to re-enter a treated patch. When the patches are near the edge, a similar phenomenon occurs because of a reflecting boundary condition. Curiously, resistance overall is lower when the two patches are exactly adjacent or at the very edge because the effects of spillover on the adjacent, untreated populations are limited. The model is available upon request.

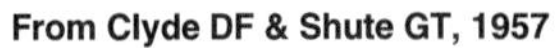

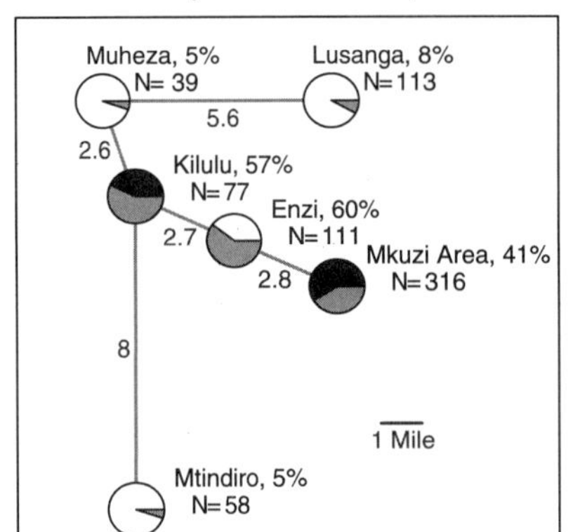

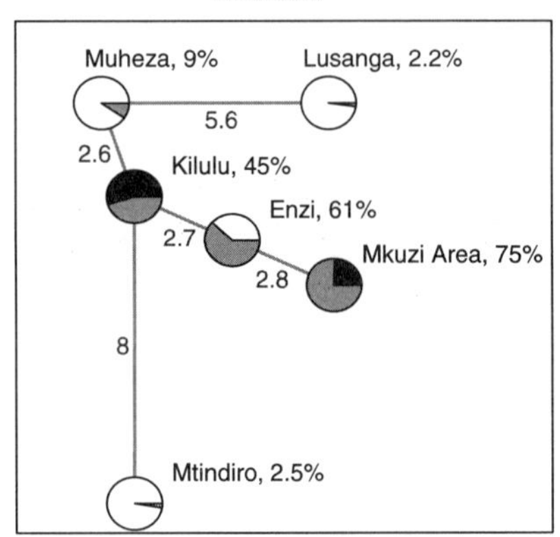

FIGURE 5. The frequency of resistance in Enzi (left) and in a simulation (right). Dosing with pyrimethamine in two cities (black background) led to high frequency of resistance (grey). Surrounding, untreated cities (black background) generally had much lower resistance, suggesting limited spread. The exception was Enzi, situated between the two treated populations, where the high frequency of resistance suggested spread of resistant phenotypes from the flanking treated populations. No resistance was found in Pongwe 12 miles east ($N = 48$), or in a northwest belt 5–9 miles away ($N = 65$). The gray lines show the roads connecting the towns at the time of the study. right) Some of these patterns can be reproduced by trial and error. The model is available upon request.

towns nearby (Figure 5, left). By trial and error, we found migration parameters and treatment frequencies that generated prevalence patterns that were close to those in the study (Figure 5, right). Try as we might, we could not find parameters that made the frequency of resistance for untreated Enzi higher than the treated cities flanking it. Mkuzi is a large and heterogeneous area. It is possible that treatment rates were very high in those parts of Mkuzi that were near Enzi but lower further away, but resistance was reported for the aggregated Mkuzi population. This is entirely speculative, but a relatively simple structured model does approximately reproduce the observed patterns.

The same principles apply to more complicated networks of interacting populations, including the flows of people on landscapes, the flow of patients among health-care institutions, and the flow of patients within a hospital.

4. Multi-drug Resistance in Structured Populations

The lessons from structured populations have an important applications antimicrobial drug policies. The previous results suggest the decision of what drug to recommend will depend, in part, on the frequency of resistance and antimicrobial resistance in neighboring populations. A primary concern here is the evolution of multi-drug resistance (MDR). Recent models for the evolution of MDR have focused on developing strategies to prevent the rapid evolution of MDR-strains [**6, 7**].

While some of these models have managed to simplify the dimensionality of such systems substantially, it is not always clear that such reductions in model size are desirable and that lower-dimensionality models are good approximations of the true models. To illustrate the complexity of MDR-models, we present a summary of the variables necessary to model a human-bacterial interaction with two antimicrobials. In the appendix, we describe some common approximations. Finally, we present some results on antimicrobial prescribing strategies in a structured population.

To model the different chemo-prophylaxis scenarios with two antimicrobials, we need to break up the host population into four population classes. We call our two antimicrobials x and y, and we denote by subscripts which type of chemo–prophylaxis a host is undergoing, if any. For example, a host colonized by the wild-type and currently taking antimicrobial y will be in population class W_y. We use the letter z to denote both antimicrobials; hosts in the class W_z are prophylaxed by antimicrobials x and y simultaneously. Subscript n indicates no prophylaxis.

To determine infection status, we need a further sub-division into five population classes: one class for uncolonized individuals, and four classes for the four possible types of resistance in the bacteria. Hosts in the class U are currently uncolonized; hosts in W are colonized by the wild-type strain; hosts in X and Y are colonized by a strain resistant to antimicrobial x or y, respectively; and, hosts in the population class Z are colonized by a bacterial strain resistant to both antimicrobials x and y. Five classes for infection status and four classes for prophylaxis status result in an unwieldy 20 population classes (see Table 3). The full model with 20 differential equations can be seen in Appendix 3. Below we present a lower-dimensional, collapsed version of this model.

TABLE 3. Asterisks (*) denote individuals who are effectively prophylaxed. These hosts cannot shed/transmit their pathogen; their microbial populations can evolve resistance if these hosts come into contact with another host infected with a strain resistant to their antimicrobial.

col. strain / proph. state	not proph.	proph. by ab x	proph. by ab y	proph. by abs x and y
none	U_n	U_x	U_y	U_z
wildtype, ab-sensitive	W_n	W_x (*)	W_y (*)	W_z (*)
resistant to ab x	X_n	X_x	X_y (*)	X_z (*)
resistant to ab y	Y_n	Y_x (*)	Y_y	Y_z (*)
resistant to abs x and y	Z_n	Z_x	Z_y	Z_z

We set $q = 0$, so that we do not have reversion to sensitives via a superinfection mechanism, $c = 0$ so that there is no evolution of novel resistance, and $r = 1$ so that drug-resistant strains can always invade prophylaxed hosts colonized by a sensitive population. We once again compartmentalize our host population by their treatment status. We say that a fraction ξ_n are not prophylaxed, a fraction ξ_x are prophylaxed by antimicrobial x, a fraction ξ_y are prophylaxed by antimicrobial y, and a fraction ξ_z are prophylaxed by both antimicrobials; $\xi_n + \xi_x + \xi_y + \xi_z = 1$.

This method allows us to make the same ξ-like approximation that is presented in Appendix 1.

Dimensionality reduction in our dynamical system relies on collapsing the infected classes into forces of infection and the susceptible classes into effective susceptible populations as has been done elsewhere [**13, 14, 23**]. The forces of infection are defined as

$$\begin{aligned} \Lambda_w &= \beta W_n \\ \Lambda_x &= \beta(X_n + X_x) \\ \Lambda_y &= \beta(Y_n + Y_y) \\ \Lambda_z &= \beta(Z_n + Z_x + Z_y + Z_z). \end{aligned} \tag{4.1}$$

Let Q_s be the class susceptible to the strain s. Then,

$$\begin{aligned} Q_w &= U_n \\ Q_x &= U_n + U_x + W_x + Y_x \\ Q_y &= U_n + U_y + W_y + X_y \\ Q_z &= U_n + U_x + U_y + U_z + W_x + W_y + W_z + X_y + X_z + Y_x + Y_z \end{aligned} \tag{4.2}$$

are the four effective susceptible classes as perceived by each of the four pathogenic strains. After some rearranging and approximating (Appendix 3), the 8 classes yield the closed dynamical system

$$\begin{aligned} \dot{Q}_w =& - Q_w(\Lambda_w + \Lambda_x + \Lambda_y + \Lambda_z) \\ &+ \frac{\lambda}{\beta}\left(\Lambda_w + \frac{\lambda+\phi_1}{\lambda}\frac{\xi_n}{\xi_n+\xi_x}\Lambda_x + \frac{\lambda+\phi_1}{\lambda}\frac{\xi_n}{\xi_n+\xi_y}\Lambda_y + \frac{\lambda+\phi_2}{\lambda}\xi_n\Lambda_z\right) \\ \dot{Q}_x =& - Q_x(\Lambda_x + \Lambda_z) - Q_w(\Lambda_w + \Lambda_y) \\ &+ \frac{\lambda}{\beta}\left(\Lambda_w + \frac{\lambda+\phi_1}{\lambda}\Lambda_x + \frac{\lambda+\phi_1}{\lambda}\frac{\xi_n}{\xi_n+\xi_y}\Lambda_y + \frac{\lambda+\phi_2}{\lambda}(\xi_n+\xi_x)\Lambda_z\right) \\ \dot{Q}_y =& - Q_y(\Lambda_y + \Lambda_z) - Q_w(\Lambda_w + \Lambda_x) \\ &+ \frac{\lambda}{\beta}\left(\Lambda_w + \frac{\lambda+\phi_1}{\lambda}\frac{\xi_n}{\xi_n+\xi_x}\Lambda_x + \frac{\lambda+\phi_1}{\lambda}\Lambda_y + \frac{\lambda+\phi_2}{\lambda}(\xi_n+\xi_y)\Lambda_z\right) \\ \dot{Q}_z =& - Q_z\Lambda_z - Q_y\Lambda_y - Q_x\Lambda_x - Q_w\Lambda_w \\ &+ \frac{\lambda}{\beta}\left(\Lambda_w + \frac{\lambda+\phi_1}{\lambda}\Lambda_x + \frac{\lambda+\phi_1}{\lambda}\Lambda_y + \frac{\lambda+\phi_2}{\lambda}\Lambda_z\right) \\ \dot{\Lambda}_w =& \beta Q_w\Lambda_w - \lambda\Lambda_w \\ \dot{\Lambda}_x =& \beta Q_x\Lambda_x - (\lambda+\phi_1)\Lambda_x \\ \dot{\Lambda}_y =& \beta Q_y\Lambda_y - (\lambda+\phi_1)\Lambda_y \\ \dot{\Lambda}_z =& \beta Q_z\Lambda_z - (\lambda+\phi_2)\Lambda_z, \end{aligned} \tag{4.3}$$

where ϕ_1 is the cost of resistance (i.e. higher clearance) of the singly-resistant strains X and Y, and ϕ_2 is the cost of resistant of the doubly-resistant strain Z.

These equations have six free parameters (we can scale out λ), and they can serve as a useful guide as to how the microbial population structure would respond to various antimicrobial-prescribing strategies. The equations also allow us to approximate basic reproduction ratios for the four strains, relative to that for the wild-type in an untreated population, $R_0 = \beta/\lambda$. In a different environment defined by some treatment rates, the basic reproductive numbers of the pathogens are:

$$\begin{aligned}
\mathcal{R}_{0,w} &= R_0\xi_n, \\
\mathcal{R}_{0,x} &= R_0(\xi_n + \xi_x)\frac{\lambda}{\lambda + \phi_1}, \\
\mathcal{R}_{0,y} &= R_0(\xi_n + \xi_y)\frac{\lambda}{\lambda + \phi_1}, \\
\mathcal{R}_{0,z} &= R_0(\xi_n + \xi_x + \xi_y + \xi_z)\frac{\lambda}{\lambda + \phi_2} = R_0\frac{\lambda}{\lambda + \phi_2}.
\end{aligned} \tag{4.4}$$

Notice that the basic reproductive number for each type is lower than R_0, but the highest $\mathcal{R}_{0,s}$ varies, depending on the amount and type of antimicrobial being used. The biological cost of resistance is assumed to increase with the number of drugs to which the pathogen is resistant. Countering this cost, the population that is susceptible to infection increases with the number of antimicrobials to which a strain is resistant. Since MDR strains are resistant to every antimicrobial, they can infect any individual. These values underline the important effect prophylaxis and treatment can have on the rates of spread of the antimicrobial-sensitive pathogens.

MDR Dynamics in Space. We now consider the spread of resistance in two equally sized patches, or spatial locations (named 1 and 2) to illustrate how antimicrobial prescribing strategies and resistance patterns vary in the simplest spatial model. Instead of eight state variables we now need sixteen; we call them $Q_{w,1}, \Lambda_{y,2}$, etc. Our Q-equations will now look like

$$\dot{Q}_{w,1} = -Q_{w,1}(\Lambda_{w,1} + \Lambda_{x,1} + \Lambda_{y,1} + \Lambda_{z,1}) + \frac{\lambda}{\beta}(\cdots) + \sigma(Q_{w,2} - Q_{w,1}). \tag{4.5}$$

Similarly, our Λ-equations will be

$$\dot{\Lambda}_{y,1} = \beta Q_{y,1}\Lambda_{y,1} - (\lambda + \phi_1)\Lambda_{y,1} + \sigma(\Lambda_{y,2} - \Lambda_{y,1}). \tag{4.6}$$

Note that now we will have eight parameters (six free parameters) describing prescribing frequencies: $\xi_{n1}, \xi_{x1}, \xi_{y1}, \xi_{z1}$ in patch 1 and $\xi_{n2}, \xi_{x2}, \xi_{y2}, \xi_{z2}$ in patch 2.

Treatment strategies for MDR-models specify what fraction of the population is treated, as well as how the different antimicrobials will be distributed among treated hosts. Some common multi-drug treatment strategies are (1) load balancing (also called 50-50 treatment [**7**] and mixing [**6**]), where half of the treated hosts are given antimicrobial x and the other half are given antimicrobial y; (2) combination therapy, where all treated hosts are given both antimicrobials simultaneously; and (3) sequential treatment or antimicrobial cycling, where hospitals treat all hosts with one antimicrobial for a given period of time, then switch to the second for some time, switching back and forth between two or cycling through three or more. Single-drug treatment is of course also an option. In the two-patch model, we will consider the case where antimicrobial x is used at one location and antimicrobial y in the other; this can be thought of as "load balancing in space".

Sequential treatment is believed to be the poorest strategy [**6, 7**], in that it drives the evolution of double-resistants the most quickly. Load-balancing and combination therapy are slightly better, though combination therapy puts more "direct" favorable selection pressure on the double-resistants, while a load-balancing strategy puts indirect selection pressure on the double-resistants. Load-balancing in space has not yet been studied. We present two simple examples of spatial treatment regimes and their resulting bacterial strain structures.

Load Balancing in Space. We consider a scenario of two hospitals where patients sometimes get transferred from one to the other, or two cities between which individuals frequently migrate. Patch 1 chooses to use antimicrobial x to treat all its patients, while patch 2 uses antimicrobial y.

When there is no migration between the two patches (hospitals), single-drug use will drive the evolution of single-resistants. If treatment levels are high enough, strain X will fix in patch 1 and strain Y will fix in patch 2. In this scenario, when we allow individuals to migrate between patches, these fixation dynamics can change. From the perspective of hosts in the W - and Z-classes, the patch these hosts occupy is irrelevant since treatment in both patches is effective against hosts infected with the wild type, and ineffective against hosts infected with the double-resistant. However, hosts infected with a strain resistant to only one antimicrobial see the two patches quite differently. In patch 2, for example, strain Y will dominate and eventually fix; if a host from patch 2 carrying a strain resistant to antimicrobial y migrates to patch 1, his potential susceptible pool changes from all hosts in patch 2 to a fraction ξ_{n1} of the hosts in patch 1. The fraction ξ_{x1} of hosts in patch 1 who are prophylaxed by antimicrobial x will not be able to contract an infection from the new immigrant since his strain is susceptible to the antimicrobial x. This means that migration is detrimental to the single-resistant strains.

This effect can be seen in Figure 6. Here we chose $\beta = 2$, $\lambda = 1$, $\zeta = 0$, $\phi_1 = 1/4$, and $\phi_2 = 1/2$, so that in the absence of antimicrobial treatment, the strains' basic reproduction ratios are $\mathcal{R}_{0,w} = 2$, $\mathcal{R}_{0,x} = \mathcal{R}_{0,y} = 8/5$, and $\mathcal{R}_{0,z} = 4/3$. In patch 1, we designate the single-drug treatment regime via $\xi_{n1} = 2/3$ and $\xi_{x1} = 1/3$; in patch 2, we have $\xi_{n2} = 2/3$, and $\xi_{y2} = 1/3$, so that 2/3 of all hosts remain non-prophylaxed while 1/3 receive single-drug treatment with the drug depending on their location. Under these treatment frequencies, the replacement numbers is patch 1 are

$$\mathcal{R}_{0,x} = 8/5 \quad > \quad \mathcal{R}_{0,w} = \mathcal{R}_{0,z} = 4/3 \quad > \quad \mathcal{R}_{0,y} = 16/15,$$

and in patch 2,

$$\mathcal{R}_{0,y} = 8/5 \quad > \quad \mathcal{R}_{0,w} = \mathcal{R}_{0,z} = 4/3 \quad > \quad \mathcal{R}_{0,x} = 16/15.$$

When the $\mathcal{R}_0$-values are ordered in this way, it becomes clear why migration is unfavorable to the single-resistants. They have the highest basic reproduction ratio in one patch and thus increase in relative frequency in this patch, but upon migration they have the lowest reproduction ratio in the other patch and are out-competed by the other three strains.

In Figure 6, we see the slow increase and then decrease in frequency of the single-resistant strains. The remarkable result in this simple model setup is that the addition of spatial structure causes a population-wide reversion to the wild-type strain, a demonstration that load balancing in space can reduce resistance. Without spatial structure, these parameter values would allow coexistence of all strains.

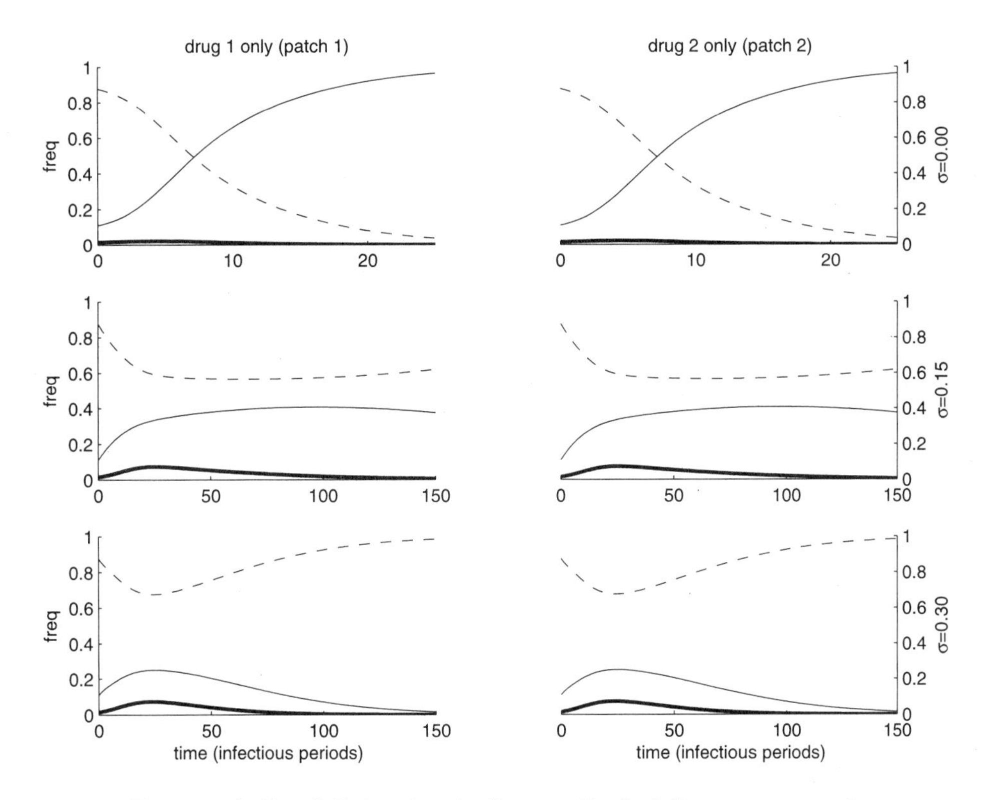

FIGURE 6. Load Balancing in Space. Dashed line represent the frequency of the wild type, solid lines the frequency of single-resistants, and the thick line represents the frequency of double-resistant strains (strength of line indicates strength of resistance). In all figures, $\beta = 2, \lambda = 1, \phi_1 = 1/4, \phi_2 = 1/2, \xi_{n1} = \xi_{n2} = 2/3, \xi_{x1} = \xi_{y2} = 1/3$. In the first row, there is no migration between patches and each spatial location undergoes selection for a particular single resistant: the single-resistant to antimicrobial x flourishes in patch 1, while the single-resistant to antimicrobial y flourishes in patch 2. In the second and third rows, the patches are coupled via a migration parameter, and the single-resistants can no longer flourish since upon migration they observe a higher prophylaxed population. Wild type strains fix for both $\sigma = 0.15$ and $\sigma = 0.30$.

Combination Therapy and Load Balancing. The previous example seems like a rational choice of prescribing strategies if each hospital has the luxury of choosing either drug for its patients. However, if each patch harbors both types of resistant hosts, as well as doubly-resistant hosts, treatment decisions will have to be made on a per-patient basis and both drugs will have to be used in each patch. In this scenario, combination therapy and within-patch load balancing are better options [**6, 7**]. We examine the consequences of implementing a combination-therapy strategy in one patch and a load-balancing strategy in the other.

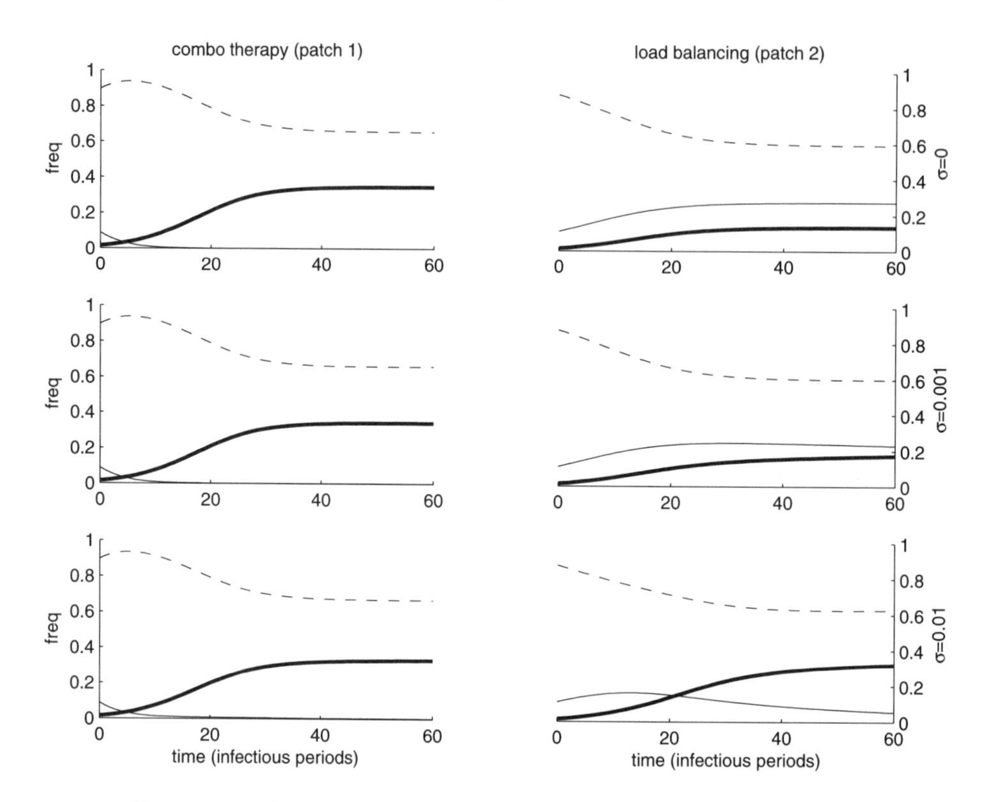

FIGURE 7. Combination Therapy vs. Load Balancing in space. Dashed line represent the frequency of the wild-type, solid lines the frequency of single-resistants, and the thick line represents the frequency of double-resistant strains. In all figures, $\beta = 2, \lambda = 1, \phi_1 = 1/4, \phi_2 = 1/2, \xi_{n1} = \xi_{n2} = 2/3$. In patch 1 (left column) we have $\xi_{x1} = \xi_{y1} = 0, \xi_{z1} = 1/3$. In patch 2 (right column) we have $\xi_{x1} = \xi_{y1} = 1/6, \xi_{z1} = 0$. We see that as coupling between the patches is increased, the combination-therapy scheme is unaffected, but the microbial population structure under the load-balancing regime does in fact change, and it begins to resemble the population structure under combination therapy. In all plots, for both patches, 25% of hosts are infected at equilibrium.

We use the same parameter values for transmission, recovery, and costs of resistance as in the first example. In patch 1, we represent combination therapy by $\xi_{n1} = 2/3$ and $\xi_{z1} = 1/3$. In patch 2, we have load balancing: $\xi_{n2} = 2/3$ and $\xi_{x2} = \xi_{y2} = 1/6$. Under combination therapy in patch 1, we have $\mathcal{R}_{0,w} = \mathcal{R}_{0,z} = 4/3$ and $\mathcal{R}_{0,x} = \mathcal{R}_{0,y} = 16/15$, so the wild types and double-resistants coexist. Under load balancing in patch 2, all strains have an $\mathcal{R}_0 = 4/3$ and all coexist. The first row of Figure 7 shows the strain frequencies in each patch when there is no migration.

Migration homogenizes the population structures in each patch. Because there are more double-resistants in patch 1, there is a net migration of hosts infected with the double-resistant from patch 1 to patch 2. Likewise, there is a net migration of hosts infected with a single-resistant from patch 2 to patch 1.

Double-resistants moving into a load-balancing scheme thrive and provide the resident single-resistants with more opportunity to acquire a second resistance mechanism. As migration increases, most of the single-resistants in patch 2 evolve into double-resistants. Single-resistants moving from load balancing to combination therapy undergo a reduction in $\mathcal{R}_0$; they migrate to a new environment where they are the least fit and cannot increase in frequency.

As coupling between the patches becomes stronger, the direct selection pressure for double-resistants under combination therapy overwhelms the balance achieved by using only single-drug treatments in patch 2. The bacterial strain structure under combination therapy is unaffected by immigration. The strain structure under load balancing, however, becomes destabilized by the arrival of double-resistants, and it begins to resemble the strain structure under combination therapy.

5. Discussion

Here, we have presented and explored some new mathematical models for the spread of resistance in a well-mixed population that explicitly consider the window-of-opportunity for resistance created by chemo-prophylaxis. The use of antimicrobials selects for resistance within a host, but if the rate of use is high enough, a resistant strain can sustain a chain of transmission within a population. Antimicrobial drug use does this by either reducing the proportion colonized by sensitive bacteria or by making it possible for resistance to invade sensitive strains directly through superinfection or the transfer resistance elements. The among-host component of selection and associated transmission are generally considered to be more important than the novel resistance due to within-host component of selection in setting the frequency of resistance within a population.

In most places, the average rate of antimicrobial drug use is too low to favor resistance, but since the rate of drug use is heterogeneous, resistance is favored in some places that are effectively sources for resistance. Thus, from the perspective of resistance, population is structured into a set of sources and sinks. Source-sink dynamics favor coexistence because some people spend more time in sources than others [**30**].

Sources create a local spillover effect, where the prevalence of resistance in nearby populations is always higher. Two sources that are near one another amplify each other. The corollary is that it is better to use different antimicrobials than your neighbors. At the very least, management decisions for antimicrobial resistant pathogens should consider what nearby populations do. Even better, the decisions should be planned and coordinated at regional scales with the purpose of making antimicrobial use as heterogeneous as possible.

For MDR, the implications are more serious. Other papers have emphasized the importance of making drug use as heterogeneous as possible [**6, 7, 19**]. Here, we have shown that load balancing can be effectively done by using different antimicrobials in different hospitals or cities that interact through migration. Moreover, because of chemo-prophylaxis, load balancing has certain advantages over combination therapy, at least when it comes to selection for resistance at the population level. In combination therapy, the total levels of both antimicrobials increase and select for both single-resistant mutants as well as the MDR mutant. In load balancing, total selection pressure is reduced, and the single-mutants are more common, guaranteeing that at least one treatment option is available.

Our models are meant to illustrate some general principles and point out some intriguing possibilities, and should not be interpreted directly as policy recommendations. Antimicrobial policy must also be determined by other concerns, including the interests of the patient, the different efficacies of the antimicrobials, whether antimicrobials have a broad or narrow spectrum, economic considerations, and other factors.

6. Acknowledgements

The authors thank Ben Cooper and an anonymous reviewer for their helpful comments.

7. Appendices

Appendix 1. Let ρ denote the rate that antimicrobial prescription begins, and let $1/\delta$ denote the duration of the protective effects. We subdivide a population into the states of chemo-prophylaxed or not, denoted with subscripts x or n. We further subdivide the population into those who are uncolonized, infected by sensitive phenotypes, or infected by resistant phenotypes. Also, let $U = U_n + U_x$, $W = W_n + W_x$ and $X = X_n + X_x$. The following six equations describe the dynamics for well-mixed populations:

(7.1)

$$\begin{array}{rlll}
\dot{U}_n = & -\beta W_n U_n - \beta X U_n & +\lambda W_n + \lambda X_n + \phi X_n & -\rho U_n + \delta U_x \\
\dot{U}_x = & -\beta X U_x & +\lambda W_x + \lambda X_x + \phi X_x + \nu W_x & +\rho U_n - \delta U_x \\
\dot{W}_n = & \beta W_n U_n + q\beta W_n X_n & -\lambda W_n & -\rho W_n + \delta W_x \\
\dot{W}_x = & -r\beta X W_x & -\lambda W_x - \nu W_x & +\rho W_n - \delta W_x \\
\dot{X}_n = & \beta X U_n - q\beta W_n X_n & -\lambda X_n - \phi X_n & -\rho X_n + \delta X_x \\
\dot{X}_x = & \beta X U_x + r\beta X W_x & -\lambda X_n - \phi X_x & +\rho X_n - \delta X_x
\end{array}$$

In the variables U, X, and W, the dynamics are

$$\begin{array}{rl}
\dot{U} = & -\beta W_n U_n - \beta X U + \lambda(W + X) + \phi X + \nu W_x \\
\dot{W} = & \beta W_n U_n + q\beta W_n X_n - r\beta X W_x - \lambda W - \nu W_x \\
\dot{X} = & \beta X U - q\beta W_n X_n + r\beta X W_x - \lambda X - \phi X
\end{array}, \tag{7.2}$$

Assuming that the rate of drug use is constant and that the pharmacodynamics are fast, the proportion of some class that is protected is given by the equation

$$\dot{y}_x = \rho(y - y_x) - \delta y_x. \tag{7.3}$$

And we use the equilibrium, assuming $\dot{y}_x$ is fast relative to $\dot{y}$:

$$\bar{y}_x = \frac{\rho}{\rho + \delta} y. \tag{7.4}$$

We let $\xi = \rho/(\rho + \delta)$, and $W_x = \xi W$, $W_n = (1 - \xi)W$, $U_n = (1 - \xi)U$, and $X_n = (1 - \xi)X$. We substitute these into (7.2), and write down the much simpler system of equations (2.3). We leave it to others to show when this approximation fails.

Appendix 2. The equations describe a closed population where local population sizes are also constant, but not necessarily equal. Let N_i denote the size of the ith population; the net emigration is $\sigma_i N_i$.

Let $\psi_{i,j}$ denote the fraction of immigrants to the ith population that come from population j, and $\sum_j \psi_{i,j} = 1$. Let $\omega_{i,j}$ denote the fraction of emigrants from population i that go to population j. To balance migration:

$$\psi_{i,j}\sigma_i N_i = \omega_{i,j}\sigma_j N_j. \tag{7.5}$$

Appendix 3. The full model with 2 antimicrobials, 4 prophylaxed states, and 5 colonization states is described below via 20 differential equations describing the dynamics of the 20 population classes described in Table 2. In the equations below, $\lambda_{s,e}$ is the recovery rate of strain s in environment e. These will be simplified later.

$$\begin{aligned}
\dot{U}_n &= -\beta W_n U_n - \beta(X_n + X_x)U_n - \beta(Y_n + Y_y)U_n - \beta(Z_n + Z_x + Z_y + Z_z)U_n \\
&\quad +\lambda_{w,n}W_n + \lambda_{x,n}X_n + \lambda_{y,n}Y_n + \lambda_{z,n}Z_n \\
\dot{U}_x &= -\beta(X_n + X_x)U_x - \beta(Z_n + Z_x + Z_y + Z_z)U_x \\
&\quad +\lambda_{w,x}W_x + \lambda_{x,x}X_x + \lambda_{y,x}Y_x + \lambda_{z,x}Z_x \\
\dot{U}_y &= -\beta(Y_n + Y_y)U_y - \beta(Z_n + Z_x + Z_y + Z_z)U_y \\
&\quad +\lambda_{w,y}W_y + \lambda_{x,y}X_y + \lambda_{y,y}Y_y + \lambda_{z,y}Z_y \\
\dot{U}_z &= -\beta(Z_n + Z_x + Z_y + Z_z)U_z \\
&\quad +\lambda_{w,z}W_z + \lambda_{x,z}X_z + \lambda_{y,z}Y_z + \lambda_{z,z}Z_z \\
\dot{W}_n &= +\beta W_n U_n - \lambda_{w,n}W_n \\
\dot{W}_x &= -\beta(X_n + X_x)W_x - \beta(Z_n + Z_x + Z_y + Z_z)W_x - \lambda_{w,x}W_x \\
\dot{W}_y &= -\beta(Y_n + Y_y)W_y - \beta(Z_n + Z_x + Z_y + Z_z)W_y - \lambda_{w,y}W_y \\
\dot{W}_z &= -\beta(Z_n + Z_x + Z_y + Z_z)W_z - \lambda_{w,z}W_y \\
\dot{X}_n &= +\beta(X_n + X_x)U_n - \lambda_{x,n}X_n \\
\dot{X}_x &= +\beta(X_n + X_x)(U_x + W_x) - \lambda_{x,x}X_x \\
\dot{X}_y &= -\beta(Y_n + Y_y)X_y - \beta(Z_n + Z_x + Z_y + Z_z)X_y - \lambda_{x,y}X_y \\
\dot{X}_z &= -\beta(Z_n + Z_x + Z_y + Z_z)X_z - \lambda_{x,z}X_z \\
\dot{Y}_n &= +\beta(Y_n + Y_y)U_n - \lambda_{y,n}Y_n \\
\dot{Y}_x &= -\beta(X_n + X_x)Y_x - \beta(Z_n + Z_x + Z_y + Z_z)Y_x - \lambda_{y,x}Y_x \\
\dot{Y}_y &= +\beta(Y_n + Y_y)(U_y + W_y) - \lambda_{y,y}Y_y \\
\dot{Y}_z &= -\beta(Z_n + Z_x + Z_y + Z_z)Y_z - \lambda_{y,z}Y_z \\
\dot{Z}_n &= +\beta(Z_n + Z_x + Z_y + Z_z)U_n - \lambda_{z,n}Z_n \\
\dot{Z}_x &= +\beta(Z_n + Z_x + Z_y + Z_z)(U_x + W_x + Y_x) + \beta(X_n + X_x)Y_x - \lambda_{z,x}Z_x \\
\dot{Z}_y &= +\beta(Z_n + Z_x + Z_y + Z_z)(U_y + W_y + X_y) + \beta(Y_n + Y_y)X_y - \lambda_{z,y}Z_y \\
\dot{Z}_z &= +\beta(Z_n + Z_x + Z_y + Z_z)(U_z + W_z + X_z + Y_z) - \lambda_{z,z}Z_z
\end{aligned}$$

In these classes we have eliminated the flow of hosts to prophylaxis (ρ), and from prophylaxis (δ). We will simply assume that the host population is divided into four sub-groups undergoing varying degrees of prophylaxis: a fraction ξ_n is not prophylaxed, a fraction ξ_x is prophylaxed by antimicrobial x, a fraction ξ_y is prophylaxed by antimicrobial y, a fraction ξ_z is prophylaxed by antimicrobials x and y.

Ignoring the recovery dynamics momentarily, these 8 classes defined by (4.1) and (4.2) allow us to write our system down in 10 equations:

$$\begin{aligned}
\dot{Q}_w &= -Q_w(\Lambda_w + \Lambda_x + \Lambda_y + \Lambda_z) \\
\dot{Q}_x &= -Q_x(\Lambda_x + \Lambda_z) - Q_w(\Lambda_w + \Lambda_y) \\
\dot{Q}_y &= -Q_y(\Lambda_y + \Lambda_z) - Q_w(\Lambda_w + \Lambda_x) \\
\dot{Q}_z &= -Q_z\Lambda_z - Q_y\Lambda_y - Q_x\Lambda_x - Q_x\Lambda_x \\
\dot{\Lambda}_w &= \beta Q_w\Lambda_w \\
\dot{\Lambda}_x &= \beta(Q_x - Y_x)\Lambda_x \\
\dot{\Lambda}_y &= \beta(Q_y - X_y)\Lambda_y \\
\dot{\Lambda}_z &= \beta Q_z\Lambda_z + \beta X_y\Lambda_y + \beta Y_x\Lambda_x \\
\dot{X}_y &= -X_y(\Lambda_y + \Lambda_z) \\
\dot{Y}_x &= -Y_x(\Lambda_x + \Lambda_z).
\end{aligned} \tag{7.6}$$

If we add in the recovery terms, we can no longer express the system in 10-dimensions, unless we make a similar approximation as in the one drug case, namely that the fractions ξ_i express the relative frequencies in the 5 types of disease classes (uncolonized, infected with wild type, infected with resistant to x, infected with resistant to y, infected with resistant to x and y). Making this approximation, we see that Z_x, for example, can be expressed as $\xi_x\,\beta^{-1}\Lambda_z$. And,

$$X_n \approx \frac{\xi_n}{\xi_n + \xi_x}\,\frac{1}{\beta}\,\Lambda_x.$$

Then, our approximation yields the differential equations:

$$
\begin{aligned}
\dot{Q}_w &= -Q_w(\Lambda_w + \Lambda_x + \Lambda_y + \Lambda_z) \\
&\quad + \frac{\lambda}{\beta}\left(\Lambda_w + \frac{\lambda_S}{\lambda}\frac{\xi_n}{\xi_n + \xi_x}\Lambda_x + \frac{\lambda_S}{\lambda}\frac{\xi_n}{\xi_n + \xi_y}\Lambda_y + \frac{\lambda_D}{\lambda}\xi_n\Lambda_z\right) \\
\dot{Q}_x &= -Q_x(\Lambda_x + \Lambda_z) - Q_w(\Lambda_w + \Lambda_y) \\
&\quad + \frac{\lambda}{\beta}\left(\Lambda_w + \frac{\lambda_S}{\lambda}\Lambda_x + \frac{\lambda_S}{\lambda}\frac{\xi_n}{\xi_n + \xi_y}\Lambda_y + \frac{\lambda_D}{\lambda}(\xi_n + \xi_x)\Lambda_z\right) \\
\dot{Q}_y &= -Q_y(\Lambda_y + \Lambda_z) - Q_w(\Lambda_w + \Lambda_x) \\
&\quad + \frac{\lambda}{\beta}\left(\Lambda_w + \frac{\lambda_S}{\lambda}\frac{\xi_n}{\xi_n + \xi_x}\Lambda_x + \frac{\lambda_S}{\lambda}\Lambda_y + \frac{\lambda_D}{\lambda}(\xi_n + \xi_y)\Lambda_z\right) \\
\dot{Q}_z &= -Q_z\Lambda_z - Q_y\Lambda_y - Q_x\Lambda_x - Q_x\Lambda_x \\
&\quad + \frac{\lambda}{\beta}\left(\Lambda_w + \frac{\lambda_S}{\lambda}\Lambda_x + \frac{\lambda_S}{\lambda}\Lambda_y + \frac{\lambda_D}{\lambda}\Lambda_z\right) \\
\dot{\Lambda}_w &= \beta Q_w\Lambda_w - \lambda\Lambda_w \\
\dot{\Lambda}_x &= \beta(Q_x - Y_x)\Lambda_x - \lambda_S\Lambda_x \\
\dot{\Lambda}_y &= \beta(Q_y - X_y)\Lambda_y - \lambda_S\Lambda_y \\
\dot{\Lambda}_z &= \beta Q_z\Lambda_z + \beta X_y\Lambda_y + \beta Y_x\Lambda_x - \lambda_D\Lambda_z \\
\dot{X}_y &= -X_y(\Lambda_y + \Lambda_z) - \lambda_A X_y \\
\dot{Y}_x &= -Y_x(\Lambda_x + \Lambda_z) - \lambda_A Y_x,
\end{aligned}
\tag{7.7}
$$

where $\lambda_S = \lambda + \phi_1$ and $\lambda_D = \lambda + \phi_2$ are the recovery rates (in all environments where antimicrobials have no effect) for the single-resistant and for the double-resistant, respectively. λ_A is the recovery rate for a host undergoing effective antimicrobial treatment. Since $\dot{X}_y < 0$ and $\dot{Y}_x < 0$, we simply say that these two classes are zero, and we approximate the full system with the remaining eight equations; this is system (4.3) in the text.

References

[1] D. I. Andersson and B. R. Levin, The biological cost of antibiotic resistance, *Current Opinions in Microbiology* **2** (1999), 489–493.

[2] D. J. Austin, M. Kakehashi, and R. M. Anderson, The transmission dynamics of antibiotic-resistant bacteria: the relationship between resistance in commensal organisms and antibiotic consumption, *Proc. R. Soc. Lond. B* **264** (1997), 1629–1638.

[3] D. J. Austin and R. M. Anderson, Transmission dynamics of epidemic methicillin-resistant *Staphylococcus aureus* and vancomycin-resistant enterococci in England and Wales, *J. Infect. Dis.* **179** (1999), 883–891.

[4] D. J. Austin, M. J. M. Bonten, R. A. Weinstein, S. Slaughter, and R. M. Anderson, Vancomycin-resistant enterococci in intensive-care hospital settings: transmission dynamics, persistence, and the impact of infection control programs, *Proc. Natl. Acad. Sci. USA* **96** (1999), 6908–6913.

[5] D. J. Austin, K. G. Kristinsson, and R. M. Anderson, The relationship between the volume of antimicrobial consumption in human communities and the frequency of resistance, *Proc. Natl. Acad. Sci. USA* **96** (1999), 1152–1156.

[6] C. T. Bergstrom, M. Lo, and M. Lipsitch, Ecological theory suggests that antimicrobial cycling will not reduce antimicrobial resistance in hospitals, *Proc. Natl. Acad. Sci. USA* **101** (2004), 13285–13290.

[7] S. Bonhoeffer, M. Lipsitch, and B. R. Levin, Evaluating treatment protocols to prevent antibiotic resistance, *Proc. Natl. Acad. Sci. USA* **94** (1997), 12106–12111.

[8] M. F. Boni and M. W. Feldman, Evolution of antibiotic resistance by human and bacterial niche construction, *Evolution* **59**(3) (2005), 477–491.

[9] M. J. Bonten, R. Willems, and R. A. Weinstein, Vancomycin-resistant enterococci: Why are they here, and where do they come from? *Lancet Infect. Dis.* **1** (2001), 314–325.

[10] M. J. Bonten, D. J. Austin, and M. Lipsitch, Understanding the spread of antibiotic resistant pathogens in hospitals: Mathematical models as tools for control, *Clin. Infect. Dis.* **33** (2001), 1739–1746.

[11] CDC, *Staphylococcus aureus* resistant to vancomycin — United States, 2002, *MMWR* **51** (2002), 565–567.

[12] D. F. Clyde and G. T. Shute, Resistance of *Plasmodium falciparum* in Tanganyika to pyrimethamine administered at weekly intervals, *Trans. R. Soc. Trop. Med. Hyg.* **51** (1957), 505–513.

[13] J. R. Gog and J. Swinton, A status-based approach to multiple strain dynamics, *J. Math. Biol.* **44** (2002), 169–184.

[14] M. G. M. Gomes and G. F. Medley, Dynamics of multiple strains of infectious agents coupled by cross-immunity: A comparison of models, in (S. Blower, C. Castillo-Chavez, K. L. Cooke, D. Kirschner, and P. van der Driessche, eds.), *Mathematical Approaches for Emerging and Reemerging Infections: Models, Methods and Theory*, IMA Volumes in Mathematics and its Applications, Springer-Verlag, New York, 2001, 171–191.

[15] I. M. Hastings and U. D'Allesandro, Modelling a predictable disaster: the rise and spread of drug-resistant malaria, *Parasitol. Today* **16** (2000), 340–347.

[16] I. M. Hastings, W. M. Watkins, and N. J. White, The evolution of drug-resistant malaria: the role of drug elimination half-life, *Philos. Trans. R. Soc. Lond. B* **29** (2002), 505–519.

[17] M. B. Hoshen, W. D. Stein, and H. Ginsburg, Mathematical modelling of malaria chemotherapy: combining artesunate and mefloquine, *Parasitology* **124** (2002), 9–15.

[18] J. C. Koella and R. Antia, Epidemiological models for the spread of anti-malarial resistance, *Malaria J.* **2** (2003), 3.

[19] R. Laxminarayan and M. L. Weitzman, On the implications of endogenous resistance to medications, *J. Health Econ.* **21** (2002), 709–718.

[20] B. R. Levin and C. T. Bergstrom, Bacteria are different: observations, interpretations, speculations, and opinions about the mechanisms of adaptive evolution in prokaryotes, *Proc. Natl. Acad. Sci. USA* **97** (2000), 6981–6985.

[21] B. R. Levin, V. Perrot, and N. Walker, Compensatory mutations, antibiotic resistance and the population genetics of adaptive evolution in bacteria, *Genetics* **154** (2000), 985–997.

[22] B. R. Levin, Minimizing potential resistance: a population dynamics view, *Clin. Infect. Dis.* **33**(Supple 3) (2001), S161–S169.

[23] S. A. Levin, V. Andreasen, and J. Lin, The dynamics of cocirculating influenza strains conferring partial cross-immunity, *J. Math. Biol.* **35** (1997), 825–842.

[24] M. Lipsitch, C. T. Bergstrom, and B. R. Levin, The epidemiology of antibiotic resistance in hospitals: paradoxes and prescriptions, *Proc. Natl. Acad. Sci. USA* **97** (2000), 1938–1943.

[25] M. Lipsitch, Measuring and interpreting associations between antibiotic use and penicillin resistance in *Streptococcus pneumoniae*, *Clin. Infect. Dis.* **32** (2001), 1044–1054.

[26] M. Lipsitch and M. H. Samore, Antimicrobial use and antimicrobial resistance: a population perspective, *Emerg. Infect. Dis.* **8** (2002), 347–354.

[27] National Nosocomial Infections Surveillance System, National nosocomial infections surveillance (nnis) system report, data summary from january 1992 through june 2004, issued october 2004, *Am. J. Infect. Control.* **32** (2004), 470–485.

[28] R. Ndyomugyenyi and P. Magnussen, Trends in malaria-attributable morbidity and mortality among young children admitted to Ugandan hospitals, for the period 1990–2001, *Ann. Trop. Med. Parasitol.* **98** (2004), 315–327.

[29] L. B. Rice, Emergence of vancomycin-resistant enterococci, *Emerg. Infect. Dis.* **7** (2001), 183–187.

[30] D. L. Smith, A. D. Harris, J. A. Johnson, E. K. Silbergeld, and J. G. Morris, Jr., Animal antibiotic use has an early but important impact on the emergence of antibiotic resistance in human commensal bacteria, *Proc. Natl. Acad. Sci. USA* **99** (2002), 6434–6439.

[31] D. L. Smith, J. Dushoff, E. N. Perencevich, A. D. Harris, and S. A. Levin, Persistent colonization and the spread of antibiotic resistance in nosocomial pathogens: resistance is a regional problem, *Proc. Natl. Acad. Sci. USA* **101** (2004), 3709–3714.

[32] D. Schellenberg, C. Menendez, E. Kahigwa, J. Aponte, J. Vidal, M. Tanner, H. Mshinda, and P. Alonso, Intermittent treatment for malaria and anaemia control at time of routine vaccinations in Tanzanian infants: a randomised, placebo-controlled trial, *Lancet* **357** (2000), 1471–1477.

[33] C. E. Shulman, E. K. Dorman, F. Cutts, K. Kawuondo, J. N. Bulmer, N. Peshu, and K. Marsh, Intermittent sulphadoxine-pyrimethamine to prevent severe anaemia secondary to malaria in pregnancy: a randomised placebo-controlled trial, *Lancet* **353** (1999), 632–636.

[34] B. Spellberg, J. H. Powers, E. P. Brass, L. G. Miller, and J. E. Edwards, Jr., Trends in antimicrobial drug development: implications for the future, *Clin. Infect. Dis.* **38** (2004), 1279–1286.

[35] J. F. Trape, The public health impact of chloroquine resistance in Africa, *Am. J. Trop. Med. Hyg.* **64**(1–2 Suppl) (2001), 12–17.

[36] S. Yeung, W. Pongtavornpinyo, I. M. Hastings, A. J. Mills, and N. J. White, Antimalarial drug resistance, artemisinin-based combination therapy, and the contribution of modeling to elucidating policy choices, *Am. J.Trop. Med. Hyg.* **71** (2004), 179–186.

Fogarty International Center, National Institutes of Health, Bethesda, MD 20892
E-mail address: `smitdave@mail.nih.gov`

Department of Biological Sciences, Stanford University
E-mail address: `maciek@charles.stanford.edu`

Resources for the Future, Washington DC
E-mail address: `ramanan@rff.org`

Titles in This Series

71 **Zhilan Feng, Ulf Dieckmann, and Simon Levin, Editors,** Disease Evolution: Models, Concepts, and Data Analyses

70 **James Abello and Graham Cormode, Editors,** Discrete Methods in Epidemiology

69 **Siemion Fajtlowicz, Patrick W. Fowler, Pierre Hansen, Melvin F. Janowitz, and Fred S. Roberts, Editors,** Graphs and Discovery

68 **A. Ashikhmin and A. Barg, Editors,** Algebraic Coding Theory and Information Theory

67 **Ravi Janardan, Michiel Smid, and Debasish Dutta, Editors,** Geometric and Algorithmic Aspects of Computer-Aided Design and Manufacturing

66 **Piyush Gupta, Gerhard Kramer, and Adriaan J. van Wijngaarden, Editors,** Advances in Network Information Theory

65 **Santosh S. Vempala,** The Random Projection Method

64 **Melvyn B. Nathanson, Editor,** Unusual Applications of Number Theory

63 **J. Nešetřil and P. Winkler, Editors,** Graphs, Morphisms and Statistical Physics

62 **Gerard J. Foschini and Sergio Verdú, Editors,** Multiantenna Channels: Capacity, Coding and Signal Processing

61 **M. F. Janowitz, F.-J. Lapointe, F. R. McMorris, B. Mirkin, and F. S. Roberts, Editors,** Bioconsensus

60 **Saugata Basu and Laureano Gonzalez-Vega, Editors,** Algorithmic and Quantitative Real Algebraic Geometry

59 **Michael H. Goldwasser, David S. Johnson, and Catherine C. McGeoch, Editors,** Data Structures, Near Neighbor Searches, and Methodology: Fifth and Sixth DIMACS Implementation Challenges

58 **Simon Thomas, Editor,** Set Theory: The Hajnal Conference

57 **Eugene C. Freuder and Richard J. Wallace, Editors,** Constraint Programming and Large Scale Discrete Optimization

56 **Alexander Barg and Simon Litsyn, Editors,** Codes and Association Schemes

55 **Ding-Zhu Du, Panos M. Pardalos, and Jie Wang, Editors,** Discrete Mathematical Problems with Medical Applications

54 **Erik Winfree and David K. Gifford, Editors,** DNA Based Computers V

53 **Nathaniel Dean, D. Frank Hsu, and R. Ravi, Editors,** Robust Communication Networks: Interconnection and Survivability

52 **Sanguthevar Rajasekaran, Panos Pardalos, and D. Frank Hsu, Editors,** Mobile Networks and Computing

51 **Pierre Hansen, Patrick Fowler, and Maolin Zheng, Editors,** Discrete Mathematical Chemistry

50 **James M. Abello and Jeffrey Scott Vitter, Editors,** External Memory Algorithms

49 **Ronald L. Graham, Jan Kratochvíl, Jaroslav Nešetřil, and Fred S. Roberts, Editors,** Contemporary Trends in Discrete Mathematics

48 **Harvey Rubin and David Harlan Wood, Editors,** DNA Based Computers III

47 **Martin Farach-Colton, Fred S. Roberts, Martin Vingron, and Michael Waterman, Editors,** Mathematical Support for Molecular Biology

46 **Peng-Jun Wan, Ding-Zhu Du, and Panos M. Pardalos, Editors,** Multichannel Optical Networks: Theory and Practice

45 **Marios Mavronicolas, Michael Merritt, and Nir Shavit, Editors,** Networks in Distributed Computing

44 **Laura F. Landweber and Eric B. Baum, Editors,** DNA Based Computers II

43 **Panos Pardalos, Sanguthevar Rajasekaran, and José Rolim, Editors,** Randomization Methods in Algorithm Design

42 **Ding-Zhu Du and Frank K. Hwang, Editors,** Advances in Switching Networks

TITLES IN THIS SERIES

41 **David Aldous and James Propp, Editors,** Microsurveys in Discrete Probability

40 **Panos M. Pardalos and Dingzhu Du, Editors,** Network Design: Connectivity and Facilities Location

39 **Paul W. Beame and Samuel R Buss, Editors,** Proof Complexity and Feasible Arithmetics

38 **Rebecca N. Wright and Peter G. Neumann, Editors,** Network Threats

37 **Boris Mirkin, F. R. McMorris, Fred S. Roberts, and Andrey Rzhetsky, Editors,** Mathematical Hierarchies and Biology

36 **Joseph G. Rosenstein, Deborah S. Franzblau, and Fred S. Roberts, Editors,** Discrete Mathematics in the Schools

35 **Dingzhu Du, Jun Gu, and Panos M. Pardalos, Editors,** Satisfiability Problem: Theory and Applications

34 **Nathaniel Dean, Editor,** African Americans in Mathematics

33 **Ravi B. Boppana and James F. Lynch, Editors,** Logic and Random Structures

32 **Jean-Charles Grégoire, Gerard J. Holzmann, and Doron A. Peled, Editors,** The SPIN Verification System

31 **Neil Immerman and Phokion G. Kolaitis, Editors,** Descriptive Complexity and Finite Models

30 **Sandeep N. Bhatt, Editor,** Parallel Algorithms: Third DIMACS Implementation Challenge

29 **Doron A. Peled, Vaughan R. Pratt, and Gerard J. Holzmann, Editors,** Partial Order Methods in Verification

28 **Larry Finkelstein and William M. Kantor, Editors,** Groups and Computation II

27 **Richard J. Lipton and Eric B. Baum, Editors,** DNA Based Computers

26 **David S. Johnson and Michael A. Trick, Editors,** Cliques, Coloring, and Satisfiability: Second DIMACS Implementation Challenge

25 **Gilbert Baumslag, David Epstein, Robert Gilman, Hamish Short, and Charles Sims, Editors,** Geometric and Computational Perspectives on Infinite Groups

24 **Louis J. Billera, Curtis Greene, Rodica Simion, and Richard P. Stanley, Editors,** Formal Power Series and Algebraic Combinatorics/Séries Formelles et Combinatoire Algébrique, 1994

23 **Panos M. Pardalos, David I. Shalloway, and Guoliang Xue, Editors,** Global Minimization of Nonconvex Energy Functions: Molecular Conformation and Protein Folding

22 **Panos M. Pardalos, Mauricio G. C. Resende, and K. G. Ramakrishnan, Editors,** Parallel Processing of Discrete Optimization Problems

21 **D. Frank Hsu, Arnold L. Rosenberg, and Dominique Sotteau, Editors,** Interconnection Networks and Mapping and Scheduling Parallel Computations

20 **William Cook, László Lovász, and Paul Seymour, Editors,** Combinatorial Optimization

19 **Ingemar J. Cox, Pierre Hansen, and Bela Julesz, Editors,** Partitioning Data Sets

18 **Guy E. Blelloch, K. Mani Chandy, and Suresh Jagannathan, Editors,** Specification of Parallel Algorithms

17 **Eric Sven Ristad, Editor,** Language Computations

16 **Panos M. Pardalos and Henry Wolkowicz, Editors,** Quadratic Assignment and Related Problems

For a complete list of titles in this series, visit the
AMS Bookstore at **www.ams.org/bookstore/**.